D0299250

Tolley's Workplace Accident Handbook

Tolley's Workplace Accident Handbook

Second edition

Editor

Mark Tyler

Dr Olivia Carlton, Greg Gordon, Alexander Green, Amanda Jenkins,
Kajal Sharma, Lauren Thomas and Hannah Wilson

AMSTERDAM • BOSTON • HEIDELBERG • LONDON • NEW YORK • OXFORD
PARIS • SAN DIEGO • SAN FRANCISCO • SINGAPORE • SYDNEY • TOKYO
Butterworth-Heinemann is an imprint of Elsevier

Butterworth-Heinemann is an imprint of Elsevier
Linacre House, Jordan Hill, Oxford OX2 8DP, UK
30 Corporate Drive, Suite 400, Burlington, MA 01803, USA

First published in 2003

Second edition 2007

Copyright © 2003, 2007 Elsevier Ltd. All rights reserved

No part of this publication may be reproduced, stored in a retrieval system or transmitted in any form or by any means electronic, mechanical, photocopying, recording or otherwise without the prior written permission of the publisher

Crown copyright material is reproduced with the permission of the Controller of HMSO and the Queen's printer for Scotland. Any European material in this work which has been reproduced for EUR-lex, the official European Communities legislation web site, is European Communities copyright.

Permissions may be sought directly from Elsevier's Science & Technology Rights Department in Oxford, UK; phone: (+44) (0) 1865 843830; fax: (+44) (0) 1865 853333; e-mail: permissions@elsevier.com. Alternatively you can submit your request online by visiting the Elsevier web site at http://elsevier.com/locate/permissions, and selecting *Obtaining permission to use Elsevier material*

Notice
No responsibility is assumed by the publisher for any injury and/or damage to persons or properly as a matter of products liability, negligence or otherwise, or from any use or operation of any methods, products, instructions or ideas contained in the material herein. Because of rapid advances in the medical sciences, in particular, independent verification of diagnoses and drug dosages should be made

British Library Cataloguing in Publication Data
A catalogue record for this book is available from the British Library

Library of Congress Cataloging-in-Publication Data
A catalog record for this book is available from the Library of Congress

ISBN: 978-0-75-068151-3

For information on all Butterworth-Heinemann publications visit our web site at books.elsevier.com

Typeset by Charon Tec Ltd (A Macmillan Company), Chennai, India
www.charontec.com
Printed and bound in Great Britain

07 08 09 10   10 9 8 7 6 5 4 3 2 1

Working together to grow
libraries in developing countries

www.elsevier.com | www.bookaid.org | www.sabre.org

ELSEVIER   BOOK AID International   Sabre Foundation

Contents

Foreword

Health and safety is an issue of vital importance to all employers. Not only is it the employer's responsibility under the law to protect staff, contractors and visitors, but effective health and safety management systems save time and money and increase the efficiency of the organisation.

It remains essential to understand the factors in the workplace that can contribute to accidents, and to be well-prepared with first aid and emergency planning. We also see much scope for putting greater focus on the rehabilitation of the victims of accidents.

The CBI supports the investigation of accidents, incidents and near-misses as an integral part of effective health and safety management. It enables operators to identify what went wrong, ascertain potential gaps in their risk assessment, review current performance and put in place improved systems to prevent other incidents. New corporate manslaughter legislation and plans to increase the range of penalties for health and safety offences generally could herald a significant change in attitudes to accident investigation. Organisations will have to anticipate a more assertive response from the enforcing authorities. In the tragic event of fatalities, the involvement of the police with their powers of arrest and focus on senior management will require that companies understand the enforcing authorities' procedures and powers and their managers' and their own corporate rights and obligations.

This book deals with these and many other issues relating to workplace accidents in a practical way, and it is a welcome guide for who all work in this important area.

Janet Asherson
Head of Health and Safety, CBI

Foreword

Acknowledgements

This second edition is much more than just an update of the first; it has seen a substantial amount of revision and the introduction of several entirely new chapters. I am grateful to Doris Funke of Elsevier for her thoughtful guidance on these changes, and of course to my fellow contributors for their efforts.

I greatly appreciated support I had from Lawrence Bamber, Roger Bibbings, Jessica Burt, David Carney, Rachel Czernobay, Mark Morsman, Alison Newstead and Ian Schofield who in a variety of helped me research and obtain new material for the second edition. My thanks also go to RoSPA and Associated British Ports for kindly permitting the re-publication of their material in this book.

Last and by no means least, Sarah Tylee deserves credit for preparing and correcting a long series of manuscripts and proofs with great efficiency, and for creating many of the tables and figures found throughout the book.

Mark Tyler
London

Contributors

Dr Olivia Carlton FRCP, FFOM
Head of Occupational Health, London Underground Ltd

Greg Gordon LLB (Hons), LLM, Dip LP
Lecturer, University of Aberdeen

Alex Green M Theol, LLB, LLM
Solicitor

Amanda Jenkins
British Red Cross

Kajal Sharma BSc (Hons)
Solicitor (non-practising)

Lauren Thomas BA MSc CPsychol FHEA
Chartered Occupational Psychologist

Mark Tyler MA, LLM, CMIOSH
Solicitor and Chartered Safety and Health Practitioner

Hannah Wilson MA (Hons) (Cantab), MPhil
Barrister, Henderson Chambers

1 Introduction

In this chapter:

Rationales for management focus on accidents

1.1 The pain, suffering and sometimes grief that work-related injuries and ill health can lead to, the disruption to family life, the financial and other pressures that can ensue, underline the moral case for organisations and individuals who work in them to prevent accidents throughout their daily activities. Conditions minimising the potential for accidents need to be maintained, and – in planning for the worst – steps must be taken to mitigate the consequences when accidents do occur by providing aid and support.

The moral case for accident prevention is obvious, but the case for management action to minimise their occurrence and effects is not difficult to make either. There is the simple utilitarian case that organisations perform better, protect their reputations and reduce their costs through the application of sound risk management principles, factors which now increasingly influence the related areas of corporate governance and corporate social responsibility. Increasingly organisations' own health as businesses or public bodies is being judged by standards that include the care taken of their employees' and the public's safety and welfare.

There is the wider economic case too, which applies to the economy as a whole. A Health and Safety Executive report (HSE, 2004) assessed the costs to British economy in 2001/2002 as being between £20 and £31.8 billion. That was between 1.9 and 3 per cent of the gross domestic product (GDP) for 2002, or around half of annual growth in GDP in the last decade. The International Labour Organization (ILO) has estimated the global costs of workplace accidents to be at least $1,250 billion.

Legal compliance is one of the strongest drivers for accident prevention. Statutes and regulations – increasingly emanating from the EU – together with the common law comprise a set of norms which are non-delegable and non-negotiable. These norms exist both as *minimum standards* which society expects all organisations to adhere to, and as a framework for *penalising* non-compliance or *compensating* victims. Legal requirements may sometimes lag behind public expectations of safety standards and good industry practice, but in many instances they can also set standards which are more demanding than managers might consider to be "reasonable" or justifiable based on their experience and their perceptions of low risks. Risk management concepts such as tolerability of risk and *ALARP* (ensuring risks are as low as is reasonable practicable) are flexible, and have their roots in the *Health and Safety at Work Act 1974* (*HSWA*) scheme of duties which are summarised in Chapter 3. Health and safety legislation is closely intertwined with best practice in risk assessment, emergency planning, fire protection, first aid and employers' liability insurance. The legal consequences which can follow for non-compliance – criminal proceedings and claims for damages – are described in their various stages leading ultimately to the courtroom are explained in Chapter 8.

In addition to these important areas this book provides guidance on aspects of accidents which are much less regulated, but which may have come to the fore as management tools, in particular the thorough investigation of accidents to learn lessons, and optional rehabilitation approach to enable employees to return to work.

Trends and targets

1.2 The last two decades have seen steadily falling numbers of fatal accidents in the workplace, and a similar but less pronounced decline in major accidents and over-3-day injuries (as classified by the *Reporting of Injuries, Diseases and Dangerous Occurrences Regulations 1995*). Tables 1.1–1.3 use statistics published by the HSE to demonstrate the numbers of accidents to employees reported from 1981 to 2006, and also to show these numbers adjusted to take into account the fluctuating size of the workforce by reference to the incidence rate per 100,000 employees.

The data on non-fatal accidents can be hard to interpret because the relevant reporting regulations and accident criteria have changed twice during the period, which explains the "jumps" in accident rates after 1985 and 1995. After each change, the downward trend has continued, but the rate of improvement is now slowing markedly.

Some of the improvement in accident statistics is due to changing patterns of employment, most notably the decline in mining and heavy engineering. There is some evidence that the incidence rates for non-fatal accidents to *self-employed* workers

Table 1.1: Fatal accidents to employees and incidence rates per 100,000

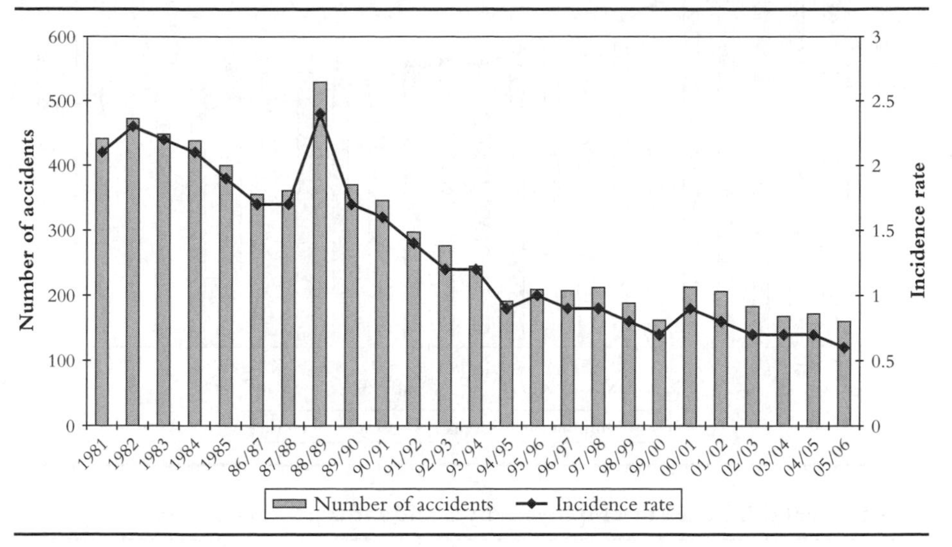

Table 1.2: Non-fatal major accidents to employees and incidence rates per 100,000

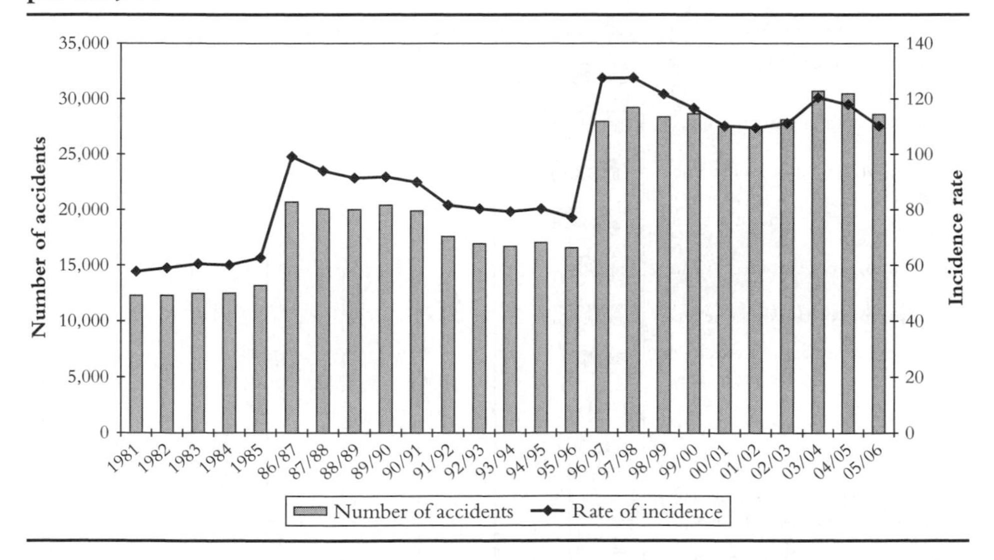

have begun to increase after 2000, possibly reflecting the shifts in the workforce away from permanent jobs and towards more fluid employment relationships.

These accident rates nevertheless compare very favourably with the situation in other developed countries. The fatal accident rate is about half the EU average

Table 1.3: Over-3-day injuries to employees and incidence rates per 100,000

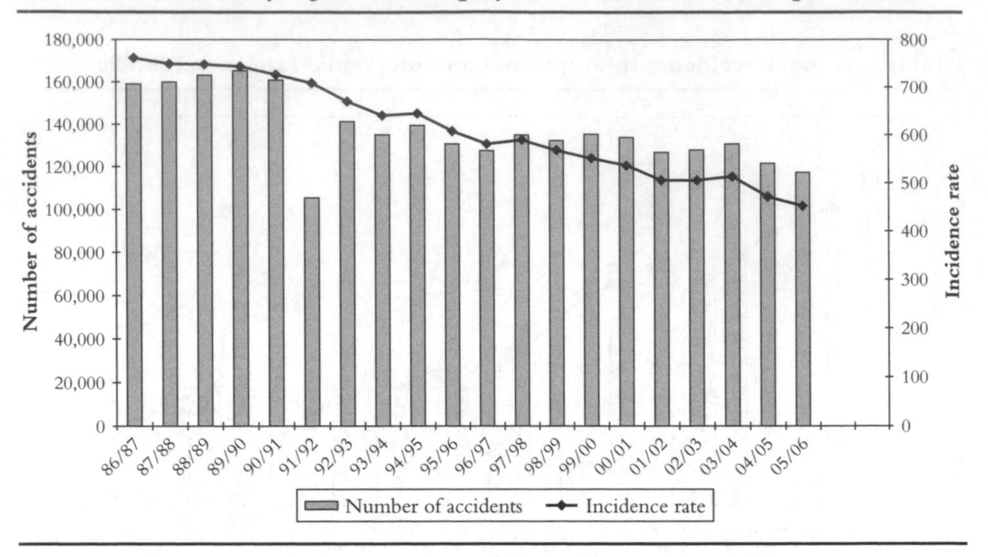

Table 1.4: Fatal injuries to employees by accident type, 2004/2005

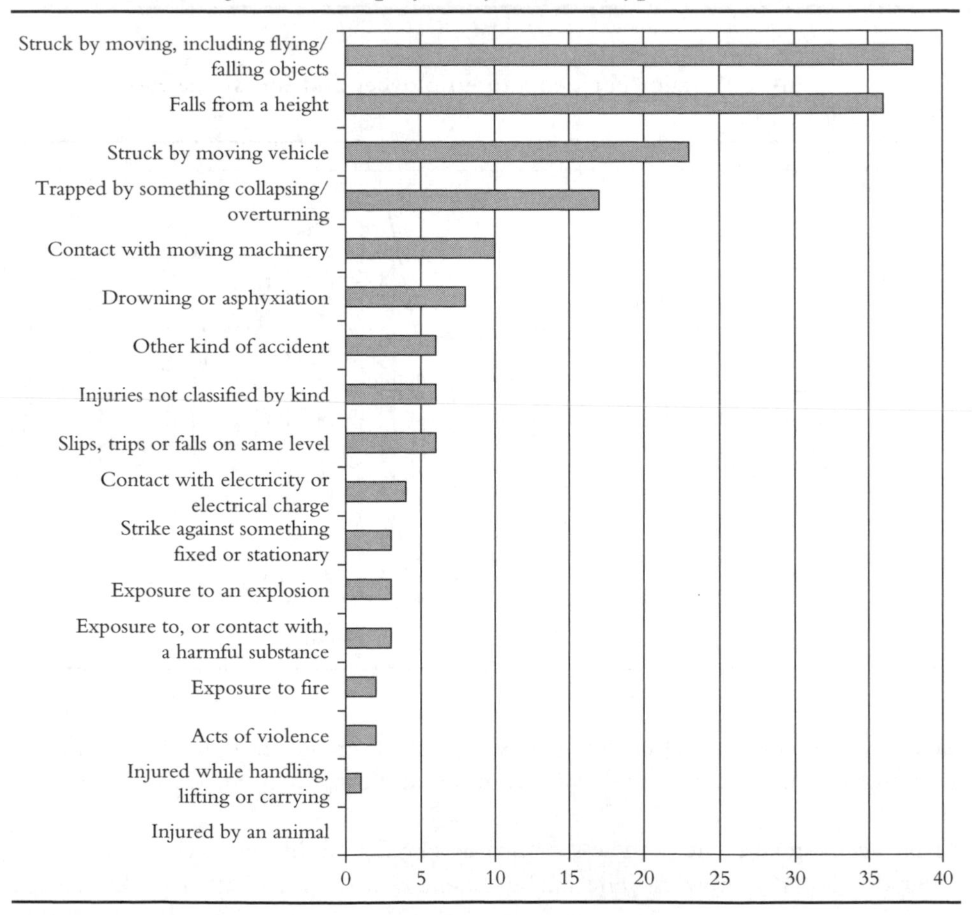

(better than any other Member State) and around three-quarters that of the USA. Measured in terms of over-3-day accidents in the EU only Holland, Ireland and Sweden have (slightly) lower incidence rates than the UK when comparisons were made on the basis of 2003 data.

Tables 1.4–1.6, also based on HSE-published statistics, show the most frequent types of accidents causing death, major injuries or over-3-day injuries in 2004/2005. (This data was provisional at the time of writing.)

In June 2000 the HSC and the Department for Environment Transport and the Regions (as it then was) launched their *Revitalising Health and Safety* strategy (DETR, 2000). This acknowledged that though there had been steady progress since the passing of the *HSWA* the reduction in levels of accidents was slowing.

Table 1.5: Non-fatal major injuries to employees by accident type, 2004/2005

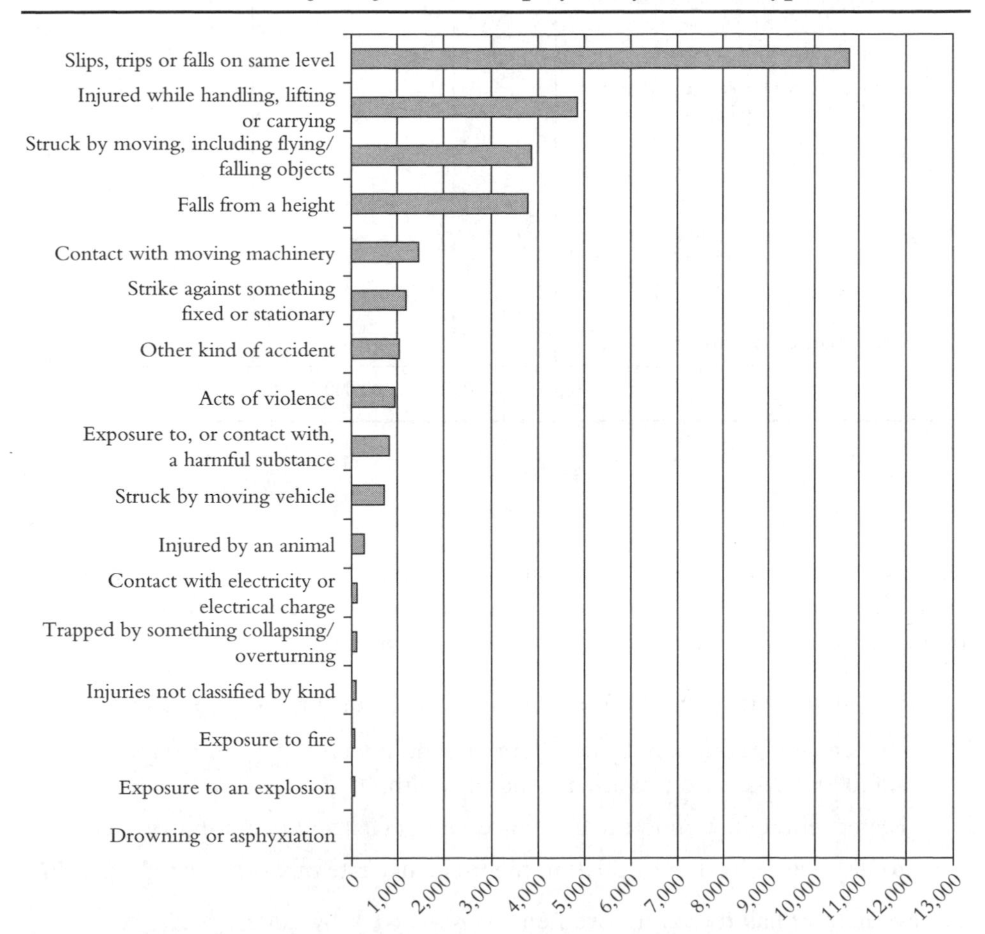

Table 1.6: Over-3-day accidents to employees by accident type, 2004/2005

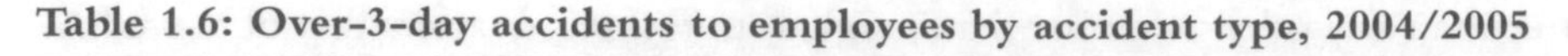

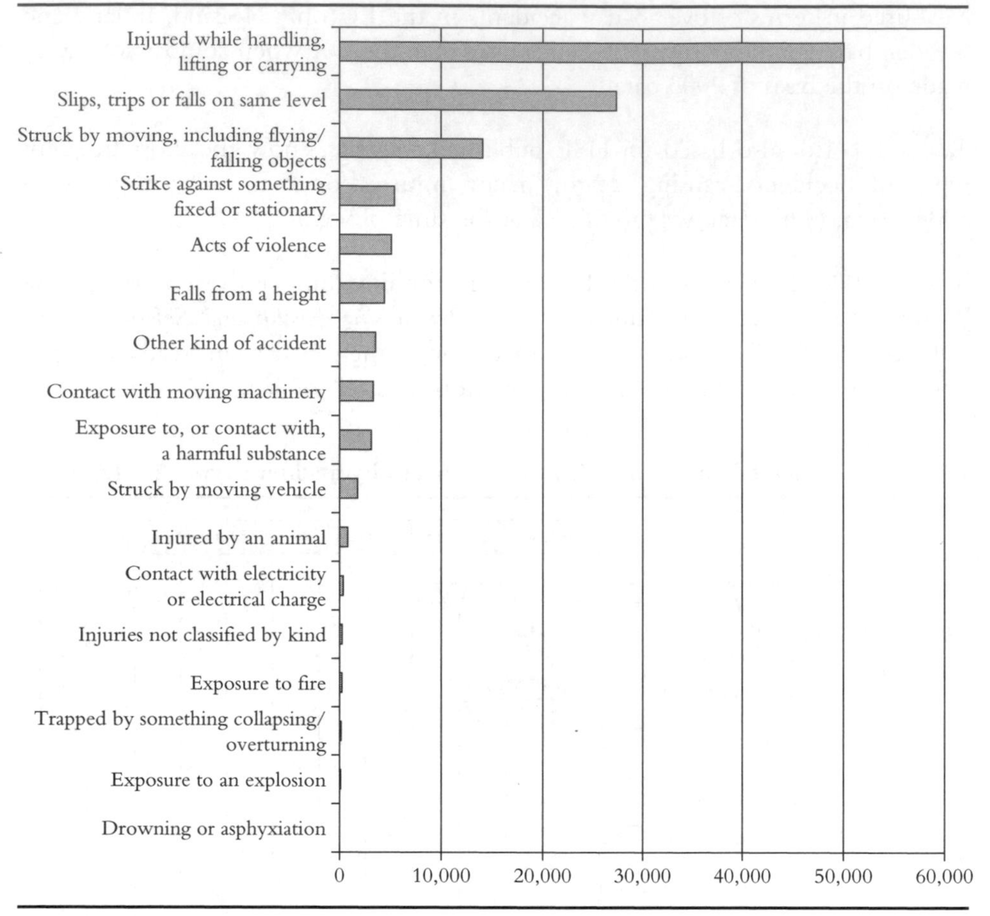

The reasons for this are complex; in part it is because the "easy gains" have already been taken with the shift in economy away from manufacturing to services and the regulatory environment having reached a mature state.

The *Revitalising* strategy in 2000 set national targets for three key improvements:

1. 30 per cent reduction by 2010 in the number of working days lost per 100,000 workers from work-related ill health.
2. 10 per cent reduction in the incidence rate of fatal and major injury accidents.
3. 20 per cent reduction by 2010 in the incidence rate of work-related ill health.

Achievement of half these improvements was aimed at by 2004.

In 2004/2005 – the mid-point of the strategy – progress was disappointing. Taking into account year on year fluctuations there was a downward trend in working days lost, but no clear change for all three targets since 1999/2000 (HSE, 2005). Nevertheless, as will be seen in Chapter 10 the strategy has been successful in part in making performance targeting more a prominent feature of health and safety policy and driving safety up the agenda for senior executives. The largest companies are now routinely reporting publicly on their own individual targets for safety improvements and accident records.

Risk management

1.3 It is now widely accepted that management controls need to be supported by a defined policy set by management at the highest level, and directors of companies (and equivalent officers of other organisations) as corporate manslaughter and new duties have increasingly been in the spotlight are well advised to take a close interest in the implementation of the policy. The published guidance from the HSE emphasises their roles of providing leadership and being accountable (HSE, 2002).

The recommended structures for managing the controls are those of the HSE's guidance document *Successful Health and Safety Management* (HSE, 1997) or the closely related model given as an alternative in BS 8800 (BSI, 2004) and OHSAS 18001 (BSI, 1999) (both produced by the British Standards Institution). The aspects of the safety management system which need to be developed specifically for accident management purposes are:

- emergency planning;
- first aid provision;
- accident investigation;
- appropriate record keeping;
- active and reactive monitoring;
- provision of rehabilitation programmes;
- reviewing and auditing performance.

Most of these requirements (an important exception being rehabilitation) are subject to legal requirements. It is important that employers are aware of the relevant HSC Approved Codes of Practice (ACoPs), HSE publications and other authoritative sources of guidance on this legislation. These can be influential because they provide evidence of how the goals can be met, and they may affect what it is "reasonably practicable" for an employer to do in cases where this is the way in which a relevant statutory duty as framed. As these are priced publications there is little alternative to filling the bookshelves with this important material unless one signs up to on-line subscription services such as *HSE Direct* which in a few years have become very comprehensive in the information they can provide.

The management tools which can support these controls include legal compliance reviews, auditing performance and benchmarking, and different forms of monitoring aimed at discerning trends and underlying causes which can be used to inform the process of planning, revising risk assessments and reviewing the adequacy of the control measures.

Risk management is attracting greater interest generally as a result of changes in attitudes for corporate governance and director's duties. The guidance for directors on the *Combined Code on Corporate Governance* for companies listed on the stock exchange (usually referred to as *Turnbull*) issued in 2003 but now updated (FRC, 2005) stated that boards of directors should understand properly the nature and extent of risks facing companies, and review systems of control at least annually including financial, operational and compliance controls, and should make an annual statement on internal control. Health and safety risks are expressly included in the guidance as being relevant for those purposes. As we will see in Chapter 10 company law and best practice are increasingly embracing these principles.

The cost of accidents

1.4 Another of the objectives of the HSC and the HSE has been to raise awareness among managers of the costs of accidents, not just the immediate costs of damages, replacement of equipment and criminal fines but the term economic cost to an organisation. For example, even a quite minor accident involving for example a person slipping on a loose stair carpet can involve substantial management time being incurred. This could involve:

- first aid provision;
- arranging transport to a doctor or hospital for treatment;
- engaging a temporary replacement worker;
- interviewing the injured person and colleagues to establish the facts;
- checking security video's records and making sure any film is not deleted;
- completing accident book and statutory accident reporting tasks;
- preparing an accident investigation report;

- payroll action to deal properly with sick pay;
- notifying insurers and/or regulatory agencies, completing their questionnaires and dealing with related correspondence;
- reviewing maintenance procedures and devising new instructions; dealing with telephone inquiry from local health and safety regulatory authority to ascertain the current situation;
- briefing the legal department/solicitor;
- meeting regulatory authority inspectors, accompanying them on an inspection of the office and providing details of the steps taken to avoid similar accidents.

At the other end of the scale, the costs of dealing with a major disaster can be enormous. For example, after the Southall rail crash of September 1997 where an express train passed a signal at danger and collided with freight vehicles in its path, the train company involved, Great Western Trains, was subject to investigation for over two years by the police and HM Railways Inspectorate. A major public inquiry was held into the causes of the accident and the train protection systems which should be available for avoiding collisions. Numerous parties from across the rail industry took part who, along with victims and the bereaved, had to be legally represented. Approximately 100 employees and managers of Great Western and Railtrack were required to give evidence to the public inquiry. Later, in criminal proceedings, Great Western and the train driver were acquitted of manslaughter but the company was nevertheless fined £1.5 million for contraventions of the *Health and Safety at Work etc Act 1974*. In addition there were massive claims for loss of life, serious injury, lost rolling stock, damage to track and infrastructure and disruption of the network. The total cost of the tragedy to all involved has never been calculated.

Accidents are reported by the HSE to cost British employees somewhere between £3.7 and £6.4 billion per year in lost wages in 2001/2002, and losses to employers of £3.9 to £7.8 billion (HSE, 2004). An example given by the HSE is an injury caused by working with an unguarded drill which cost a small engineering company £45,000, not counting the fine and costs of being prosecuted subsequently. As a rationale for accident prevention being on the management agenda, the cost of accidents has obvious attractions for promotion by the health and safety authorities: it is a rare example of them being able to point to positive economic gains which can be attained in a short-time frame.

As will be seen in later chapters there are certain tensions between on the one hand the demands of health and safety management and on the other, commercial drivers and incentives and liability considerations. The economic case is seen to ultimately appeal to shareholders and other funders and stakeholders who exert influence on directors and senior managers. One of the latest contributions in this direction has been the HSE's on its "Ready Reckoner" homesite www.hse.gov.uk/costs/index.asp for illustrating accident and incident costs. There is no substitute however for studying one's own costs of injuries, ill health and other accident outcomes to gauge the full impact on the bottom line.

Learning from bad experiences

1.5 Accident investigations in particular are a subject of increasing importance. In 2001 the HSE undertook a consultation on proposals for a new statutory duty to investigate accidents, as there is no explicit statutory duty to do this and it was seen by some as an important gap in the existing legislation (HSE, 2001a). Although previous consultation had seemed to indicate widespread support for a duty to carry out investigations which are proportionate to the scale or complexity of an incident, the HSE eventually announced in 2003 that new regulations would not be taken forward. Instead, new HSE guidelines were drawn up. The Chairman of the HSC was quoted as saying:

> "We want people to learn the lessons from work-related incidents with the potential to cause injury and ill health so that they can prevent similar occurrences in the future. We recognise that some employers need help to tackle this issue, so we are preparing a range of guidance material. We will monitor the effectiveness of this guidance closely – and if there is no improvement in incident investigation then we may consider the possibility of recommending new legislation."

From a risk management perspective good accident investigation techniques are highly desirable. Careful gathering of information from a variety of sources about accidents and "near-misses" builds an understanding of the underlying as well as immediate causes of accidents, helps prevent recurrences and can reduce future legal liabilities and damage to employment relations. It is also important for monitoring trends, reviewing the effectiveness of safety policy and prioritising the commitment of resources and effort in accident prevention. The process of investigation does not always though sit easily with concerns for liability or disciplinary procedures which may face individuals or the organisation collectively after an accident, particularly a serious one. Investigation ideally needs to be carried out with total objectivity, active participation and without seeking to identify or apportion blame. The threat of criminal penalties or dismissal for misconduct is far from being conducive to these goods and it is another example of non-aligned, or even competing, policy objectives in accident prevention.

Insurance and benefits

1.6 Insurance arrangements and social security are used to spread the financial risks involved in accidents. It might be thought that the insurance market promotes accident prevention policies, since better risk management should make it easier for organisations to obtain cover at the most competitive premium rates, and conversely organisations with poorer records and safety management systems would be penalised. In practice the market has not always been able to operate this efficiently.

Employers' liability insurance has been compulsory in the UK for over thirty years, it has historically been loss-making for insurers. 2003 saw a crisis in this area of the insurance market and others (such as public liability) as capacity shrank and premiums rose dramatically after five years of underwriting losses totalling £761 million. There were various reasons for this: among them the weak balance sheets of insurers after the decline in stock markets after 2001 and the long-tail liabilities for asbestos. Even employers with good claims records struggled to obtain cover. In response to a government consultation exercise in 2003 on the case for reforming employers liability insurance the Association of British Insurers argued that no one benefited from the present system and that "claimants, employers, insurers and the public interest are all being short-changed". The fundamental problem however is that the availability of universal insurance and benefits provides little financial incentive to the creator or "owner" of the risk to guard against large financial losses. Unless there are means of rewarding those who manage risk effectively (something akin to the motor insurance no-claim bonus) and means of re-distributing the financial risks more effectively to those who are responsible for the risks to health and safety (insurers' health and safety audits and individually assessed premiums) insurance may have little impact as a driver.

The Government began to address these issues in the 1990s when social security benefits paid to accident victims were clawed back from defendants to personal injury claims. Even so, continuing benefits would be payable in cases of long-term incapacity of accident victims, and in the present context, the re-distribution of cost is still largely borne by insurance.

The "*Revitalising*" strategy in 2000 identified the need to re-design benefits and insurance systems to give positive motivation for improved health and safety performance. In practice this is likely to mean closer examination in future of ways in which reduced premiums or other credit can be given to organisations on the basis of their individual risk profiles and accident records, particularly where they are able to offer assurance that they might recognise minimum standards of risk management and safety management. Some notice may be taken of new legislation introduced by Belgium in 2006 which will require businesses with accident rates and severities which are above certain thresholds (at least five times the national average) to pay a *cotisation de prevention forfitaire* or fixed preventative contribution to their insurance companies of somewhere between €3,000 and €15,000. The insurer will be required to spend the contribution on accident prevention measures by the employer concerned.

Rehabilitation

1.7 Another fertile area for development which employers and the employers' liability system historically not supported as fully as it might is in the provision of services and support for the treatment and rehabilitation of accident victims.

The TUC has called for a National Rehabilitation Committee to be set up to oversee the broadening of rehabilitation provision and this has been welcomed by insurers. Insurers and organisations representing lawyers who deal with personal injury claims have also publicly supported the *Rehabilitation Code* (BICMA, 2002) which is designed to encourage early assessment of whether an insured person would benefit, and the production of an independent rehabilitation needs report which it is agreed neither side will seek to use in any subsequent litigation. While there has been recognition of the value of rehabilitation, and large employers increasingly view this as a natural part of occupational health services they provide, it is fair to say that most personal injury claims still proceed without much serious consideration rehabilitation the full benefits are not yet being seen (IUA, 2002).

As will be seen in Chapter 7 the issue of rehabilitation has begun to come to the fore as the Disability Discrimination Act 1995 has affected employment law. The need to retain and if necessary make reasonable adjustments for employees suffering from work-related injury or ill health now needs to be considered very carefully to avoid breaching this Act. Reasonable adjustments could comprise:

- allowing an employee time off for rehabilitation or treatment;
- making adjustments to premises;
- making equipment modifications;
- changing the employee's normal working hours or allowing home working;
- allocating some of an employee's duties to another colleague;
- transferring the employee to fill a vacancy.

Accountability or compensation culture?

1.8 Pressure has been growing for more and more accountability for employers, especially in the wake of major transport accidents in the UK. This is by no means a modern phenomenon. The course of the industrial revolution can be charted by the gradual creation of new legislation and the adoption of the common law aimed at providing basic protection under the law for employees and the public. Health and safety legislation has extended since to issues such as openness with employees and safety representatives, disclosure of information to the authorities (often without safeguards against self-incrimination), obligations to notify the authorities soon after an accident has occurred, and other obligations to co-operate with inquiries and investigations. Nowadays calls for more accountability are in reality targeted much more focused agendas of personal in terms of accountability of senior managers, and calls for harsher penalties which inflict lasting effects on companies' finances and reputations. In 2006 these emerged as a major policy issue with the publication of a major consultation on proposed new penalty regimes (Macrory, 2006) considering with the introduction of the corporate manslaughter Bill into Parliament which many MPs wished to see create new offence for directors too.

The notion of accountability – and how it can be increased and made more visible – underlines much of the "*Revitalising*" strategy. This has led for example to the expansion of the HSE's "name and shame" website so that it publicises details of not just conviction and fines, but also improvement notice and prohibition notices served by the HSE. Another development with less punitive objectives has been the guidance published and actively promoted to large companies by the HSC (HSE, 2001b). This guidance advises publication by companies on:

- health and safety policy details;
- significant risks and the systems in place to control them;
- health and safety goals, with reasonable targets;
- a report on progress towards the goals in the reporting period;
- data on health and safety performance;
- details of any fatalities and the steps taken to prevent any recurrence;
- numbers of work-related health problems first reported in the reporting period;
- details of enforcement notices and convictions, and remedial action taken;
- the total cost of work-related illness in the reporting period.

Some other significant changes taking place in the litigation system have however caused widespread concern about accountability being taken to extremes, with someone having to be found at a fault for every misfortune which life brings. Measures directed towards reforming and making the civil justice system more accessible have been beneficial to many people, but have come at a cost. Pre-action protocols which require defendants to incur costs investigating claims that may be of little value, the relaxation of restrictions on lawyers, advertising by unregulated claims companies, having a stake in the success or failure of their customers cases, and the introduction of lawyers' success fees as a new litigation funding mechanism have given rise to fears of the appearance of US-style speculative litigation practices starting to take root.

Views have polarised on whether these changes have in reality led to a "compensation culture". The TUC points to data which indicate that surprisingly less than only around 10 per cent of people who are injured at work actually receive compensation. The government points to numbers of claims reaching court or being notified to the Compensation Recovery Unit (which is meant to be obligatory) which went down by 5.3 per cent between 2000 and 2005.

Prime Minister Tony Blair recognised the elusive nature of the truth of the problem in a speech in May 2005 where he pointed out the UK's favourable cost of claims compared to most countries and said:

> "... But the facts too often do not prevail. You may recall the stories of the girl who sued the Girl Guides Association because she burnt her leg on a sausage or the man who was injured when he failed to apply the brake on a toboggan run in an amusement park. Neither of these cases produced big compensation awards in the courts. But this is not the impression that is left. The headlines have an after-life. They leave behind the sense that, not only are such cases being brought all the time, but that huge sums of money are being wasted. This impression, in turn, has genuine effects. Public bodies, in fear of litigation, act in highly risk-averse and peculiar ways. We have had a local authority removing hanging baskets for fear that they might fall on someone's head, even though no such accident had occurred in the 18 years they had been hanging there. A village in the Cotswolds was required to pull up a seesaw because it was judged a danger under an EU Directive on Playground Equipment for Outside Use. This was despite the fact that no accidents had occurred on it."

Perceptions of more active enforcement activity and frequent prosecutions – again not entirely borne out again by the facts as shown in published statistics – have also fuelled disquiet and popular criticisms of the "elfansafety brigade". In 2006 the

Chairman of the HSC Bill Callaghan was driven to join the debate, when in launching a revised risk management guide he stated publicly that:

> "we must, and will, promote the sensible management of risks that protects people from real harm and suffering, but avoids bureaucratic back covering. My clear message is that if you are using health and safety to stop everyday activities – get a life and let others get on with theirs."

References

BICMA (2002) *The Rehabilitation Code, Early Intervention and Medical Treatment in Personal Injury Claims*. The Bodily Injury Claims Management Association. www.bicma.org.uk/rehabilitation

BSI (1999) *Occupational Health and Safety Management Systems – Specification*, OHSAS 18001.

BSI (2004) *Guide to Occupational Health and Safety Management Systems*, BS 8800.

DETR (2000) *Revitalising Health and Safety*. www.hse.gov.uk/revitalising/strategy.pdf

FRC (2005) *Internal control – Revised Guidance for Directors on the Contained Code (Financial Reporting Council)*. www.frc.org.uk

HSE (1997) *Successful Health and Safety Management*. HS(G)65. HSE Books.

HSE (2001a) *Consultation Document – Proposals for a New Duty to Investigate Accidents, Dangerous Occurrences and Diseases*. www.hse.gov.uk/consult/condocs/cd169.htm

HSE (2001b) *Revitalising Health and Safety: Health and Safety in Annual Reports*. www.hse.gov.uk/revitalising/annual.htm

HSE (2002) *Directors' Responsibilities for Health and Safety (INDG 343)*. www.hse.gov.uk/pubns/indg343.pdf

HSE (2004) *Interim Update of the "Costs to Britain of Workplace Accidents and Work-Related Ill Health*". HSE Economic Advisors Unit. www.hse.gov.uk/statistics/pdf/costs.pdf

HSE (2005) *Achieving the Revitalising Health and Safety targets*. Statistical Progress Report, November 2005. www.hse.gov.uk/statistics/pdf/prog2005.pdf

IUA (2002) *Report of the Rehabilitation Working Party*. www.iua.co.uk

Macrory (2006) *Regulatory Justice: Sanctioning in a Post-Hampton World*. www.cabinetoffice.gov.uk/regulation/documents/pdf/macrory060524.pdf

2 Accident causation

In this chapter:

Definitions and terminology

Dictionary and technical definitions

2.1 Definitions of incidents and accidents vary widely according to industry and sector: an incident in one industry may be regarded as an accident in another. Further, definitions of incidents and accidents vary greatly according to whether the terms are used in everyday life, or whether the terms are used in a technical sense by health and safety practitioners or managers, reliability engineers, lawyers or safety psychologists. Before considering what causes accidents, it is necessary to spend some time considering what they are.

The *Concise Oxford English Dictionary* defines an *incident* as "an event or occurrence; an instance of something happening" (2004). This definition says only that an incident is a circumstance in time, and nothing about whether the incident is noteworthy or insignificant, unexpected or anticipated, how the event came about, and whether there was the potential for serious harm.

The same source defines an *accident* as "an unfortunate incident that happens unexpectedly and unintentionally; something that happens by chance or without apparent cause". In this case, the definition indicates that an accident is a random and unexpected negative event.

Neither of these definitions would be accepted by a technical specialist, because they do not provide sufficient information on the true nature of incidents and accidents.

In a technical sense, and especially within safety critical industries and/or high reliability organisations, these terms have more specific meanings. As such, the definitions used often incorporate detailed criteria which a given event must meet in order for the term to apply. For example, the International Civil Aviation Organization (ICAO) provides an internationally agreed convention for investigating civil aviation accidents (2001). The ICAO convention contains a whole chapter on definitions, and the definition of an aviation accident is as follows:

> "An occurrence associated with the operation of an aircraft which takes place between the time any person boards the aircraft with the intention of flight until such time as all such persons have disembarked, in which:
>
> (a) a person is fatally or seriously injured as a result of
> - being in the aircraft, or
> - direct contact with any part of the aircraft, including parts which have become detached from the aircraft, or
> - direct exposure to jet blast,
>
> **except** when the injuries are from natural causes, self inflicted or inflicted by other persons, or when the injuries are to stowaways hiding outside the areas normally available to the passengers and crew: or
>
> (b) the aircraft sustains damage or structural failure which:
> - adversely affects the structural strength, performance or flight characteristics of the aircraft, and
> - would normally require major repair or replacement of the affected component,
>
> **except** for engine failure or damage, when the damage is limited to the engine, its cowlings or accessories: or for damage limited to propellers, wing tips, antennas, tires, brakes, fairings, small dents or puncture holes in the aircraft skin: or
>
> (c) the aircraft is missing or is completely inaccessible"
>
> ICAO (2001, p. 1-1)

Similarly, the aviation industry defines an incident as "an occurrence, other than an accident, associated with the operation of an aircraft which affects or could affect the safety of operation" (ICAO, 2001, p. 1-1). This definition of an incident suggests that it is an event which may have had the *potential* to be more serious.

In transport, the Air Accident Investigation Branch (AAIB), Marine Accident Investigation Branch (MAIB) and recently established Rail Accident Investigation Branch (RAIB) draw distinctions between incidents and accidents in order that investigative resources can be allocated accordingly. Accidents are certainly investigated by the relevant authority, while incidents *may* be investigated by the appropriate authority if the occurrence is relevant to the entire sector. Incidents are more likely to be investigated locally, probably by the airline, shipping company or train operating company concerned. Near misses, or "close shaves", are often unreported in many industries, but in the aviation, marine and rail industries there is the option of reporting them to a confidential or anonymous reporting scheme. Such schemes can be useful where personnel want to report a safety concern or near-miss experience, but do not want to draw the attention of line management to their actions, for understandable reasons.

Health and Safety Executive definitions

2.2 The Health and Safety Executive (HSE) also uses specific definitions of incidents and accidents. The HSE uses the term "adverse event" to include both accidents and incidents. An accident is defined as "an event that results in injury or ill-health" (HSE, 2004, p. 2), while an incident is classified as a near miss or an undesired circumstance. A near miss is "an event that, while not causing harm, has the potential to cause injury or ill health" (HSE, 2004, p. 2). Reportable dangerous occurrences under RIDDOR are defined as near misses in this context. An undesired circumstance is defined as "a set of conditions or circumstances that have the potential to cause injury or ill-health" (HSE, 2004, p. 2). The HSE definitions focus on the potential of an event to cause ill health, injury or death.

Near misses, incidents and accidents

2.3 It is clearly reasonable to provide definitions based on drawing a distinction between an event that harms a person, or a number of people, and one that damages only plant or equipment. It is possible that events may occur which cause significant damage to plant and equipment, and yet injure no one. There may still be advantages for an organisation in classifying such events as accidents, and investigating them thoroughly; it may be possible to take preventative action and avoid similar failures or damage in the future.

Investigations of more minor incidents and "near misses" – events that may be regarded as a "close shave" – may also be useful. Both incidents and near misses may provide valuable information if investigated, because the aetiology of near misses, incidents and accidents is believed to be similar. That is, these events are often believed to have similar causes. However, minor events occur much more frequently than major ones, and therefore investigation of near misses and incidents may allow for corrective action to be taken *before* a similar set of circumstances leads to a more serious accident. Some authors would suggest that such events provide "free lessons" in organisational safety and safety deficiencies. This is why confidential reporting schemes are used in several industries to encourage reporting of near misses and incidents. Such schemes do have their limitations, but they may help organisations to capture this type of safety information in an anonymous format, without reliance on the formal management reporting schemes which employees can be reluctant to engage with.

As an example of the relative frequencies of near misses and incidents to accidents, consider the data in Figure 2.1. This shows the ratio of major injury accidents to minor injury accidents and non-injury accidents, from HSE statistics.

Figure 2.1: Relationship between major injuries, minor injuries and non-injury accidents (adapted from HSE, 1997, p. 8)

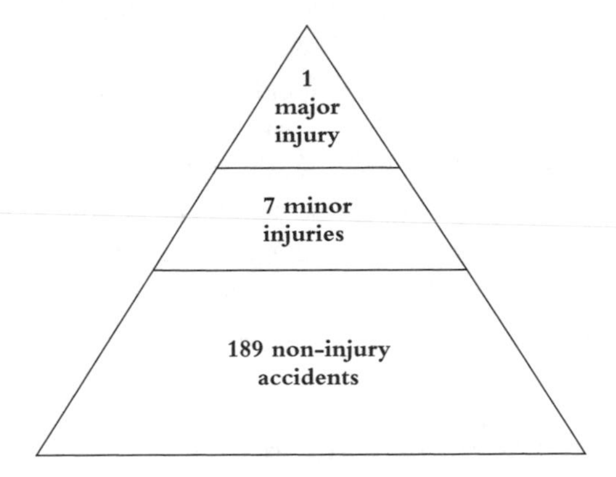

In reality, whatever definition is selected, it is clear that accidents are events with undesired outcomes, which could include injury or fatality to persons, and/or damage to plant and/or equipment. So, in this chapter, the term "accident" will apply to events which meet this criteria, whether they might be known as incidents or accidents in other settings, and whether they may be called individual, organisational or system accidents in other texts.

Accident outcomes

2.4 Defining accidents in terms of direct outcomes such as injuries, fatalities and losses to equipment and assets oversimplifies the "result" of an accident. It must always be remembered that, in addition to the injuries, fatalities, and loss and damage of physical assets, there are a wide range of additional outcomes associated with accidents. Generally, the more serious the accident, the wider the repercussions, but even relatively "minor" accidents can have profound personal effects.

Following accident involvement, an employee will suffer loss of confidence, and his or her relationship with his or her colleagues and supervisor will shift. If an injury has been sustained, then there may be long-term workplace consequences. Economically, injury and ill health can place additional financial burdens on individuals and families, at a time when these resources may already be stretched. Changes in mobility, fitness or health resulting from the accident will also influence life satisfaction and well-being. Behavioural and psychological changes resulting from the accident may be of long-term consequence. The nature of family relationships and relationships with friends may change (Adams et al., 2002; Cormack et al., 2006; HSE, 2006).

For the organisation, additional negative outcomes include the direct economic costs of loss of working time and production; the loss of future contracts and possible threats to the continuity of the business; increased insurance risks and premiums; and loss of reputation and goodwill, all of which can be difficult to quantify. In addition, and depending on the severity of the accident, there may also be far wider social and environmental costs which result. With very serious accidents, the social and environmental consequences can be extreme, and may persist for many years, or even decades.

Once an event has occurred, and depending on the timescales involved, there may be time to mitigate for an unfolding accident. Fires can be put out, plant can be shut down and people can be evacuated. These actions may reduce the severity of the outcome of the accident, but they do not necessarily downgrade it to a more minor event – the *potential* for very serious outcomes was always present. This is another reason why defining an accident based solely on its outcome can be misleading, because doing so fails to take into account any deliberate interventions taken to minimise the impact of the accident. Primary safety involves reducing the probability of accidents occurring in the first place. Secondary safety initiatives are concerned with reducing the probability of harm and damage from accidents once they have occurred. Both primary and secondary safety are vital in minimising overall losses from accidents, and hence accident management and emergency planning strategies are also important considerations.

Accidents and chance

2.5 More seriously, lay definitions of the term "accident" imply that these events are somehow caused by blind or random chance; as if they were an act of God, a total fluke, a freak occurrence. In fact, this is often far removed from the truth. Although every accident is unique, the study of accidents in a wide range of sectors has shown that there are certain commonalities, and it is therefore possible to formulate some clear conclusions about organisational and system safety.

Table 2.1: Chance, outcome and public perceptions

A specialist radiation contractor was engaged to transport decommissioned equipment from a cancer treatment unit in Leeds to Sellafield for disposal. A specially constructed 2.5-tonne container was used to carry the contaminated material on a lorry for 130 miles. Travelling by road for 3 hours, the material reached Sellafield, where it was discovered that radiation levels were up to 1,000 times greater than what would normally be considered a very high dose rate.

Investigation revealed that a shield plug – a vital part of the approved packaging – was missing from the flask. The radiation fortunately took the form of a narrow beam, which had been directed towards the ground throughout the trip. Had it been emitted horizontally, dangerous radiation would have been emitted for 980 feet from the flask.

It was "pure good fortune" that no one was exposed to dangerous radiation levels (BBC, 2006a). The HSE and Department for Transport brought a prosecution because they regarded it as a serious incident. Clearly, the outcome could have been much worse – and would have "upgraded" the event to an accident. However, the *potential* for harm was always present, and only chance prevented this situation from being much more serious.

Public perceptions of nuclear risks are typically influenced by the "unknown factor" and the "dread factor". The unknown factor describes the extent to which radioactive effects are unobservable and delayed, and the extent to which the risks are unknown to science. The dread factor relates to the belief that nuclear risks are fatal, uncontrollable, catastrophic, and will remain involuntarily present for future generations. Incidents and accidents in the nuclear industry are regarded as particularly "doom-laden" by the general public (Slovic, 1987). Hence, reports that contractor employees had a "relaxed and cavalier" approach would have been of particular concern to the general public.

One of the key findings has been that there is very rarely a *single* cause of an accident. A single act, event, omission or failure very rarely "causes" an accident. Accidents are complex combinations of multiple causal and contributory factors. A causal factor is a factor that has a direct, one-to-one link to the accident. If a causal factor is removed from a sequence of events, the accident does not occur. A contributory factor is a factor with a probabilistic relationship to the accident – removing a contributory factor reduces the chances that an accident will occur, but does not directly prevent it. Hence, accidents are not random, but are the culmination of a complex series of causal and contributory acts, events, omissions and failures.

Some of the failures which cause an accident will be active failures, meaning failures at the front end of operations, which often have an immediate effect. Others will be latent failures, "resident pathogens" (Reason, 1990), which lie dormant, unnoticed and unrecognised, until an unforeseen circumstance reveals the gap in the safety defence. Typically, the unforeseen and unpredictable interactions between events and circumstances mean that it simply would not have been possible to predict the accident in advance (Perrow, 1997).

In reality, if it were possible to predict the nature, outcome, location and timing of an accident with any degree of accuracy, then organisations and systems, and the people who work within them, would never have them in the first place. Unfortunately, the advantage of hindsight, which is conferred automatically on all accident investigators by virtue of their role, sometimes obscures this very simple fact.

In addition, luck does play a role in determining the outcomes of adverse events. It may be a cliché, but luck comes in two flavours: good luck and bad luck. The causes of an incident and a more serious accident may be identical, but perhaps because of a stroke of good luck, the more serious accident outcomes were averted. As an example, consider a failure to the main rotor gearbox on a helicopter. If a helicopter's main rotor gearbox fails a few seconds after take-off, then the outcome will probably be a premature landing, with minimal damage and minor injury. The same failure, with absolutely identical causes, but occurring at 6,000 feet above the North Sea during winter conditions, will have very different and far more serious outcomes.

If it is merely providence which has prevented an incident from becoming an accident, then there is clearly a case for conducting an investigation. "Incidents that by chance fall short of developing into major accidents should attract an equal intensity of investigation if they are to serve as sources of insight into causes and allow future accidents to be prevented that may not benefit from the same fortuitous chance" (RAE, 2005, p. 7).

Accident phases

2.6 Finally, it is also clear from the study of many accidents that they move through a common set of phases. Although the details will vary depending on the regulatory context, sector, organisation and individuals involved, accidents begin with an initiating event(s). This event is what starts the accident, and it can be a technical failure or an unsafe act of a person. Sometimes, a series of technical failures and unsafe acts are the immediate causes. These are the issues which will typically occupy lawyers for months after the event, prosecuting and defending, and determining fault and liability to address compensation and disciplinary issues.

As already mentioned, there is sometimes a short period after the initiating event(s) in which the situation could potentially be redeemed. This is known as the "amelioration phase". If this opportunity is available, it depends critically on the operator or supervisor recognising that something is amiss, making the correct diagnosis and instigating the correct recovery processes. Given that the time available may be limited, and that the situation confronting the person may be highly unfamiliar, this may be highly unlikely. The danger release phase of the accident is where the accident "happens", which may or may not be the same moment in time as the harm release phase. The harm release phase is where damage to people, plant and assets actually occurs – this may occur shortly after the danger release phase, or it may begin (and continue for) some time later.

Early theories of accident causation

Domino theories of accident causation

2.7 One of the earliest formal theories of accident causation was developed by Heinrich in 1931. He analysed 75,000 accident reports from companies insured with Travelers Insurance Company and developed the "domino" theory of accident causation (cited in Bamber, 2003). Heinrich concluded that 88 per cent of all accidents were caused by unsafe acts of people, 10 per cent by unsafe conditions and that 2 per cent were caused by acts of God (e.g. "natural" accidents).

Based on his findings, Heinrich identified five factors in the accident sequence, shown in Figure 2.2. These are social environment and ancestry, fault of the person (carelessness), unsafe act and/or mechanical/physical conditions, the accident, and finally, the injury itself. Social environment and ancestry included the social learning of custom and practice in the workplace, as might be evidenced in an apprentice learning from his master. The "carelessness" factor included negative personal

Figure 2.2: Heinrich's domino theory of accidents

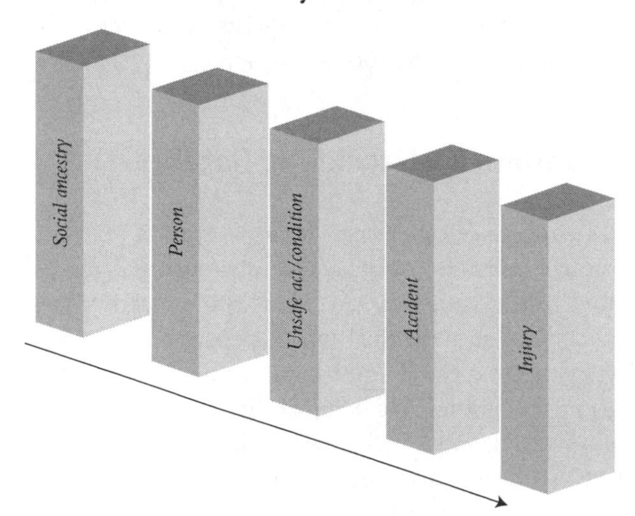

characteristics of the individual; however they might have been acquired. Unsafe acts or physical conditions were the errors or technical failures, which led directly to the accident which resulted in the injury. In essence, removal of the unsafe act or condition would prevent the accident.

Heinrich's theory of accident causation is a simple linear sequence of events. It explains "what happened", but it doesn't provide much information on why the accident occurred. In addition, the model essentially lays accidents firmly at the door of the unsafe act or mechanical condition – without examining any underlying or contributory factors. This early theory has been superseded by more sophisticated theories of accident causation, but Heinrich's terms "unsafe act" and "unsafe condition" are still very much in use.

A more complex domino theory suggests that adverse events have immediate causes, underlying causes and root causes (HSE, 2004). The immediate causes are the actual agents of injury or ill health – the blade, the substance or the fumes. The underlying causes of a workplace accident are the unsafe acts and unsafe conditions which gave rise to the immediate cause, such as where a guard is removed, the window closed or the ventilation switched off. The root causes of the adverse event are the failures from which all other failures stem. For example, a failure to adequately assess risk, incomplete analysis of training needs, lack of monitoring and control. Elimination of underlying and root causes potentially prevents a whole series of adverse events, while addressing immediate causes would only prevent recurrence of the specific adverse event which arose. This model is more sophisticated than that proposed by Heinrich, but many safety and accident theorists would nevertheless dispute the existence of a "root cause" as being too simplistic.

Accident proneness and accident liability

2.8 Heinrich's concept of carelessness almost "blames" the operator for having the injury. The more modern terms of "accident proneness" and "accident liability" are sometimes used in the same way. Accident proneness is defined as a relatively enduring characteristic of an individual which predisposes those possessing it to be involved in a higher than expected number of accidents, compared to other people in the same situation. In contrast, accident liability is the propensity for any individual to be accident involved, and it includes situational, personal and task-related factors. A great deal of research has been conducted into individual differences in accident proneness and accident liability. However, there are many methodological difficulties in conducting research in this area.

One of the main difficulties is in ensuring that comparison groups are equally matched for their exposure to the risk. For example, in examining accident rates among occupational drivers, how can a researcher be sure that an individual's higher accident rate involvement is not simply a result of spending more miles or hours behind the wheel, or of driving different types of vehicle on different types of road? This distinction would be necessary to distinguish accident proneness from accident liability. Further, if accident proneness is a relatively enduring characteristic, then people with high levels of accident involvement at work should also experience higher accident rates in other areas of their life – and controlling for exposures to risks outside work makes research of this nature even less reliable.

A comprehensive review of the research literature on accident proneness recently concluded that "there is little evidence to support a grand theory of accident proneness" (Lawton & Parker, 1998, p. 41). And while accident liability does appear to differ between individuals, this could of course be due to many factors other than "accident proneness": factors associated with the task and the situation will play a role. Research has shown that individual differences in accident involvement fluctuate over time – which has led one author to conclude that the "accident liability" club has very few members, and an ever-changing membership list (Reason, 1990, p. 199).

Because of these issues, strategies which aim to reduce the risk or frequency of accidents by identifying those who are particularly prone or liable to accidents are unlikely to be very effective. If it were possible to reliably and consistently identify "accident repeaters", then organisations would be able to take appropriate action. Possibilities could include selection screening (such as psychometric testing) to avoid offering employment to higher-risk individuals, or retraining (and maybe ultimately redeploying) those personnel with poor accident involvement records. This is known as the "bad apple" theory of accidents, and although intuitively appealing, the reality is not so simple. An example provided by Hopkins (2006) will serve to illustrate this point.

Imagine that a company has 1,000 employees. In any 12-month period, most will have no workplace accidents resulting in injuries, although a small proportion may have one injury (say, 16.4 per cent). Another 1.6 per cent may have two injuries and 0.1 per cent may have three injuries. So in total, 819 employees will have had no injuries (although a high proportion of these may have had near misses or close shaves), 164 will have had one injury, sixteen will have had two and one employee will have had three. Disciplining this one individual as an "accident repeater" or "bad apple" does not take into account the fact that some of the injuries sustained may be due to organisational factors which are beyond the individual's control, or even due to chance. How many of the 819 uninjured employees escaped accidents through luck alone? A much longer period of time would be required to identify "accident repeaters" with any degree of confidence, since the relative rarity of injuries and accidents means that the data are limited, and conclusions drawn from small amounts of data may well be spurious.

Taken to the extreme, this approach would mean that organisations would only eliminate "bad apples" if it could be shown that no organisational or chance factors contributed to their own injuries or accidents, and to the absence of injuries and accidents among their colleagues. In reality, because multiple factors cause accidents, including organisational issues and chance, this is extremely unlikely to be the case. However, it is certainly true that organisations can screen out individuals who report a higher likelihood of risky behaviour. For example, people who report enjoying driving at speed may not be the best recruits to driving vacancies in a road haulage company, since they may have a higher frequency of optimising violations (see page 33) compared to other drivers.

Nevertheless, the *overall* level of risk to an organisation which can be attributed to recruiting such individuals is likely to be small for two reasons. Firstly, not all instances of speeding result in accidents. This means that there is no direct, 1:1 correspondence between speeding and road accidents. If there were, then very quickly everyone would learn never to speed because of the absolute certainty of accident involvement. Secondly, although speeding reduces the time available to the driver to perceive, identify and appropriately respond to hazards, and also potentially worsens the consequences of any accident that does occur, it is often not the sole cause of occupational driving accidents. Additional factors could include poor weather conditions, which could decrease adherence and lead to reduced handling capability, or fatigue and tiredness, which may be related to hours worked, distance driven, time of day and driver scheduling. Further, some drivers may fail to perceive, identify and appropriately respond to hazards even when driving within the speed limits. It is likely that all of us have done this at some point or another, although we may normally attribute the problem to the other driver and not ourselves (see Table 2.2).

Table 2.2: Attribution theory

Scenario 1: Imagine that you are driving in the middle lane on the motorway, overtaking some traffic in the left-hand lane. However, the car in front of you is moving very slowly, and it looks as if he is just sitting in the middle lane. You prepare to pull out, into the outside lane, to overtake him. As you do so, you glance over your shoulder and spot another car in your blind spot, driving very quickly in the outside lane. The driver beeps his horn at you to warn you that he is there, and you quickly pull back into the middle lane, swerving slightly. What is your first reaction? Probably that you don't know where he came from, that he appeared from nowhere, and that you can't believe that he has the nerve to honk at you!

Scenario 2: Imagine that you are driving down the middle lane of the motorway, with a fair distance between you and the car in front. You suddenly notice the car in front pull out, and then swerve violently back into the middle lane as a car in the outside lane speeds past, beeping his horn. What is your first reaction? Probably that the driver in front was an idiot not to check his blind spot, and that there could have been a collision.

In the first scenario, you are the driver who pulled out without checking your blind spot, but you attributed the near miss to the situation you were in. In the second scenario, which describes an identical situation from a different perspective, you probably attributed the near miss to the driver in front. This is a common phenomenon – we attribute our own near misses and failures to the situation we were in. However, the near misses and failures of others we almost invariably attribute to them personally. The reality is that a combination of factors will determine how we behave in a given situation – but we tend to judge other people quickly, without placing enough emphasis on the situation they were in. This is worth remembering when you find yourself thinking that someone is "accident prone".

Human error

To err is human

2.9 Humans make errors very frequently, and there is much truth in the ancient Latin phrase "errare humanum est", translated as "to err is human". Human performance is variable, and sometimes, some of the variations in our performance fall outside some limit of tolerance. No human can be 100 per cent reliable in thought or deed all of the time; it is part of the human condition to mess things up occasionally. After all, if we did not make errors, then we could not learn from our experiences. Any parent will tell you that their children take great pains to ignore all well-intentioned advice, precisely so that they can go off and make their own mistakes.

We make errors in every field of human endeavour, and not just in novel and unexpected situations. We frequently make errors in simple, everyday tasks that we have completed hundreds, if not thousands, of times before. The consequence of most of our errors is minimal, and in many cases, perhaps the worst outcome is that we might actually have been observed, that someone may have seen our failure. Even then, the consequence may be nothing more than mild embarrassment: perhaps a collapsed soufflé at a dinner party, or stalling the car at a set of traffic lights. However, some human errors become rather more public: Table 2.3 provides some examples of everyday human errors that made the news.

Types of human error

2.10 If you drive and have ever hired a car, the chances are that that you have switched on the windscreen wipers when you really intended to use the indicator to signal your intention to make a left or right turn. The first time you do this, assuming that you did not read the user's manual before driving off in the hire car, you have made a *mistake*. The action was *intended*, because although you did not intend to make the mistake, you incorrectly thought that the location of the indicator control on the hire car would be the same as on your own vehicle. You will probably make the same error several times during the hire period, but the second and subsequent times you make the error, it will be a *slip*. This is because you now *know* where the correct control for the indicator is located in the hire car, but your experience driving your own vehicle is stronger than your limited experience with the hire car. The more familiar habit will intrude on your intended actions: this is known as a *double-capture slip*.

Slips differ from mistakes because the actions involved are *unintended*. Observation alone will not tell them apart, because in each case the behaviour *looks* identical, and

Table 2.3: Human error in the news

The Swedish state broadcast channel accidentally showed explicit footage of an adult film on a monitor in the background of a news broadcast. The incident occurred because a technician had been watching sport on a cable channel on the monitor and forgot to turn the channel back over. The adult film began broadcasting at midnight and was clearly visible in the background during the news. A spokesperson was "shocked and dismayed" at the "huge blunder" (BBC, 2006b).

A Tokyo trader cost his employers at least £128 million after he mistakenly sold 600,000 shares for 1 yen each, instead of selling 1 share at 600,000 yen (about £3,000). Another trader at a Swiss investment bank lost £71 million in seconds after trying to sell 16 shares for 600,000 yen each – he actually sold 610,000 shares at 6 yen each. Errors such as these are known in financial circles as "fat-finger syndrome", a memorable, if inaccurate, term (Lewis, 2005).

In a letter to 15,000 holders of Individual Savings Accounts (ISAs), a financial institution provided the wrong contact details. Those customers who rang the number listened to disco music, and were then told that they had reached "Britain's best place to meet other men". A spokesperson for the company said "A letter was recently issued to a number of customers which, due to human error, contained an incorrect telephone number. We have put this right and have written to customers to apologise for the error" (BBC, 2002).

A businessman made a return trip to Italy using his wife's passport, which he had picked up accidentally as both passports were kept in the same drawer at their home. In spite of the passport being checked several times during the trip, and twice by UK and Italian immigration officials, no one noticed that he had the wrong passport. The passenger said that he was extremely surprised, as he did not "bear any resemblance to my 5′7″ brown-haired wife" (BBC, 2004).

An electricity company sent an account statement to a domestic customer showing that she owed £2,131,474,163. The £2 billion debt was incurred in spite of the fact that the customer had a prepaid meter. The letter said "Because of this, your meter will be re-set in the next few days so that you can gradually pay off this debt". The company were relieved that the customer saw the funny side of the incident, and blamed the erroneous account statement on "human error" (BBC, 2001).

it is only the intention which differs. It is the intent of the act which will determine whether a given error is a slip or a mistake. Slips are most frequently associated with highly skilled or automatic behaviours, and for this reason they are sometimes known as the "absent-minded professor" syndrome, or more colloquially as having a "senior moment"; they are failures of attention. They are normally made by experienced and competent people, and as such, they are usually detected and corrected quickly, with minimal consequence, although of course there are exceptions.

As well as the double-capture slip, other slips include *behavioural reversals*, where the correct action is performed, but to an inappropriate object. An example would be where you make a hot drink, and throw the teaspoon in the bin and the used tea bag in the sink. These slips are also called *behavioural spoonerisms*, named not after the teaspoon, but after Reverend Archibald Spooner, Warden of New College, Oxford, who made similar verbal reversals. One apocryphal example is his toast to Queen Victoria: "Three cheers for our queer old dean". *Place-losing slips* occur where you can't remember how far you have come in a sequence of events: How many sugars have I put in this cup of tea? *Lost-goal slips* are where the intention itself has been forgotten: I came upstairs to my study to get something, but what on earth am I here for?

In contrast to slips, mistakes are usually associated with highly conscious and deliberate performance on a task. They are more often made by novices than experienced people. Mistakes are essentially planning failures, resulting from an incomplete knowledge of the situation (a knowledge-based mistake), or misapplication of an "if-then" rule (a rule-based mistake). To continue with the previous driving example, the first time you drove the hire car, and turned on the windscreen wipers instead of the indicator, you made a knowledge-based mistake because you did not know where the indicator control was on the hire car. Knowledge-based mistakes occur during the assessment of a situation. They are errors which occur in diagnosing and interpreting a situation because the operator has incomplete or insufficient information to be fully aware of the situation.

In contrast, rule-based mistakes occur once an assessment of the situation has been made, and are normally errors made in formulating a plan to deal with the situation. An example of a rule-based mistake in driving would be misapplication of the rule that it is usually acceptable to turn the steering wheel in the direction that you want the car to move in. There are situations where this rule would not apply, such as where the car is skidding on ice. In these situations, the steering wheel should be turned in the direction of the skid in order to regain control of the car. Rule-based mistakes occur where someone has, or thinks they have, a full understanding of the situation, and invokes a plan or rule to deal with it. The error occurs because the rule itself, normally an "if-then" rule, is inappropriate in the *actual* situation.

As has been demonstrated, mistakes and slips are both errors of *commission*: errors which result from *doing*. Lapses are errors of *omission*, since they are errors which result from *failing to do*. Lapses are essentially memory failures, and they often occur when we forget to do something. Frequently occurring lapses are those associated with behaviours which have to be done in a specified order or sequence. This is the case with procedural tasks such as product assembly, completing a checklist or following a recipe. The longer the sequence of actions, the more likely it is that something will be omitted. Perhaps more worryingly, the longer the sequence the more likely it is that the operator will be distracted, and interruptions may cause slips as well as lapses. Steps which contain more complex information than preceding steps, or which do not follow on logically from them, are also more likely to induce error. In everyday terms, this helps to explain why one may have a couple of screws left over after assembling some flat-pack furniture: a step was probably left out of the procedure.

Violations are another type of human error. Like mistakes, these are intended errors, and they involve deviations from safe operating procedures and standard rules. *Exceptional violations* are the exception, by name and nature. These are wilfully malicious or malevolent extreme acts, usually committed with the intent to cause maximal harm and damage. Sabotage and terrorism are both examples of exceptional violations, and both are extremely rare. The vast majority of violations are non-malicious deviations or adaptations from specified procedures, usually committed for convenience or expedience, and normally motivated by a desire to simply get the job done.

There are three main types of non-malicious violation: the routine violation, the necessary violation and the optimising violation (Reason, 1997). To avoid the more emotive terms "violation" and "deviation", some authors and organisations use the term "adaptation". These terms are used somewhat interchangeably in the following:

- *Routine adaptations* are short-cuts which usually involve a degree of corner-cutting. They are normally made by skilled and experienced operators, usually because they make working life easier. They are essentially "work-arounds" – people will work around the specified procedures where these are perceived to be too clumsy or inconvenient. Examples include "rigging" switches and turning off auditory alarms. These adaptations are called routine because they become the norm for an individual, or for a group of individuals, over a period of time: this is known as the "normalisation of deviance" (Vaughan, 1996). The habitual nature of these adaptations develops where there is no sanction for not following procedures, and no reward for completing the task "by the book". First-line supervisors are clearly the key to ensuring that such short-cuts, work-arounds and corner-cutting do not become the norm. However, this can be very difficult to achieve if supervisors do not fully appreciate how their safety leadership roles require them to use sanctions and rewards in order to reinforce the appropriate behaviours and maintain compliance. To win compliance, the psychological trade-off between safe, inconvenient behaviours and unsafe convenient ones has to be altered.

- *Necessary adaptations* are those which are actually required to get the job done. These adaptations are typically provoked because the site, equipment or tools provided to do the job are simply ineffective given the procedures. Sometimes, necessary adaptations result from a technological lag. When people work with equipment that was designed ten or twenty years previously, it is not always fit for purpose given the current working demands. Sometimes, the problem is with the procedures – they may appear to have been written by people with no appreciation of how the job works in practice. Indeed, there are instances on record where it has been impossible to complete the task using the specified procedures. In situations where the site, tools and equipment are mismatched to the procedures, an individual has no choice but to violate the procedures in order to get the job done.

Table 2.4: Can human error be eliminated?

Some engineers may believe that human error can be eliminated from safety critical systems by introducing improved technology. The hope is that with reduced scope for human input, there will be fewer opportunities for human error. In fact, introducing new technology typically changes the type and nature of the human failure; it does not eliminate it.

A very simple example will illustrate this point: imagine a traditional alarm clock. The clock has a conventional "hammer-and-bell" mechanical alarm, set by use of a twelve-hour wheel on the back of the clock. In setting this alarm to sound at 7.23 a.m., it is likely than an error will occur. It is difficult to set the alarm accurately, so it will probably sound somewhere between 7.18 and 7.28 a.m. Switching to a digital alarm clock, the alarm can now be set with a much higher degree of accuracy, 7.23 a.m. precisely. However, a different type of error has been introduced, because the alarm is now set using the twenty four-hour clock. If the alarm does not sound at 7.23 a.m., it is probably because the operator set it for 7.23 p.m. The new error may be less likely to occur, but it is of a higher magnitude (twelve hours instead of five minutes), and probably of greater consequence (it is harder to make up for lost hours than lost minutes). Has the new technology reduced or increased the risk of waking late?

- *Optimising adaptations* are thrill-seeking adaptations, where people seek to obtain maximum satisfaction from the process of achieving a task. For example, breaking a speed limit in a company car may be against company policy, but some individuals are motivated to speed simply because they enjoy it.

Error promoting conditions

2.11 Certain conditions and factors promote human error; these are also known as *stressors* or *performance shaping factors*. Situational factors which promote human errors include working in sub-optimal conditions. Examples of sub-optimal conditions include poor weather, poor lighting, high noise or vibration levels, extremes of temperatures or humidity or working in a confined space. In such situations, the probability of human error increases. Factors relating to the task, such as how critical the task is and how much time is available to complete it, will also influence the error rate. A task that is complex, novel or unfamiliar will give rise to more errors than a task which is simple.

Completing a task under high stress conditions, such as in an emergency, will increase the probability of error. Poorly designed equipment, or equipment that is unfamiliar, will also lead to a higher incidence of error. Other performance shaping factors may be related to the job instructions and practices; to psychological factors such as current skill levels, motivation, attitude and personality; and to physiological factors such as fatigue, hunger and thirst.

An employee's ability to cope with these factors will depend on his or her level of training and experience. All other things being equal, an individual with less experience will make more errors than one with greater experience. Table 2.5 shows how the level of stressors present in the work environment will increase the error rate of

Table 2.5: Level of stressors and increases in error rates in completing routine tasks

Level of stressors	*Error rate increased by a factor of:*	
	Novice/inexperienced operator	*Skilled/experienced operator*
Very low	×2	×2
Optimal level	×1	×1
Moderately high	×4	×2
Very high	×10	×5

From Miller and Swain (1987), used with permission of the publisher routine tasks, relative to the skill level of the operator at the workface. Error rates will also increase as a function of task complexity.

As can be seen, the error rate for routine tasks increases for both skilled and novice performers when the level of stressors is very low, as well as when the level of stressors is very high. This is because some level of stress is necessary for optimal performance. If there are insufficient demands on the employee, then performance is likely to decrease. This is what happens when a task is monotonous, boring or routine – the employee may well look for other things to occupy his attention and can therefore be distracted easily. Similarly, if there are too many demands, then performance will also decrease, because there will be too much information for the employee to take in, and his attention will narrow. For example, this often happens when a number of auditory alarms sound at the same time – the employee finds it difficult to pay attention to them all, and his performance at identifying the reason(s) for the alarms may be impaired as a result.

This principle, that there is an optimal level of stress or arousal necessary for peak performance, is illustrated in the inverted-U curve shown in Figure 2.3. This principle is known as the Yerkes–Dodson law. It applies universally to human performance, although the actual point of peak performance on the arousal curve for a particular task or work situation will vary between individuals.

Figure 2.3: The relationship between performance and stress or arousal (the Yerkes–Dodson law)

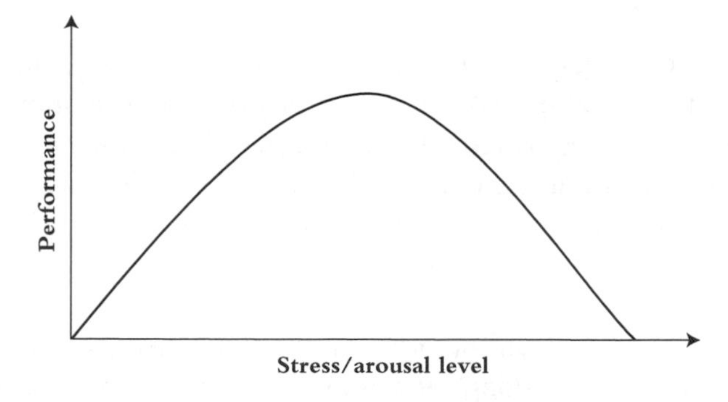

Human error in safety critical industries

2.12 Technically, an error can be defined as "an inappropriate or undesirable human decision or behaviour that reduces, or has the potential for reducing, effectiveness, safety or system performance" (Sanders & McCormick, 1987, p. 606).

This definition is suitable for all workplace errors, including mistakes, slips, lapses and violations. The definition focuses on human decisions and acts, yet still encompasses organisational and system goals. It is important to note that human error is as frequent and pervasive in working life as it is in any other area of human activity. However, as the antithesis of human reliability, it can have serious and significant workplace consequences. Indeed, some authors suggest that human error is a primary cause in as many as 85–90 per cent of accidents (Sanders & McCormick, 1987; Wickens & Hollands, 1999).

Table 2.6: Error and blame

Many safety critical and high-risk organisations have formal error management processes to manage human reliability. These strategies can ultimately reduce both the probability and consequence of errors. For example, in aviation, highly trained assessors sitting in on routine flights can observe which errors occur most frequently. Detailed debriefing sessions are conducted with the pilots to establish the type of error and to investigate why they occurred (Helmreich, 1998).

This information can be fed back to flight crew, to instructors and training personnel, to the safety department, and even to the designers of flight deck displays, controls and software. Managing human error in the workplace is a complex matter, but many organisations in high-risk sectors now take a more "error tolerant" approach than perhaps they once did.

Some organisations openly encourage reporting of mistakes, slips, lapses and associated incidents, without taking punitive action against employees. In order to be effective, such an approach must incorporate a zero tolerance approach to violations, since violations cannot be tolerated by a safety critical or high-risk organisation. The aim of such strategies is ultimately to shift *organisational culture*: to encourage a *just culture* as opposed to a *blame culture*.

A just culture is one in which the organisation acknowledges that human errors are as much a product of working conditions as they are the individual. Individuals are encouraged to report unsafe acts, error promoting conditions and unworkable procedures. These are quickly corrected by management, and the individuals involved are not "blamed" for any errors which resulted from them. In response, employees take individual and professional responsibility for following company procedures and policies to the letter, accepting that serious action will be taken against violations.

In spite of this, it must be remembered that human errors at the workface or operational front-line are usually one-off events. An error may even be a non-event, if it is of no consequence. Errors are occurring all the time, while accidents are extremely rare. Clearly, the vast majority of errors do not trigger accidents. However, there are situations where an error at the front-line combines with latent factors such as supervisory and management errors to cause a serious "organisational accident". This is why human reliability is of major concern to safety critical and high-risk industries: the costs and consequences of human error can be extremely high.

Organisational accidents

The "Swiss cheese" model

2.13 Reason (1990, 1997) developed a more complex linear model of organisational accidents. His theory is currently one of the most influential and widely used theories of accident causation among safety professionals. The model simply suggests that in any organisation, defences are used to prevent hazards from becoming losses. Organisational defences can be "hard" or "soft". "Hard" defences include automatic warning devices and alarms, engineered technical safety features; physical barriers and guards; and protective weak points designed in to the system (e.g. fuses). "Soft" defences are based on personnel and procedures, and include legislative and regulatory requirements; assurance, inspection and checking; standard operating procedures; training and briefing; permits to work; and of course supervisors and operators at the "front-line". Hazards are possible threats to the organisation, which could *potentially* result in a loss. A loss to the organisation is harm or damage occurring to people, equipment, plant and/or assets.

Reason argues that there is always a trade-off between production and the protection afforded to vulnerable people and assets via the organisational defences. If the organisation provides too much protection for the level of risk associated with production, the company will cease to be commercially viable. However, if there is too little protection for the risks associated with production, then the chances of suffering a catastrophic accident, and of going out of business, increase. To maintain some form of equilibrium, the organisation needs to operate at parity level, where the protection provided adequately covers the risks of production. Most rational managers would accept that it is necessary to balance these two competing demands for the long-term survival of the business. However, parity is difficult to achieve in practice. This is because while production output can be measured and therefore managed, protection is usually invisible until it is insufficient. It is impossible to measure accurately how many accidents a safety management system has prevented; only those which occurred can actually be counted.

Unsafe acts and unsafe conditions

2.14 Organisational defences against accidents can be conceptualised as barriers between hazards and losses. However, these barriers and defences have holes in them, just like slices of Swiss cheese. This is why Reason (1997) called his model, shown in Figure 2.4, the Swiss cheese model. The "unsafe acts" slice of cheese represents operations at the "sharp-end", the front-line employees in an organisation. This slice of cheese represents defences within the realm of direct and immediate human activity at the workface. The holes in this slice are unsafe acts (or failures to act), meaning human errors. According to Reason (1997), mistakes, slips, lapses and violations which occur here are active failures, and they are the immediate cause(s) of the accident. Human errors can be immediate causes either singly or in combination: a violation paired with a mistake is a common immediate cause. It should be noted that some immediate causes may be technical; not all are human failures. However, improvements in engineering reliability mean that simple technical failures are very rarely found to be the primary cause of accidents. Conversely, the proportion of accidents with a human cause has correspondingly increased.

Figure 2.4: The accident trajectory (based on Reason's Swiss cheese model, 1990)

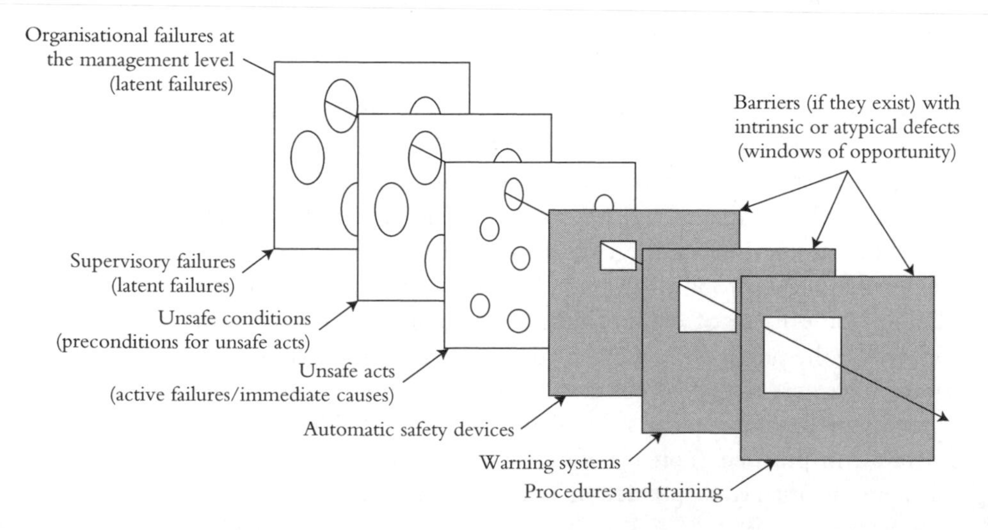

Beyond "unsafe acts", the holes in the next slice of cheese represent "unsafe conditions". These unsafe conditions are unsafe practices and conditions within the local workplace and psychological factors, all of which increase the probability of unsafe acts. These holes are latent factors which contribute to accidents, rather than active failures. In effect, the unsafe conditions are the error promoting conditions and performance shaping factors outlined above. There is a one-to-many mapping between unsafe conditions and unsafe acts, since an unsafe condition can promote many types of human error, and allow them to promulgate.

Table 2.7: Reinforcement of unsafe acts and unsafe conditions

Sometimes, unsafe acts are promoted and reinforced by working practices and working conditions. For example, consider the case of a machine operator on a night shift, who operates a cutting tool with a colleague. Both employees have three years of experience operating this equipment together. However, for several months, they have been operating the cutting tool without the guard in place.

The first time they "rigged" the guard, it was done because they were behind target, having come back late from their break. As they were required to complete a set number of pieces, it was important to make up the time. However, on subsequent shifts, they always worked without the guard because it was quicker. Without having to continually engage and disengage the guard, they saved time, and they could have an extra ten minutes on their break while still meeting their target.

On the night of the accident, they behaved exactly as they had done for the previous four months – but this time, the tool operator judged the timing slightly inaccurately, and lost the tips of four of his fingers.

The unsafe acts were violations – operating that piece of machinery without the guard was against operating procedures. The initial violation was motivated by time pressure, because the operators were behind target. Subsequent violations became routine, in order that the operators could always have some extra time on their break. However, these unsafe acts were reinforced because for four months they had successfully violated the procedures with no consequence. There were no near misses and no close shaves, and perhaps they gradually came to believe that operating the cutting tool wasn't so risky after all. Since the shift supervisor only visited their part of the line infrequently, and never made any comment to them about their failure to use the guard, the violation became habitual.

Supervisory and management errors

2.15 The next slice of cheese represents the organisational defence of supervision. The holes within this slice of cheese are again unsafe acts (or failures to act), meaning human errors. However, in this slice of cheese they are removed from the immediate workface. These are latent supervisory failures. There is a one-to-many mapping between supervisory failures and unsafe acts, since a single poor supervisory practice can allow many different types of unsafe conditions and unsafe acts to promulgate. Failures to monitor and motivate are key supervisory issues, particularly where there is a need to manage and reduce violations. Further examples of supervisory failures include planning failures, where production demands may lead to an increased temptation to allow employees to take risks, and failures to fix known problems, where a supervisor does not correct a reported unsafe condition.

Research shows that supervisors often engage in very little safety-related activity, in spite of their importance in maintaining safe systems of work. One study of 100 construction accidents found that supervisors and their operatives could rarely describe any supervisory safety behaviours, and that supervisors were often complicit in violations (Haslam et al., 2005).

Organisational failures

2.16 Finally, according to Reason's model, the ultimate causes behind the accident will be latent management failures, since it is managers who typically implement organisational strategy. Examples include cuts in organisational and department budgets, changes in the way that organisational resources are managed, alterations to the way that work is organised and scheduled, and the way that all such changes are managed. These are all management issues which could harbour latent failings. As with supervisory failings, they have a one-to-many mapping; a single management decision can influence any number of activities further down the accident trajectory: supervisory activities, unsafe conditions and workplace practices, and unsafe acts. However, the management failings here are again instances of human error. Human error is just as prevalent in the supervisory and management defences as it is at the workface – it is not just employees at the front-line who make mistakes. However, it is often very simple to point the finger at the failures which occur at the front-line or the "coal-face" of an activity, without examining the human failures which may lie behind this.

Reason (1997) argues that the difference with errors at the supervisory and management levels is that they tend to be latent failures, rather than active ones. They do not usually immediately cause accidents, because they are often further removed from the workface. However, they clearly meet the definition of a human error: "an inappropriate or undesirable human decision or behaviour that reduces, or has the potential for reducing, effectiveness, safety or system performance" (Sanders & McCormick, 1987, p. 606). Pressing the wrong key on a control panel is obviously an error, while a particular management decision is not so immediately identified as such. Identifying management errors becomes even more difficult as one moves up the organisational hierarchy, since management decisions may be made by consensus, and may of necessity be taken on the basis of incomplete or limited information. Very senior management roles cannot be proceduralised, and it is often difficult to identify management decisions as failures at this level without some degree of hindsight. However, management decisions undoubtedly influence the actions and decisions which are taken at the workface, albeit that they may be removed in time and space. A discussion of some management and supervisory errors is provided in Table 2.8.

Table 2.8: Supervisory and management errors

A company engaged a specialist cleaning contractor to clean a crude oil storage tank at one of its terminals (HSE, 1989). The floating roof tank was supported on 219 pillars, and although it had been emptied, it contained two banks of sludge, a total of approximately 1,000 tonnes, which needed to be cleared. As might be expected with a site of this nature, there were clear site regulations prohibiting smoking outside specific areas, and which required surrender of all lighters and matches at the main gate. This information was available in a booklet and clearly displayed on site entry passes. In addition, the oil company operated a permit to work system which required all contractors to minimise the health, safety and environmental risks as far as was reasonably practicable.

The tank was not vapour free, but breathing apparatus was to be used as a precaution. On arrival at the site, the contractor's supervisor gave instructions to the men on how to fit the breathing apparatus mouthpiece, how to use the regulator to obtain air from the compressor outside the tank and how to deal with air supply failure and snagged hoses. The contractor had two teams of four men working on the tank in shifts, with three men inside the tank and one outside. Although there was no written procedure for cleaning the tank, the procedures to be used had been discussed and agreed with the site management.

On one shift, the outside man looked into the tank and saw the three men surrounded by a wall of fire. Two managed to escape the flames, but the third died from asphyxiation. His body was recovered late the next day, some thirty two hours after the fire had started, still wearing what remained of his breathing apparatus. The investigation revealed that one of the men who survived had been smoking inside the tank, and the fire had started when he dropped his cigarette end onto the floor. The deceased was a non-smoker.

Was this just a simple violation? What supervisory and management failures might have contributed to the accident? The investigation revealed several latent supervisory and management failings (HSE, 1989):

- Site health and safety regulations were available in a booklet which included information on smoking prohibitions, but this was not issued to all of the contractors.
- Site entry passes which contained health and safety information and included details of the smoking prohibitions were not issued to all the contractor's personnel.
- The site gate was not manned, so although the written procedures required lighters and matches to be surrendered, there was no enforcement of this at the gate, and nobody immediately available to surrender these items to.
- The contractor's foreman stated that all employees knew that smoking was forbidden unless in specified areas and would result in disciplinary action, and yet smoking inside the tanks was commonplace amongst certain employees.
- The site operator had assumed that the contractor's team were specialists – in fact, several of the men working for the contract had extremely limited experience and almost no training.
- The contractor's supervisor provided very limited training in the use of the breathing apparatus, and some men did not appreciate that there could be toxic hazards associated with fumes in the tank, or that the concentration of fumes would be greater nearer the floor.
- Some of the men found that they could see better without the breathing apparatus masks, and since they suffered no ill effects from doing so, they concluded that there were no fumes and that it was safe to smoke.
- The contractor's supervisor only attended the tank sporadically, and had on occasion removed his own mask inside the tank to provide verbal instructions to the men, reinforcing the erroneous perception that it was safe.

Amelioration and redemption

2.17 The amelioration phase of an accident is the point in time, if it exists, where an accident situation can be redeemed. The grey slices of cheese in Reason's model represent the final accident barriers, the very last defences against the accident. These are the "last stand" – the opportunity for the human operator at the workface (or some other person supervising or controlling operations) to recognise the situation, correctly diagnose it, and correctly initiate whatever recovery processes are available (if any). The grey slices of cheese represent those defences which operate once the initiating events have taken place. However, these defences may also be imperfect, and therefore the grey slices which represent them also contain holes. The relative size of the holes represents their likely efficacy in an emergency situation.

Because stress itself increases the probability of human error, reliance on human activity to recover an accident once the initiating events have taken place is a risky strategy. Last-line defences against accidents which rely on engineered mechanisms are likely to be more reliable than human ones. The most reliable choice would be an automatic safety device that did not require any human input. The next most reliable choice would be an automatic warning system, to alert the operator and perhaps warn him to take a particular course of action. Both these solutions are likely to be more effective than relying on training and procedures. Emergency procedures and training are likely to be the least practised skills of any employee, since the situations in which they can be rehearsed are few and far between: by definition, emergencies are rare. Emergency procedural and training defences firstly require the operator to determine that they actually apply, which involves correctly perceiving that something is amiss. In addition, the operator is then required to assess and correctly diagnose the problem, and then select and execute the appropriate course of action. These actions are up to ten times more prone to human error now that the crisis has arisen than they were just seconds before, because of the stress of the situation (see Table 2.5).

Table 2.9 provides details of an air traffic control accident that highlights how procedural defences can be ineffective as the last barrier against an accident. Air traffic control (ATC) is a complex process involving the control of aircraft movements. Some air traffic controllers work at airports, controlling aircraft during ground movements, take-off and landing. Other controllers work in approach, controlling arriving aircraft shortly before landing, and departing aircraft shortly after take-off. Area controllers control the movement of aircraft through clearly defined airspace, once the aircraft is en route. However, the primary aim of all air traffic control is to avoid conflict between aircraft by maintaining separation minima between aircraft, while also maintaining an efficient traffic flow. Much of the work involved in managing and controlling air traffic is routine and even mundane, but of course there are exceptions. In common with other process industries such as nuclear power, air traffic control has been described as "hours of intolerable boredom punctuated by a few minutes of pure hell" (Wickens & Hollands, 1999, p. 517).

Table 2.9: Procedures as the last line of defence

In July 2002, two aircraft were travelling en route over Überlingen in Germany. One was a Boeing 757 freight aircraft operated by DHL with two pilots, and the other was a Tupolev 154 with nine crew and sixty passengers on board (BFU, 2004). At around 9.20 p.m., the Boeing 757 first officer received air traffic control clearance to climb to 36,000 feet. He climbed the aircraft over a period of minutes, and then he handed control of the aircraft to the captain, to go to the lavatory. Seconds later, the on-board automated traffic alert and collision avoidance system (TCAS) gave a traffic alert ("traffic, traffic"), closely followed by a resolution alert. The resolution alert commanded the Boeing 757 crew to "descend, descend". Rejoining the captain, the first officer could see the second aircraft through the cockpit window. The TCAS then gave another command to the pilots to "increase descent, increase descent". Fast to react to the TCAS commands, the captain had been descending the aircraft throughout this period, but was now encouraged to descend even more rapidly. The pilots reported the TCAS descent to the air traffic controller, but unfortunately the transmission was lost as the controller was then dealing with another aircraft.
Meanwhile, the Tupolev 154 was also at 36,000 feet with five flight crew on board. The pilot who was actually flying the aircraft was in the captain's seat, but he was under instruction from the pilot in command, who was the most senior pilot with the airline company. Another three crew members were on the flight deck: the flight navigator, the flight engineer and a non-flying pilot. At around the time that the Boeing 757 pilot went to the lavatory, the Tupolev crew were discussing an aircraft which was represented on a visual display in the cockpit, part of the TCAS.
They then sighted the aircraft, and within seconds received a traffic alert from TCAS ("traffic, traffic"). This was acknowledged by both the pilot flying and the pilot in command. Seven seconds later, air traffic control asked them to descend, as there was another aircraft crossing their path. The pilot in command requested the pilot flying to descend. Immediately following this request, the TCAS issued a resolution alert, telling the crew to climb. The pilot flying questioned whether he should climb or descend: the pilot in command again advised the ➧

pilot flying to follow the air traffic controller's instructions. Shortly afterwards, the air traffic controller again provided another instruction, requesting that the crew descend more quickly. Following this, the TCAS system advised the crew to increase the rate of climb. At this point, the pilot flying commented: "It says climb!" Eight seconds later, the two aircraft collided at 34,890 feet, with the loss of all 71 lives.

TCAS was introduced in civil aviation as an additional safety barrier, to provide an on-board warning of potential collisions in situations where ground-based human control had failed to maintain separation. At Überlingen, TCAS had functioned as intended on both aircraft. Had the crew followed the commands provided by the TCAS resolution alerts, the collision would have been avoided. However, the Tupolev crew had been trained by their airlines that the TCAS alerts were merely advisory, and their procedures did not require them to comply with the TCAS resolution command. This was in accordance with aviation regulations in Russia at that time. However, in the rest of Europe, aviation regulations required pilots to be trained to obey TCAS without question, since TCAS provides resolution alerts when separation has already been compromised. Instead, the Tupolev crew followed the commands of the air traffic controller, who was unaware of the TCAS instructions.

As with other accidents, the actual causal and contributing factors in this case were numerous and complex. However, the potential to mitigate for the consequences of the loss of separation between aircraft, and to redeem the situation, relied critically on procedural factors. In this event, they failed.

It is important to note that the actual trajectory of an accident, shown as an arrow moving through the holes in the cheese slices, can often only be constructed with hindsight. It is not always possible to predict in advance which defences will fail, or which failures will allow an accident to slip through. Reason (1997) suggested that the *actual* gaps in safety defences, represented by holes in the slices of cheese, move and fluctuate all the time, depending on local circumstances and conditions. For example, in maintaining an automatic safety device, it may be necessary to take it out of action temporarily, until the work is complete. In this instance, this whole "slice" of the safety defence is removed.

Similarly, it is not possible to predict *when* certain failures will occur. For example, meteorologists know that extreme storms occur infrequently. Some storms are so severe that they are known as "100-year storms". This means that a storm of this magnitude will occur approximately once in every 100 years. However, that does not mean that such a storm will occur like clockwork on the same date every century. Nature might decide to have two 100-year storms a week apart. However, no one could have predicted that the turn of the century would occur sometime during that seven-day period. In fact, it is more likely that having recently experienced such a severe storm, most people would estimate the date of recurrence to be far into the distant future.

While it may be possible to reduce or eliminate some of the more immediately obvious gaps in safety defences, eliminating every possible latent failure in every defence, or continuing to add layers of defences without limit, are not feasible options. An organisation that attempted to do this would find that protection would strait-jacket production. Risks are inherent to any production system or organisation. However, it is also true that some industries are higher risk than others, and may require more, and tougher, defences.

System accidents

2.18 Charles Perrow, a sociologist and organisational theorist, suggested that in high-risk and/or safety critical sectors, accidents are "normal". This is because the very nature of interactions between the components of such systems means that accidents, while rare, are inevitable (Perrow, 1997). System components include design, equipment, procedures, operators, supplies/materials and/or the environment.

According to Perrow (1997), all industries vary in the extent to which the interactions between system components are linear or complex. Linear interactions between system components are those in expected and familiar production or maintenance sequences, which are visible even if not planned. For example, the failure of a part on a production line will have predictable consequences. It is clear that upstream, products entering the line will pile up quickly, and downstream, products will be incomplete. The situation can be easily addressed: the fault can be bypassed, with the missing parts assembled later on, or the fault can be fixed, with incomplete products upstream being temporarily stored. In contrast, complex interactions involve unfamiliar sequences and are often unplanned, unforeseen and/or hidden – they are often not visible and not easy to predict or comprehend. These situations arise where a component serves two or more functions – known as common modes. The example Perrow gives is of a heater which heats gas in one tank, and also acts as a heat exchanger, absorbing excess heat from a chemical reactor. As a design, this saves energy, but it means that if it fails, it is more difficult to predict the effects. As complexity increases, the difficulties in predicting the effects of failure increase exponentially.

Further, all industries vary in the extent to which system components are coupled (Perrow, 1997). Coupling refers to the extent to which changes in one part of the system can effect change in another. In tightly coupled systems, there is a great degree of time dependency, so that changes in one part of a system quickly have knock-on effects, and there is normally only one way to reach the production goal. Further, there is little slack in the system, resources cannot be substituted and failed equipment causes a shutdown. In loosely coupled systems, there is more independence between components, a greater degree of flexibility and less time dependency. If something is not correct the first time, it can be done again, and loosely coupled systems can therefore tolerate more failures without becoming unstable.

For Perrow (1997), there is an inherently high level of risk in those industries in which the system is more complex and is also tightly coupled. The ability of any human operator to correctly predict actions and interactions, dependency and effect in such situations is greatly reduced. Such systems are opaque rather than transparent, and causes and effect are networked rather than directly related. Perrow argued that high-risk and safety critical industries such as nuclear power, air traffic control, and chemical and allied industries were both complex and tightly coupled, which increases the risks exponentially. An inability to predict, anticipate and mitigate failings in such industries is particularly critical, since the costs and consequences may well be extreme.

Organisational resilience

2.19 A relatively new concept in safety research is that of organisational resilience, sometimes known as resilience engineering (Hollnagel et al., 2006). This approach to accidents and safety is more proactive and less focused in hindsight than other models of accident causation. It is a very recent development in accident and safety research, and hence it lacks the body of practitioner knowledge which is available for other theories of accidents and methods of accident investigation. However, resilience engineering is a very promising paradigm, which focuses on finding out how people manage to produce successful results in the face of failure. Resilience engineering acknowledges that safety is simply the flip-side of risk; and that success is the flip-side of failure. Organisations operate in risky environments, and the key to operating safely and successfully is to accurately anticipate and appropriately respond to the ever-changing nature and levels of risk faced by the organisation. In a nutshell, resilience is defined as "the ability of systems to anticipate and adapt to the potential for surprise and failure" (Woods & Hollnagel, 2006, p. 4).

Some characteristics of a resilient organisation are given by Woods (2006). Firstly, organisations need to cope with the constant pressure to be "faster, better, cheaper". This management maxim used by Daniel Saul Goldin, a NASA administrator, generally requires the people within an organisation to do things more effectively, more quickly or in more complex ways. However, there is often a trade-off between initiatives intended to increase speed, improve quality and cut costs – the three are not often compatible and cannot often be achieved simultaneously. Secondly, there is often an unreasonable drive in organisations for people to be increasingly efficient and thorough without a corresponding increase in resources. This leads to a conflict in short- and long-term goals, where it is easier for short-term goals to take precedence. Longer-term goals have to be given core corporate value status, with short-term goals made subordinate. In this way, a resilient organisation can better manage the complexity involved in attempting to meet multiple goals.

Thirdly, Woods (2006) argues that resilient organisations have to accept some degree of production and efficiency sacrifice in order to encourage safer decisions. If someone suspects a problem, and the evidence is ambiguous or subtle, then he might courageously call for production to be stopped in order for the matter to be investigated. If the system is shut down, and this individual is proved to have been right in diagnosing a problem, then the management will typically praise the individual for their role in avoiding the loss, failure or injury. This is a good test of a safety culture, but not a good test of organisational resilience: that requires changing the ending of the story. If production is shut down, and the investigation reveals that nothing is amiss, what then is the management's response? Woods argues that a resilient organisation will still acknowledge the stoppage as valuable, where other organisations will not.

Conclusions

2.20 In conclusion, it can be seen that accidents, whether regarded as individual, organisation or system accidents, have multiple and complex causes. No single failure leads to an accident, and luck and chance will play a role in determining the outcome. Early theories of accident causation included simple linear models, such as that developed by Heinrich (1931, cited in Bamber, 2003). More sophisticated linear models include the domino model, which emphasised the importance of a "root cause". Complex linear models include Reason's "Swiss cheese" model, which is probably the dominant model used in current safety and accident practice and research. However, recent advances in the theory of resilience engineering (Hollnagel et al., 2006) develop the concepts of complexity and coupling originally discussed by Perrow (1997). These theories of system accidents emphasise networks of causal and contributory factors, and are more forward looking than models of accident causation which rely heavily on hindsight.

As we create ever more complex organisations and equipment, so we have also become more aware of the limitations of the people who work within these systems. The role of human factors in incidents and accidents is now increasingly acknowledged. Human factors is the study of the inter-relationships between humans, tasks, equipment and organisations in accomplishing specific tasks or missions in given environmental and commercial contexts. Given that all non-technical failures are human failures, it is clear that the remit of human factors includes all human behaviour at work, including organisational and management factors, first-line managers and supervisors, and activity at the workface.

Some organisations are reluctant to use the term "human factors", since there is a perception that it means counselling, dealing with the psychological aftermath of accidents or treating mental health problems among victims of disaster. These *can* be important considerations for accident investigators, but they do not describe the discipline of "human factors". Human factors in accident investigation simply means examining behaviour in context, and finding out why people did the things they did, said the things they said or thought the things they thought. In such answers can often be found the key "reasons" why an accident happened.

References

Adams, M., Burton, J., Butcher, F., Graham, S., McLeod, A., Rajan, R., Whatman, R., Bridge, M., Hill, R. & Johri, R. (2002) *Aftermath: The Social and Economic Consequences of Workplace Injury and Illness*. New Zealand Department of Labour, Wellington, New Zealand.

Bamber, L. (2003) Accident causation and impact. In Tyler, M. (ed.) *Tolley's Workplace Accident Handbook*. Lexis Nexis, Reed Elsevier, UK.

BBC (2001) *Shocking £2 bn Electricity Bill*. BBC News, 25 August 2001. From www.news.bbc.co.uk/1/hi/uk/1509599.stm; accessed 24 July 2006.

BBC (2002) *Savers Given Gay Service*. BBC News, 14 February 2002. From www.news.bbc.co.uk/1/hi/England/1819948.stm; accessed 24 July 2006.

BBC (2004) *Husband Flies with Wife's Passport*. BBC News, 17 February 2004. From www.news.bbc.co.uk/go/pr/fr/-/1/hi/England/Oxfordshire/3495299.stm; accessed 24 July 2006.

BBC (2006a) *Road Container "Leaked Radiation"*. BBC News, 17 February 2006. From www.news.bbc.co.uk/go/pr/fr/-/1/hi/England/4725686.stm; accessed 24 July 2006.

BBC (2006b) *Swedish Blunder Puts Porn on News*. BBC News, 22 August 2006. From www.news.bbc.co.uk/go/pr/fr/-/1/hi/entertainment/5273706.stm; accessed 22 August 2006.

BFU (2004) *Investigation Report: Accident 1 July 2002 near Überlingen/Lake Constance/South Germany*. German Federal Bureau of Aircraft Accident Investigation, Braunschweig.

Cormack, H., Cross, S. & Whittington, C. (2006) *Identifying and Evaluating the Social and Psychological Impact of Workplace Accidents and Ill-Health Incidents on Employees*. Research Report 464. Health and Safety Executive, Sudbury, Suffolk.

Haslam, R.A., Hide, S.A., Gibb, A.G.F., Gyi, D.E., Pavitt, T., Atkinson, S. & Duff, A.R. (2005) Contributing factors in construction accidents. *Applied Ergonomics*, 36, 401–415.

Helmreich, R.L. (1998) Error management as an organisational strategy. *Proceedings of the IATA Human Factors Seminar*, Bangkok, Thailand, 20–22 April 1998, pp. 1–7.

Hollnagel, E., Woods, D. & Leveson, N. (2006) *Resilience Engineering: Concepts and Precepts*. Ashgate, Aldershot.

Hopkins, A. (2006) What are we to make of safe behaviour programs? *Safety Science*, 44, 583–597.

HSE (1989) *A Report of the Investigations by the Health and Safety Executive into the Fires and Explosion at Grangemouth and Dalmeny, Scotland, 13 March, 22 March and 11 June 1987*. HSE, Sudbury, Suffolk.

HSE (1997) *Successful Health and Safety Management*. HSG 65. HSE Books, Sudbury, Suffolk.

HSE (2004) *Investigating Accidents and Incidents*. HSG 245. HSE Books, Sudbury, Suffolk.

HSE (2006) *Health and Safety Offences and Penalties 2004/2005: A Report by the Health and Safety Executive*. Published by the Health and Safety Executive. From www.hse.gov.uk/enforce/off0405/index.htm; accessed 20 July 2006.

ICAO (2001) Aircraft accident and incident investigation. *International Standards and Recommended Practices: Annex 13 to the Convention on International Civil Aviation*, Ninth Edition. ICAO, Montreal, Canada.

Lawton, R. & Parker, D. (1998) *Individual Differences in Accident Liability: A Review*. Contract Research Report 175/1999. HSE Books, Sudbury, Suffolk.

Lewis, L. (2005) *Fat Fingered Typist Costs a Trader's Bosses £128 Million*. *The Times*, 9 December 2005. From www.timesonline.co.uk; accessed 24 July 2006.

Miller, D. & Swain, A. (1987) Human reliability analysis. In Salvendy, G. (ed.) *Human Factors Handbook*. Wiley, New York.

Perrow, C. (1997) *Normal Accidents: Living with High Risk Technologies*, Second Edition. Princeton University Press, Princeton, NJ.

RAE (2005) *Accidents and Agenda: An Examination of the Processes that Follow from Accidents or Incidents of High Potential*. The Royal Academy of Engineering, London.

Reason, J. (1990) *Human Error*. Cambridge University Press, New York.

Reason, J. (1997) *Managing the Risks of Organisational Accidents*. Ashgate, Aldershot.

Sanders, M.S. & McCormick, E.J. (1987) *Human Factors in Design and Engineering*, Sixth Edition. McGraw-Hill, New York.

Slovic, P. (1987) Perception of risk. *Science*, 236(4799), 280–285.

The Concise Oxford English Dictionary (2004) Soanes, C. & Stevenson, A. (eds) *The Concise Oxford English Dictionary*. Oxford University Press, Oxford Reference Online; accessed 20 July 2006.

Vaughan, D. (1996) *The Challenger Launch Decision: Risky Technology, Culture and Deviance at NASA*. The University of Chicago Press, Chicago, IL.

Wickens, C.D. & Hollands, J.G. (1999) *Engineering Psychology and Human Performance*, Third Edition. Prentice Hall, Upper Saddle River, NJ.

Woods, D. (2006) Essential characteristics of resilience. In Hollnagel, E., Woods, D. & Leveson, N. (eds) *Resilience Engineering: Concepts and Precepts*. Ashgate, Aldershot.

Woods, D. & Hollnagel, E. (2006) Resilience engineering concepts. In Hollnagel, E., Woods, D. & Leveson, N. (eds) *Resilience Engineering: Concepts and Precepts*. Ashgate, Aldershot.

3 Compliance with legislation

In this chapter:

Introduction

3.1 This chapter summarises the basic requirements of British health and safety law and looks at a range of provisions found in legislation for the implementation of measures designed to manage the consequences of workplace accidents.

These include fire protection, making careful assessments of the arrangements needed to respond to accidents and emergencies, and of course the provision of properly trained first aid personnel. This chapter concludes with a brief checklist of legal compliance issues.

In the following chapter more detailed consideration is given to the duties to record information and to report events to the relevant authorities as a means of enabling them to undertake formal investigations under other statutory powers.

Overview of health and safety law

Legislation

3.2 Legislation provides extensively for measures to be taken for the prevention of accidents. Protection by employers of their employees (and others affected by work activities) is a fundamental requirement of the *Health and Safety at Work etc Act 1974* (*HSWA*). The penalties for non-compliance are generally fines of an unlimited amount, and in some (rare) cases imprisonment is possible (see page 254). On some occasions, the courts have chosen to characterise these offences as being not truly criminal but rather of a regulatory nature designed to protect the interests of broad segments of the public. In reality, the offences are dealt with as crimes, in the criminal courts, and sentencing of offenders is becoming increasingly stringent.

Statutory requirements aimed at accident prevention exist in the form of obligations to undertake and record suitable risk assessments, and to devise, record and implement appropriate control measures. These most basic requirements, and the "principles of prevention and protection" to be applied by all employers, are contained in the *Management of Health and Safety at Work Regulations 1999* (*MHSWR*). Many more sector-specific or risk-specific regulations supplement these 1999 regulations.

The focus of this chapter is the element of health and safety legislation which requires employers to make provision for mitigating the outcome and effects of accidents, and also to retain data about the circumstances in which accidents (and ill health) occur so that investigations can be undertaken subsequently. The obligations exist in two main forms – the so-called *general duties* under the HSWA, and health and safety regulations (made in the form of *statutory instruments* (SIs)). These statutory provisions, which are usually expressed in very general terms or which describe goals without being prescriptive as to the exact measures to be taken, are often fleshed out and interpreted by good practice guidelines of a variety of kinds.

HSWA: Health and Safety at Work Act 1974 general duties

3.3 In Part 1 of the *HSWA* are a series of obligations on all those whose activities could affect safety in workplaces in different ways:

- *Employers*:Employers owe a duty to ensure, so far as is reasonably practicable, the health, safety and welfare at work of their employees and trainees (*HSWA, s 2*). This extends to the provision of appropriate information and training, and adequate welfare facilities.
- *Self-employed*: Every self-employed person must ensure his own safety so far as is reasonably practicable (*HSWA*, *s 3(2)*).
- *Employers and self-employed*: Both owe an additional duty in relation to persons who are not their employees but who may nevertheless be adversely affected by the conduct of the employers or self-employed persons' business activities

(or "undertaking") (*HSWA, s 3(1) – (2)*). The duty is to ensure, so far as is reasonably practicable, that such persons are not exposed to risks.

- *Employees*: Every employee is obliged to take reasonable care for his own and others' safety, to co-operate with the employee on safety matters and not to interfere or misuse any safety equipment (*HSWA, ss 7–8*).
- *Owners and managers of property*: A person who controls (to any extent) non-domestic premises, plant or substances kept on premises has to take reasonable measures to ensure, so far as is reasonably practicable, the safety of the premises and the plant or substances (*HSWA, s 4*).
- *Manufacturers and suppliers*: (Those who manufacture, import or supply substances and articles for use at work and those who design articles) are required to ensure, so far as is reasonably practicable, the safety of the products (*HSWA, s 6*). This entails the provision of adequate safety information.

In each case these general duties have important implications for accident risk management. The codes of practice and good practice guidelines described elsewhere in this book will in most situations represent what it is reasonably practicable for duty holders to do to ensure that, where the risk of accidents cannot be avoided, the consequential risks to health are minimised through good accident management systems.

It is also implicit in the general duties that lessons learned from accidents (including *near misses*) are used to inform the ongoing risk management process and so reduce risks further. In fact there are more specific obligations as well to this effect (see 5.25–5.26). Failures by employers and others to implement reasonably practicable precautions (or in the case of employees, failure to follow them) can lead not just to criminal prosecutions as well as other enforcement measures designed to compel compliance with the legal obligations to ensure health and safety.

Health and safety regulations

3.4 Regulations are made by ministers in the form of Statutory Instruments ("SIs") pursuant to powers granted by the *HSWA* or other Acts of Parliament. These can vary considerably in their scope and function. For instance, the *Reporting of Injuries, Diseases and Dangerous Occurrences Regulations 1995* (*RIDDOR*) are concerned purely with administrative issues of recording and notification of occurrences to the Health and Safety Executive (HSE) or the local authority. The *Control of Major Accident Hazards Regulations 1999* (*COMAH*) lay down important safety and emergency planning procedures to sites holding significant quantities of chemicals and other dangerous materials. The *Control of Asbestos at Work Regulations 2006*, on the other hand, lay down specific control measures and comprehensive

CASE STUDY 1

3.5

Multi-party responsibility under the HSWA

Company A was engaged by Client X as a principal contractor on a project involving the construction of high-rise office buildings. Company A employed Company B to undertake the preparation of the site and to lay the foundations. Company B contracted with Company C to provide workers for these initial works.

During the preparation of the site one of Company C's workers was injured when an unsupported trench in which he was working collapsed.

The Health and Safety Executive conducted an investigation and brought charges against Company A, Company B and Company C under Sections 2 and 3 of the HSWA 1974. Company C was found to have breached its duty under Section 2(1) in that it had failed to ensure that the injured workman was not working in unsupported trenches or to have instructed him of the dangers of so doing. Companies A and B were found to have failed in conducting their undertakings in a manner which did not pose a threat to individuals working on site. The companies' breaches related to their failure to draw up and enforce adequate method statements in respect of trench work, and to their failure to perform proper risk assessment given the poor condition of the land.

risk management processes for the control of this specific hazard to health, and also include provisions for dealing with situations where accidental exposure could be imminent or has occurred. Other regulations can cover generic risks such as maintenance and use of electrical equipment, while others regulate specific industry activities such as docks, rail transport or construction. There are over a hundred SIs of these different kinds.

It follows that all those charged with health and safety responsibilities need to (a) identify both their industry-specific requirements and the other more generic regulations which apply to all their business activities, and (b) maintain a comprehensive understanding of all these provisions (Table 3.1).

Failure to comply with health and safety regulations, as well as being a criminal offence for which financial penalties are imposed if the enforcement authorities

Table 3.1: Sources of legislation

Office of Public Sector Information
http://www.opsi.gov.uk/legislation/uk.htm
This site contains the full text of Acts of Parliament from 1988 onwards and SIs (including health and safety regulations) from 1987 onwards. Where legislation has been repealed or amended this is not however always tracked and so Acts and legislations may not appear in up-to-date form. There is an advanced search facility.

The Stationery Office (formerly HMSO)
http://www.tso.co.uk
Printed copies of all UK health and safety legislation can be purchased from TSO. There are the original Acts and Regulations and are not necessarily up-to-date texts; amending legislation has to be purchased separately.

hsedirect
http://www.hsedirect.com
A subscription-based service provided by the HSE in conjunction with a commercial publisher. It provides up-to-date legislation in conjunction with a large range of other materials.

Croner-*i*
http://www.croner.co.uk
A subscription-based service which includes text of legislation (not necessarily updated with amendments) as well as extensive other sources of health and safety materials.

bring prosecutions and will also generally enable an injured person to bring a claim for compensation in reliance upon a breach of statutory duty. These issues of enforcement and compensation are the subject of Chapter 8.

Authoritative sources of guidance on legislation

3.6 Guidance on appropriate action to ensure health, safety and welfare in the workplace can be obtained from a huge variety of sources. The most authoritative are the Approved Codes of Practice (ACoP), which are issued by the Health and Safety Commission (HSC). Although compliance with ACoPs is not mandatory, it is necessary for practical purposes to adhere to their guidelines. If criminal proceedings are brought in relation to an alleged breach of a statutory requirement to which any ACoP relates a failure to observe the Code's relevant provisions will create a presumption that the statutory provision was breached. The onus will then be on the defendant to demonstrate that it had devised other measures at least as effective as those in the ACoP (*HSWA, s 17*).

Important examples of ACoPs are those which describe the means of complying with the *MHSWR* and the *Health and Safety at Work (First-Aid) Regulations 1981.*

Other forms of authoritative guidance on good practice are HSE guidance notes, British and European Standards, and trade association or professional body codes and guidelines. The HSE's own guidance has no special evidential weight as compared with those other sources of advice. However, in recent years the HSE's material has been prefaced with the following advice:

> "This guidance is issued by the Health and Safety Executive. Following the guidance is not compulsory and you are free to take other action. But if you do follow the guidance you will normally be doing enough to comply with the law. Health and safety inspectors seek to secure compliance with the law and may refer to this guidance as illustrating good practice."

It is true to say that in many enforcement situations – whether they be formal advice and warnings, enforcement notices or prosecutions – inspectors refer to or quote from the extensive volumes of guidance material produced by the HSE (and other bodies).

Therefore, proper management relating to accidents requires managers to maintain databases which correctly identify not just statutory requirements contained in Acts and Regulations but applicable ACoPs, as well as other authoritative sources of guidance (Table 3.2). Most booksellers can supply priced HSC/HSE publications.

Leadership responsibilities of directors and equivalent senior executives

3.7 Directors and other officers were not included in the description of the *HSWA* general duties above because no duties are expressly given to them. In 2005 and 2006 the HSE and ministers considered whether to introduce explicit directors' responsibilities into UK health and safety legislation but at that time no conclusive view was formed.

Nevertheless directors (or equivalent senior-level managers) have ultimate responsibility for ensuring that the risk of accidents is controlled in any organisation. This is true, not just of directors of corporations who hold office for the purposes of company law, but also for those who govern the affairs of other organisations such as

Table 3.2: Sources of guidance

HSE Books *www.hsebooks.com* This is a mail order service for all HSC/HSE publications, videos, etc. Some leaflets are free. ACoP and the main HSE guidance series are priced publications which have to be purchased.
HSE Website Many free publications and reports can be downloaded from www.hse.gov.uk/ The site is fully searchable. Specific topics (e.g. first aid) may have their own mini-sites where all the main guidance is listed, with links to further information.
HSE Infoline 0845 345 0055 This is described as a one-stop shop for information, expert advice and guidance. Call-centre staff can search databases to locate guidance publications relevant to particular activities or risks. However, priced publications still need to be purchased from HSE Books for other outlets.
Commercial providers There are a number of on-line sources which are subscription based from which virtually all these publications can be downloaded. Some provide summaries of other documents (e.g. British Standards). See hsedirect and Croner-*i* referred to in Table 3.1.

local authorities and some other types of public bodies. Such persons can be criminally liable in a personal capacity for the corporation's health and safety failure where the failure has been with their "consent" or "connivance" (tacit acceptance), or where it is attributable to any "neglect" on the part of the individual (*HSWA, s 37* – see page 241).

Similar provisions can apply in certain industry sectors to other senior managers below director level who have overall control of operations such as at offshore installations, mines and quarries, although in these situations their responsibilities are generally defined in terms of specific obligations on them in the legislation.

As the obligations of directors and other officers are not defined under health and safety legislation (other than negatively in terms of the failings described in *HSWA*

1974, s 37) it is difficult for them to obtain assurance that they are not personally at risk of prosecution. However, the HSC issued guidelines in 2002 which highlighted action points which should be addressed in terms of leadership of health and safety (HSC, 2002) (Table 3.3).

Table 3.3: Directors' responsibilities: HSE guidance

Action Point 1
The board needs to accept formally and publicly their collective role in providing health and safety leadership in their organisation.

Action Point 2
Each member of the board needs to accept their individual role in providing health and safety leadership for their organisation.

Action Point 3
The board needs to ensure that all board decisions reflect their health and safety intentions, as articulated in the health and safety policy statement.

Action Point 4
The board needs to recognise their role in engaging the active participation of their staff in improving health and safety.

Action Point 5
The board needs to ensure that it is kept informed of, and alert to, relevant health and safety risk management issues. The HSC recommends that boards appoint one of their number to be the health and safety director.

The guidelines expand on these action points, but only to a limited degree. They highlight an important issue for the director with responsibility for health and safety which is "to monitor health and safety performance and to be kept informed about any significant failures, and the outcome of any investigations into their causes". This is part and parcel of a safety management system which provides for effective monitoring and review of an organisation's performance.

Employees' duties

3.8 The burden of maintaining a safe working environment does not rest wholly with the employer, in theory at least. *HSWA Section 7* requires an employee to take reasonable care for the health and safety of himself or herself and his or her colleagues in the workplace. When a legal requirement is placed upon an employer to fulfil certain health and safety objectives the employee must co-operate with his or her employer to ensure compliance. The aim of the legislation is to ensure that health and safety is a concern at all levels of an undertaking.

An employee in breach of the duty set out in *Section* 7 may be liable for a fine upon conviction. The consequences of a breach can also extend to a dismissal if the employee is found to have been in breach of the contractual term to perform his or her duties with due care and skill, despite having received proper instruction on safety measures and being conscious that interference with safety equipment and procedure (prohibited under *Section 8 HSWA*) could lead to dismissal (*Martin v Yorkshire Imperial Metals Ltd*).

The *MHSWR* also place certain duties on employees. *Regulation 14* requires an employee to use machinery, equipment, dangerous substances, transport equipment, means of production and safety devices provided to him or her in accordance with any training and instruction he or she has received from an employer acting in compliance with the relevant statutory provisions (e.g. *HSWA*). *Regulation 14* also demands that an employee inform his or her employer, or colleague responsible for health and safety, of work situations which could pose a serious and immediate danger to health and safety, or which would represent a failure on the part of the employer to provide adequate protection for health and safety. Similar duties are placed on employees under the *Regulatory Reform (Fire Safety) Order 2005* (see 3.42).

Although these provisions place an ostensible burden on employees both to maintain health and safety in the workplace and to act as informant in respect of poor practice, in reality they are infrequently enforced. Statistics published by the HSE reveal that in the year 2004–2005 only fifteen people were charged with breaches of Section 7, of which twelve were convicted. The average fine per conviction was £679.

Common law duties

Employers' liability

3.9 The common law has been developed over many years by judges, but the law can evolve as higher courts may depart from earlier decisions and adapt the law to changing employment practices and public policy considerations. The common law is therefore flexible but at the same time it may be entirely predictable.

The common law duty of an employer can be summarised as one to take reasonable care for the health and safety of its employees, and an injured employee can sue the employer for negligence where the duty is breached. Often this is described more specifically as a set of obligations to provide a safe place of work, safe plant and equipment, competent employees and a safe system of work, though this is by no means an exhaustive list of the duty's components.

The extent of the duty has evolved significantly over time, with increasing emphasis on employers' responsibilities to identify and manage risks to a high standard, and it has gradually extended beyond just the prevention of accidents in the course of an employee's normal working conditions. For example, an employer nowadays owes an obligation to take measures to protect employees against reasonably foreseeable risks of violent attacks – even though generally an employer would not have a liability for the act or default of a third party, and there is not an equivalent duty on an employer to prevent other kinds of crime against employees such as theft.

There is also a recognised common law duty on an employer to have available arrangements for first aid or to facilitate other appropriate medical assistance in the event of accidents (*Kasapis v Laimos*), even where the employer is not responsible for causing the accident. This duty can also arise where the employee is working away from the employer's place of business, and even when the employee is working for a different undertaking having been transferred there on a temporary basis to work for the other employer (*McDermid v Nash Dredging and Reclaimation Co Ltd*).

The employer's duty of care has also been found to be sufficiently broad to cover some risks to employees which would appear to be primarily their own responsibility. Employees who have been injured after becoming drunk have successfully sued their employers for taking insufficient care of them while being put to bed in the sleeping quarters (*Barrett v Ministry of Defence*) or while being transported back from a social night (*Jebson v Ministry of Defence*). In such cases though the amount of damages awarded will generally be scaled down substantially to allow for the element of the employee's contributory negligence. In the *Jebson* case the employee was held to have been 75 per cent at fault and his damages were reduced accordingly.

Unless an employer has made a commitment to do so in the terms of an employee's contract of employment there is usually no duty to take out life insurance or critical illness cover for an employee against the adverse financial consequences of an injury or disease caused entirely accidentally or by another party. In one case an employee who had been working abroad in a country where there was no compulsory motor insurance claimed against the employer in respect of a road accident. The employer was held not to be liable (*Reid v Rush & Tomkins Group plc*). (The statutory requirements for compulsory employers' liability insurance are different – this compulsory insurance is for situations where injury and loss arises because the employer has breached a duty of care owed to an employee. See 3.13–3.15.)

Employment is a contractual relationship and consequently employees can sue for breach of express and implied terms in the contract of employment. In most situations this makes no practical difference to making a claim for common law negligence as far as proving liability for injury is concerned. However, it does mean that if

CASE STUDY 2

3.10

> ***Summoning medical assistance***
>
> Mr J was employed as a security guard by Company K, a multi-national insurance business. Mr J regularly worked a night shift at one of Company K's 16 buildings in London. He was required to monitor and control access to and from the building. He also conducted security patrols of the underground car park.
>
> During a patrol Mr J fell down a poorly lit concrete staircase, hitting his head and breaking his leg. He fell unconscious. Mr J had not been instructed to telephone the control room at regular intervals to inform his colleagues that he was safe and well, nor was he carrying a mobile phone. Consequently, Mr J remained in the stairwell for 5 hours during which his condition deteriorated. He was eventually discovered by members of staff arriving for work the following day.
>
> Company K were found to have been in breach of their duty to have proper procedures in place to protect members of staff from risks posed by working alone.

arranging insurance of any kind is a term of the contract, or if any other benefits have been incorporated (such as sick pay, periods of recuperation, health services or injury rehabilitation) compensation will be recoverable for these if the employer fails to provide them, even though a common law negligence claim for them would have failed.

Other duty holders

3.11 As a rule the common law does not impose obligations on anyone to act positively so as to prevent or mitigate harm. A common example given is that it is not negligent to fail to give assistance to a stranger who is drowning, even if the means to do so are readily available. The exception to this rule is where the law recognises a duty to act based on an assumption of responsibility which goes with a special relationship between a party and the person at risk. The employer–employee relationship is a prime example, but other recognised duty holders are schools in relation to their pupils, police in relation to persons in custody and local authorities in relation to children vulnerable persons in their care.

The position of the emergency services has been considered on a number of occasions in recent cases which have resulted in a somewhat inconsistent approach. In the case of *OLL Ltd v Secretary of State for Transport*, the coastguard was held not liable for the late arrival of rescue helicopters where it was claimed that inadequate location details were given to pilots. In the case of *Kent v Griffiths* where an ambulance failed to attend an asthmatic person having an attack within a reasonable time after a 999 call from her doctor, the Court of Appeal held the London Ambulance Service liable for the further injury caused by the delay. The reasoning here is that the ambulance service, by accepting the call, had undergone the requisite assumption of responsibility for a duty of care to be recognised, and there was no good reason the ambulance service had been able to provide for the delay. (By implication, declining to attend because of the unavailability of an ambulance, or even failing to take the 999 call, might not have rendered the service liable.)

On the other hand, a fire brigade was held not to have a duty to answer an emergency call on the basis of a ruling that the duty of the fire services (and the police) is owed to the public at large but does not extend to being a duty of care to anyone individual in danger. As it was graphically put by the Court of Appeal: "if they fail to turn up, or fail to turn up in time, because they have carelessly misunderstood the message, got lost on the way, or run into a tree, they are not liable".

In contrast, in other situations where firemen turned off sprinkler systems negligently a claim was successful (*Capital & Counties plc v Hampshire County Council*).

CASE STUDY 3

3.12

> ***The liability of the emergency services***
>
> A fire ignited in the boiler room of School D. The sprinkler system was activated, but Mr E, the caretaker, remained in the building, attempting to minimise any damage. The fire brigade were eventually summoned. The fire officer on the scene ordered that the sprinklers be switched off, although the seat of the fire had not been located, and although Mr E's presence in the building had not been ascertained. The blaze intensified and spread. Mr E was placed in grave danger and suffered severe smoke inhalation before he was rescued by the fire brigade. School D and Mr E brought proceedings against the fire authority for damages. They successfully argued that the fire officer had made a decision that no reasonably well-informed fireman could have made. This amounted to negligence as the extensive damage and serious injury to Mr E would have been avoided if the sprinklers had remained on.

Liability insurance

When insurance is required

3.13 The *Employers' Liability (Compulsory Insurance) Act 1969* (*EL(CI)A 1969*) makes it mandatory for most employers to take out insurance for liability to employees for injury or disease. It is an offence to fail to maintain an insurance policy complying with the Act's requirements, or to fail to display the certificate of insurance where it can easily be seen and read at each place of business. Employers may be fined up to £2,500 for each day there are without suitable insurance. Failure to display a certificate of insurance may result in a fine of up to £1,000.

There are no compulsory insurance requirements for the self-employed to insure themselves against injury.

Certain employers are exempt from the Act, principally nationalised industries and various state bodies such as the NHS and others for whom the government in effect provides an ultimate indemnity against liability, but who in practice have to meet the cost of claims out of their budgets. Small family businesses are also partially exempted in that there is no requirement to insure an employer's liability to a spouse, parent, son or daughter or other close relative (*EL(CI)A 1969, s 2(2)(a)*). Following extensive consultation, the Government enacted the *Employers' Liability (Compulsory Insurance) (Amendment) Regulations 2004*. These Regulations remove the need for employers' liability insurance where a company employs only its owner and where that owner/employee owns 50 per cent of the issued share capital.

The obligations to insure also apply only in relation to employees ordinarily resident in Great Britain or who are temporarily in Great Britain for work purposes of fourteen days or longer (or seven days in the case of offshore installations and their associated structures). Employees working abroad are not covered by the Act. It is essential therefore to check when employees work abroad that the organisation's employers liability cover extends to the particular territory and that local insurance requirements are also met.

Insurance cover and prohibited conditions

3.14 The basic requirement is that the insurance must cover body injury and disease arising out of and in the course of employment. The level of cover must be of at least £5 million, either for a single employer, or where a company has subsidiaries for that group (*Employers Liability (Compulsory Insurance) Regulations 1998, Regulation 3*). Most insurance companies have provided policies with up to £10 million of cover per incident since 1995, having previously offered effectively unlimited cover.

Cover under such policies is generally restricted to claims by those who are employees of the insured, which could exclude directors and other officers if they do not also hold contracts of employment, persons who are engaged as independent contractors or freelancers, or consultants retained on a specific project. Because the flexible nature of modern employment has produced much uncertainty about exactly where the line is drawn between a true employee and a person who provides services on a contract basis, it is common for insurers now to extend the scope cover under employer's liability policies to cover liability to self-employed people and labour only sub-contractors which arises during periods when they work for the insured employer.

The insurance cover obtained for these purposes will indemnify the employer against the compensation for the injury, including the consequential financial losses incurred by the claimant and the employer's liability to reimburse the Department for Work and Pensions the recoverable amount of benefits paid to the claimant on account of the recoupment scheme for compensation under the *Social Security (Recovery of Benefits) Act 1997*. The policy will generally also provide for payment by the insurer of the costs of legal representation incurred with the insurer's consent in investigating and defending a claim, or conducting any related criminal proceedings, inquest or fatal accident inquiry.

The policy underlying the *Employers' Liability (Compulsory Insurance) Act 1969* is that employees should always be entitled to compensation for work-related accidents and ill health. Consequently prohibitions have been placed on policy wording which would normally allow insurers to avoid liability in situations where an insured employer fails to comply with conditions designed to protect the insurer (*Employers' Liability (Compulsory Insurance) Regulations 1998, Regulation 2*).

Employers' liability insurers are not permitted to make their liability conditional upon prompt notification of any claim to them by an employer, or a term that an employer shall have exercised reasonable care to protect employees from harm or complied with health and safety legislation, or the proper retention by the employer of records of work activities, accidents or health monitoring. Insurers are also not permitted to rely – as against a claiming employee – on the agreed excess level of cover whereby the employer agreed to pay the first part of any claim. (In practice in this situation the insurer is required to meet the full amount of the employee's claim and to recover any excess from the employer.)

Transfers of undertakings

3.15 In situations where an employee has been transferred with the business to a new employer the employee's rights in relation to any claim for injury or disease will transfer as well. Under the *Transfer of Undertakings (Protection of Employment) Regulations 2006* ("TUPE") the original employer's rights to insurance cover under an employer's liability policy are transferred by law to the transferee employer (see *Bernadone v Pall Mall Services Group*). In contrast where the original company is sold (i.e. its shares are assigned to a new owner) the same corporate entity retains liability throughout for accidents.

Employers who have ceased trading

3.16 A company or individual employer which has ceased trading can still be sued, though if insolvent the permission of the court may be required. In effect the insurers will be required to defend or settle the claim on its behalf.

It is possible to make basic checks on the trading status of a company and whether it has been dissolved with the Registrar of Companies (www.companieshouse.gov.uk).

Where an employer company has been wound up since the accident or other events giving rise to a claim it may still be possible to pursue an action for damages directly against an insurer if one can be identified and traced. Under its Code of Practice for tracing lost and forgotten employers' liability policies the Association of British Insurers (ABI) will garner company information from its members at no cost to the enquirer (see Table 3.4). What is then required is for an application to the Court to be made to revoke the dissolution of the company under the *Companies Act 1985, Section 651* and this enables an action to proceed subsequently against that company. If judgement is obtained for the claimant, further action can be taken against the insurer to enforce the original insurance indemnity by virtue of the *Third Parties (Rights Against Insurers) Act 1930* (without the claimant having an obligation to pay any arrears of insurance premium).

Records of insurance policies

3.17 It is a legal requirement to retain copies of certificates of employers' liability insurance for forty years. This requirement only applies to policies in force on or after 31 December 1998.

Records of employers' liability insurance policies should be retained for as long as there is a possibility of a claim being made. A claim for personal injury can generally be made at any time with three years of an individual suffering an injury or disease and being aware of that he/she has a claim. There is no finite period that can be set for retention of policy details. Some ex-employees might bring claims in their

Table 3.4: Sources of information on previous insurance policies

- *The business insurance brokers*: Brokers will generally retain files of their principals' historic insurance cover.
- *Former brokers*: It may be necessary to check if they are still in business. Enquiries can be made by contacting:

Association of British Insurance Brokers	Tel: 0870 950 1790
14 Bevis Marks	Fax: 020 7626 9676
London	E-mail: enquiries@biba.org.uk
EC3A 7NT	*www.biba.org.uk*

- *Employers Liability Enquiry Unit*: This is an online search service from under the insurance industry's voluntary *Code of Practice for Tracing Employers' Liability Insurance Policies* (ABI, 1999).

To seek assistance contact:

Employers' Liability Enquiry Unit	Tel: 020 7216 7546
Association of British Insurers	Fax: 020 7216 8612
51 Gresham Street	E-mail: info@abi.org.uk
London	*www.abi.org.uk*
EC2V 7HQ	

The online enquiry form is at *www.abi.org.uk/ELCode/Admin/AEDUpdate.asp*

seventies or eighties or claims might be brought by estates of deceased employees. With the ease of storing information electronically it is therefore prudent to retain insurance records indefinitely.

Lost insurance policies are a common problem encountered in asbestos and other diseases with a long period of onset, either because of inadequate archiving procedures or because a business has a history or re-organisations, acquisitions or transfers of employees from other organisations subject to the provisions of TUPE.

Accident and emergency management

Management of Health and Safety at Work Regulations 1999 (*MHSWR*)

3.18 Legislation dealing with this aspect of risk management are a relatively recent phenomenon and have been driven mainly by legislation emanating from the European Union. The most important of these was the health and safety framework Directive *89/391/EEC*, adopted in 1989, which required all Member States to align their laws so that they contained provisions dealing explicitly, amongst other things, with emergency planning.

The Directive *89/391/EEC* was implemented in Britain by the *Management of Health and Safety at Work Regulations 1992*, which were later revised and replaced by regulations of the same name in 1999. These regulations are often abbreviated to *MHSWR*.

The *MHSWR* lay down fundamental legal rules which need to underpin systems for safety management in all organisations. They also dovetail with the provisions of the *HSWA* (Section 2) that require all employers with five or more employees to prepare and maintain a written health and safety policy describing their organisational and management arrangements for health and safety.

The Framework Directive sets out nine "General Principles of Prevention", the importance of which should not be underestimated (see Table 3.5). The term "principle" is slightly deceptive, in that the requirements contained are in fact substantive elements of employers' duties; *MHSWR, Regulation 4* requires that they form the basis of any preventive and protective measures which employers take. A safety management system needs to incorporate these principles, as well as methods for their implementation and ways of checking that they are being consistently applied.

The chief requirements of the *MHSWR* relevant to accident management are as follows.

Competent persons

3.19 An employer has to appoint one or more designated persons to assist in carrying out measures related to the prevention of occupational risks. This is both an advisory role and one which can involve all aspects of the safety management system. Usually it will be fulfilled by a designated health and safety manager, full time in larger organisations or as part of wider management responsibilities in smaller organisations. There is a presumption in the regulations that the competent persons are appointed from within the organisation and that a company should only enlist external services where there is no appropriate competent member of staff available. This appointment is distinct from other positions that required separate regulatory requirements for other specialist competent persons, for example those who provide occupational health services, or who undertake inspections of high-risk equipment such as lifts and escalators or pressure systems, or radiation advisers.

Table 3.5: The general principles of prevention

The following are the nine general principles of prevention set out in the Framework Directive (*89/391/EEC*) and the *MHSWR*:

Principle 1
Avoiding risks: Examples would be to re-design tasks to eliminate a hazardous operation, or to avoid use of a dangerous substance. This overlaps with principle 2, since it presupposes identification of risks.

Principle 2
Evaluating the risks which cannot be avoided: This stresses the need for formalised risk assessment and recording.

Principle 3
Combating the risks at source: This principle entails eliminating contact with a risk, as opposed to trying to reduce the effect on the person at risk.

Principle 4
Adapting the work to the individual: Specific issues mentioned in the directive are workplace design, choice and equipment and methods of work, the aim being to maximise comfort and concentration, and reduce scope for improvising dangerous work methods.

Principle 5
Adapting to technical progress: This involves use of new technology to develop safer working methods.

Principle 6
Replacing the dangerous by the non-dangerous or the less dangerous: This principle of "substitution" can be applied to chemicals, equipment, places of work and methods of working.

Principle 7
Developing an overall prevention policy which covers technology, organisation of work, working conditions and social factors: This overarching principle requires consideration of high-level safety policies and Safety Management Systems, and development of a company's "safety culture".

Principle 8
Giving collective protective measures priority over individual protective measures: This entails maximising safety benefits (e.g. through engineering measures) to all potentially affected, rather than reliance on personal protection, such as protective clothing.

Principle 9
Giving appropriate instructions to workers: Training, instruction and communication about risks and safety measures come under this principle.

Risk assessments

3.20 All employers are required to make a suitable and sufficient assessment of (a) the risks to the health and safety of their employees to which they are exposed whilst they are at work and (b) the risks to the health and safety of persons not in their employment arising out of or in connection with the conduct by of their undertakings. Where five or more employees are employed the employer must record the significant findings of the assessment; and any group of his employees identified by it as being especially at risk (*MHSWR, Regulation 3*).

Extensive practical advice is available from a range of organisations on risk assessment techniques, ranging from very basic investigation of simple hazards, to extremely sophisticated quantified risk assessment approaches designed for estimating the probability and consequences of major disasters. For further information see www.hse.gov.uk/risk/index.htm

Retention of risk assessment documentation by companies is prudent, sometimes even risk assessments which have been replaced on review as being out of date. Apart from applicable requirements of national law for recording and retention of the documents, there are a number of reasons why this may be necessary:

- As part of formulating a properly devised safety management system, it will be necessary to carry out the most extensive possible reviews and audits of safety performance within the company, for which prior risk assessment records may be essential for comparison.
- The company may be disadvantaged commercially if it cannot provide detailed evidence of risk assessment procedures: for example, in some business sectors it is necessary for regulatory purposes for businesses to undertake investigations of the competence and safety documentation systems of their suppliers. Even when this is not a regulatory requirement, a number of companies now subject their suppliers to a form of audit or questionnaire seeking evidence of adequacy of their risk assessment procedures.
- Historical risk assessment documentation may be the only significant source of defence evidence in the event of enforcement action taken under the *HSWA* or liability claims and may be used to demonstrate retrospectively the reasonableness of the company's decision-making or safety practices (and, for that matter, that the claimant received appropriate information or warning about the risks).

Documented health and safety procedures

3.21 This is a more extensive requirement than documenting and retaining risk assessments. *MHSWR, Regulation 4* requires that every employer shall make and give effect to such arrangements as are appropriate, having regard to the nature of his activities and the size of his undertaking, for the effective planning, organisation, control, monitoring and review of the preventive and protective measures identified as being necessary from risk assessments which have been carried out.

Typically these arrangements would include areas such as lines of management responsibility, practical control measures, safe systems of work, training and supervision, and other features of health and safety practice such as monitoring compliance and maintaining records of appropriate standards, codes and advisory literature. The essential requirement here is to define management obligations and to establish the structure which will underpin the basic statutory and common law risk control duties. To do this requires a very thorough investigation of the organisation's operations and management responsibilities at all levels of the business, and planning for how the relevant health and safety obligations will be dealt with within this structure.

Training and information for employees

3.22 The overarching requirement of *MHSWR, Regulation 13* is that in entrusting work to an employee, an employer has to take into account his capabilities. Employees (including incidentally senior managers) have to be provided with adequate health and safety training on being recruited and before being exposed to new or increased risks (e.g. because of a change in their responsibilities or on the introduction of new technology). They also have to be given comprehensible relevant information on risks, safety procedures and emergency arrangements. Training should not be at the workers' expense and must be during normal working hours.

The majority of safety management systems are designed on a model set out in the HSE's seminal guidance *Successful health and safety management* (HS(G)65) (HSE, 1997) which is taken further in BS 8800: 2004: *Occupational Health and Safety Management Systems Guide* (BSI, 2004). This offers two approaches: the HS(G)65 model, and a similar model based on ISO 14001, which is an environmental management standard that is slightly more detailed. The HS(G)65 system entails establishing a policy with targets and goals, organising to implement it, setting forth practical plans to achieve the targets or goals, measuring performance against targets or goals and reviewing performance. Auditing overlies the whole process. This approach is often shortened to the pneumonic – POPIMAR (Figure 3.1):

Policy: Effective health and safety policies set a clear direction for the organisation.

Organising: An effective management structure and arrangements are in place for delivery of the policy.

Figure 3.1: Key elements of successful health and safety management

Policy
Organising
Planning and implementing
Measuring performance
Reviewing performance
Auditing
Policy development
Organisation development
Developing techniques of planning, measuring and reviewing
Feedback loop to improve performance
Information reporting
Control link

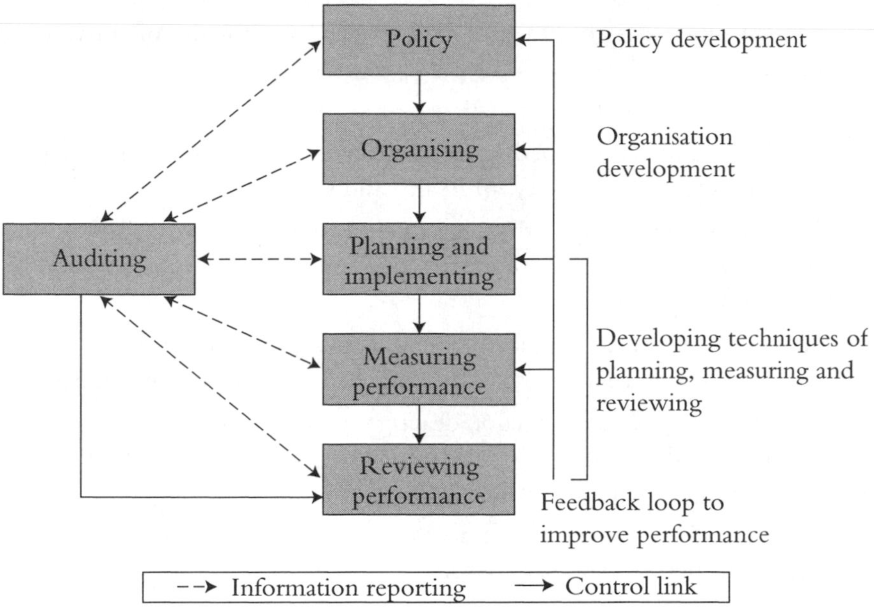

Source: HSE (1997)

Planning and **I**mplementing: There is a planned and systematic approach to implementing the health and safety policy through an effective health and safety risk management system.

Measuring/monitoring: Performance is measured against agreed standards to reveal when and where improvement is needed.

Auditing and **R**eviewing Performance: The organisation learns from all relevant experience and applies the lessons.

These key elements are linked in the form of a continual improvement loop.

BS 8800 itself is merely a guide; it is not a certifiable standard. However OHSAS 18001: 1999 *Occupational Health and Safety Management Systems Specification* (BSI, 1999) is a form of standard for which independent third party certification of conformity is available.

BS 8800 builds on the H(S)G65 approach – outlined above – but starts off the loop with an initial status review in order to establish "Where are we now?" BS 8800 also includes an alternative approach aimed at those organisations wishing to base their OSH management system not on HS(G)65 but on BS EN ISO 14001, the environmental management systems standards which again incorporates an initial status review (Table 3.6).

Table 3.6: Health and safety policy checklist

- ☐ Does an up-to-date health and safety policy exist for the organisation or location?
- ☐ Is the policy regularly reviewed?
- ☐ Has the policy been signed (and dated) by a director or senior manager who has (site) responsibility for health and safety?
- ☐ Does the policy commit the organisation or location to achieve a high level of health and safety performance, with legal compliance seen as a minimum standard?
- ☐ Does the policy provide that adequate and appropriate resources – time, money, people – will be provided to ensure effective policy implementation?
- ☐ Does the policy contain a section dealing with the organisational framework – people and their duties – so as to facilitate effective implementation?
- ☐ Does the policy allow the setting and publishing of objectives for the organisation or location and for individual directors, managers, competent persons and supervisors?
- ☐ Are the high-level responsibilities of directors (or the equivalent top tier of management in the organisation) set out in a manner which is consistent with the HSC's guidelines *Directors' Responsibilities*?
- ☐ Does the policy clearly place the prime responsibility for the management of health and safety on to line management?
- ☐ Are the statutory principles of prevention embedded in the policy?
- ☐ Does the policy ensure that employee involvement and consultation take place in order to gain their commitment to it and its implementation?
- ☐ Does the policy contain a section dealing with the need for risk assessments to be undertaken – both general and specific – their significant findings acted upon, and an assessment record-keeping system maintained?
- ☐ Does the policy deal with the need to document the arrangements, systems and procedures by which the policy will be implemented on a day-to-day basis?
- ☐ Is the need to have accident and emergency arrangements addressed?
- ☐ Does the policy require that all employees – including agency staff and temporary employees – at all levels receive appropriate training and information to ensure that they are competent to carry out their duties?
- ☐ Does the policy contain a section dealing with the monitoring and measurement of health and safety performance?

- ☐ Does the policy require that audit systems are in place?
- ☐ Has the policy been effectively brought to the attention of all employees and agency staff and temporary employees?
- ☐ Are copies of the policy on display throughout the organisation or location?
- ☐ Is the policy understood, implemented and maintained at all levels within the organisation or location via the use of suitable arrangements?

Legal requirement for procedures for serious and imminent danger and danger areas (Table 3.7)

3.23 This aspect of the Framework Directive *89/391/EEC* was given effect principally by the *MHSWR Regulations 8–10*, which deal with the planning and implementation of procedures to deal with serious imminent dangers and danger areas, and the communication issues which need to be addressed in these situation.

Table 3.7. MHSWR, Regulation 8

MHSWR, Regulation 8 consists of the following requirements:

1. Every employer shall:
   (a) establish and where necessary give effect to appropriate procedures to be followed in the event of serious and imminent danger to persons at work in his undertaking;
   (b) nominate a sufficient number of competent persons to implement those procedures in so far as they relate to the evacuation from premises of persons at work in his undertaking;
   (c) ensure that none of his employees has access to any area occupied by him to which it is necessary to restrict access on grounds of health and safety unless the employee concerned has received adequate health and safety instruction.

➧

2. Without prejudice to the generality of paragraph (1)(a), the procedures referred to in that sub-paragraph shall:
   (a) so far as is practicable, require any persons at work who are exposed to serious and imminent danger be informed of the nature of the hazard and of the steps taken or to be taken to protect them from it;
   (b) enable the persons concerned (if necessary by taking appropriate steps in the absence of guidance or instruction and in the light of their knowledge and the technical means at their disposal) to stop work and immediately proceed to a place of safety in the event of their being exposed to serious, imminent and unavoidable danger;
   (c) save in exceptional cases for reasons duly substantiated (which cases and reasons shall be specified in those procedures), require the persons concerned to be prevented from resuming work in any situation where there is still a serious and imminent danger.
3. A person shall be regarded as competent for the purposes of paragraph (1)(b) where he has sufficient training and experience or knowledge and other qualities to enable him properly to implement the evacuation procedures referred to in that sub-paragraph.

Sub-paragraphs (1) (a) and (b) of *MHSWR, Regulation 8* are expressed as being duties which an employer owes to anyone at work in his undertaking, which would include contractors and visiting clients but also others such as emergency services personnel who might be present only where the risk of danger is actually encountered. This is a wider obligation than the common law duty of an employer or occupier and, as will be seen below, it extends to a positive obligation to communicate information on the procedures to other employers and self-employed persons who could be affected.

Identification of relevant dangers

3.24 The requirements of *MHSWR, Regulation 8* are closely connected with the other parts of the *MHSWR,* which require employers to undertake and record risk assessment of the activities, to set out effective arrangements for risk management and to appoint competent persons to assist in ensuring compliance with health and safety legislation. In order to comply properly with *Regulation 8* an employer first needs to:

- identify risks that are of a magnitude that an emergency response could be required;

- determine the nature of the appropriate emergency responses; plan the emergency measures;
- identify the internal personnel and external services who may be needed to implement the measures, and determine what information needs to be provided to those;
- identify parts of the workplace which are danger areas (described in HSE guidance as work environments "where the level of risk is unacceptable without taking special precautions").

This process is necessary because the *MHSWR* wording is notably unspecific about the circumstances in which the duty to devise procedures for services and imminent dangers will arise. There are various additional health and safety regulations which address specific hazardous workplaces, and other provisions contain a separate regime imposing requirements for fire precautions. These are outlined further below.

As a minimum virtually all employers, even those whose operations might be considered very low risk, for example because they are purely office based, will need to have evacuation arrangements in the event of fire or a total power failure, and provision for first aid. Most businesses will also require some procedures to deal with bomb threats in their own premises or the immediate locality. Threats posed by the local environment, for example serious flooding, will need to be considered, as will dangers posed by catastrophic control failures at neighbouring premises such as ones used for gas storage.

More complex businesses involving processes and plant face risks of the same character as those above but added to, and possibly combined with, the risks which flow from process or equipment failures. Arrangement for effective evacuation will necessarily be a foremost consideration, but the regulations also require employers to restrict access to danger areas. The corollary of this is that locations or activities which give risk to serious dangers (and which are nevertheless justified in being used by appropriate risk assessment) should generally be isolated, with access restricted to authorised persons only and with containment measures employed where this is practicable. This will in turn require safe systems of work to be devised for those who work in the areas, taking into account all the other relevant health and safety regulations for the tasks involved.

Establishment of procedures

3.25 To meet the obligation to establish procedures they should normally be in written form. Electronic storage is permissible but as the procedure must always be retrievable for use in an emergency physical copies should be available.

Inevitably the nature of the procedures required will depend on the nature of employer's business and its circumstances. The ACoP to the *MHSWR* lays down a number of general guidelines for what should be included (HSE, 2004):

- The nature of the risk and how to respond to it.
- Procedures geared to the nature of the danger, as far as is practicable.
- Responsibilities of employees or groups of employees who may have specific tasks to perform, or who have been trained to deal with specific emergencies. Appropriate preventative and protective measures should be in place for those employees.
- The role, responsibilities and authority of the nominated competent persons, and arrangements for ensuring other employees are aware who the relevant competent persons are and understand their roles.
- Any specific obligations the employer needs to comply with in dealing with emergencies.
- Details of when and how the procedures are to be activated, allowing for the fact that if events develop rapidly there may not be opportunities to give instruction before people need to escape to places of safety. There may be a need to continence evacuations before or during attempts to bring an emergency under control.

The *MHSWR* ACoP also highlights the need to control carefully the return to any dangerous areas or the resumption of work after an emergency before they are made safe, and indicates that only in exceptional circumstances (principally where there is a threat to human life) would re-entry be justified by emergency services workers. *MHSWR, Regulation 8(2)(c)* requires that such exceptional circumstances be specified in the written procedures as a measure of competence rather than any specific qualifications and it is likely to require reinforcement of competence through regular refresher training, drills or simulations. (An employer who permits untrained employees to attempt a rescue mission for a colleague who is still in grave danger would probably be contravening these regulations.)

These guidelines omit any reference to practising emergency arrangements, in contrast with standard conditions for drills which are for example a feature of compliance with fire legislation. It is implicit in *Regulation 8* that the effectiveness of emergency procedures is subject to an assurance regime, and in order to comply fully with the provisions it is suggested that the procedures are tested once at the absolute minimum by drills/simulated exercises; further periodic testing is likely to be required however in most businesses.

CASE STUDY 4

3.26

> ***Hidden dangers to rescuers operating in confined spaces***
>
> Two men were engaged in laying pipes beneath old office buildings. They were using a petrol-powered generator to power a drill. Worker A left the vault in which the pipes were being laid for a few minutes to check his instructions. Worker B continued to operate the drill. Upon his return, Worker A found Worker B unconscious and fitting. Worker A called for assistance from his supervisor, but they were unable to effect a rescue as both became increasingly dizzy and nauseous. The emergency services were called, and paramedics extricated Worker B from the vault, although not after one of the paramedics was forced to leave the rescue site, overcome with dizziness. The vault was later found to have been contaminated with carbon monoxide.

Competent persons

3.27 A competent person for these purposes differs from the competent person appointed under *MHSWR, Regulation* 7 in that the former's competence needs to relate specifically to the carrying out of evacuation procedures, while the latter's competence is of a more general nature relating to the compliance requirements of the employer. This will generally require bespoke training.

Contact with external services

3.28

Table 3.8: MHSWR Regulation 9

Every employer shall ensure that any necessary contacts with external services are arranged, particularly as regards first aid, emergency medical care and rescue work.

The obligation under the *MHSWR*, to arrange for necessary contacts with external services is often viewed as one to have immediate access to appropriate hospital and other emergency services, but it is more extensive than this. Appropriate external services may well comprise private services such as providers of decontamination and clean-up facilities, or specialist engineers needed to disconnect or re-connect critical equipment, and training and refresher courses for first aid personnel. Again, compliance with this regulation requires careful consideration of the information derived from thorough risk assessment processes.

In some situations it may be difficult to persuade an inspector that appropriate arrangements are in place if there is no formal retainer for the services of a particular organisation. A formal retainer agreement is not legally required for these purposes but it is advisable for evidential purposes to at least maintain a record of recent correspondence with an external service provider confirming their call-out arrangements and the terms upon which services will be provided.

Regulation 9 can also apply to the actual means of making contact with external services. For example, provisions may need to be made for emergency contact by lone workers and others in remote locations where the employers will not be able to have direct control over communications. For high-risk activities it may also be appropriate to include in the procedures back-up communications facilities equipped with all necessary contact information in the event that the employer's principal communication systems are rendered inoperable.

Specific regulations requiring emergency planning

Major accident hazard sites

3.29 The *Control of Major Accident Hazards Regulations 1999* (*COMAH*) concern organisations which keep quantities of dangerous substances above certain designated thresholds. Duties apply differently in relation to top tier and lower tier levels, the details of which are beyond the scope of this chapter. Central to *COMAH* lower tier duties is the Major Accident Prevention Policy (MAPP) which describes the safety management system for preventing major accidents and, in particular, the plans for dealing with emergencies and for testing and reviewing responses. Top tier duty holders are required to go further and also produce on-site and off-site emergency plans for submission to the competent authority for consideration before operations are permitted on site.

Detailed guidance on emergency planning is found in the HSE guidance note *Emergency Planning for Major Accidents* (HSE, 1999). The local authority for a top tier establishment is in turn required to develop an emergency plan using information supplied by the operation. *COMAH Regulation 11* requires review and if necessary revision, of emergency plans, at least every three years.

Radioactive materials

3.30 The *Radiation (Emergency Preparedness and Public Information) Regulations 2001* require operators of sites holding or transporting radioactive substances in quantities above certain threshold amounts to observe a number of procedures to provide the HSE with information, including plans to deal with foreseeable radiation emergencies, and to review and test these plans at least every three years. Further information can be found in HSE guidance note *Guide to the Radiation (Emergency Preparedness and Public Information) Regulations 2001.*

For smaller scale operations involving work covered by the *Ionising Radiations Regulations 1999* there is a requirement to prepare contingency plans for foreseeable radiation accidents to restrict exposures and to undertake rehearsals at suitable intervals.

Road and rail transport

3.31 The *Carriage of Dangerous Goods and Use of Transportable Pressure Equipment Regulations 2004* place obligations on operators relating to information about and action to be taken in the event of any emergency.

Ports and harbours

3.32 A harbour authority has an obligation to prepare and keep up to date an emergency plan for emergencies which involve or could affect dangerous substances which are brought in to or handled in the harbour, and is required to give notice of the plan to those responsible for putting it into effect. See the *Dangerous Substances in Harbour Areas Regulations 1987.*

Air transport

3.33 The *Air Navigation Order 2000* requires most public transport aircraft to have an operations manual, one of the components of which should be a description of procedures for making and carrying dangerous goods and the action to be taken in the event of an emergency involving dangerous goods.

Shipping

3.34 The operator and master of a vessel are required to ensure all employees are familiar with action to be taken in an emergency involving packaged dangerous goods or marine pollutants (*Merchant Shipping (Dangerous Goods and Marine Pollutants) Regulations 1997*).

Construction work

3.35 The *Construction (Health, Safety and Welfare) Regulations 1996* (SI 1996 No. 1592), Regulation 20 provides that:

> "Where necessary in the interests of the health and safety of any person on a construction site, there shall be prepared and, when necessary implemented suitable and sufficient arrangements for dealing with any emergency, which arrangements shall include procedures for any necessary evacuation of the site or any part thereof."

These Regulations (which are replaced from April 2007 by new *Construction Design and Management Regulations 2007*) make similar provision to the *MHSWR* for emergency contacts and person nominated to implement arrangements and also provide for emergency routes and fire-fighting arrangements.

Asbestos

3.36 Where asbestos is being used, removed or repaired the *Control of Asbestos at Work Regulations 2006* place an obligation on an employer to ensure that procedures for accidents, incidents and emergencies are prepared, including relevant safety drills which are tested at regular intervals. Information on emergency arrangements must be provided and suitable warning and communication systems must be in place to enable remedial action or rescue operations to be activated immediately when an event occurs.

The Regulations also positively require an employer to:

- mitigate the effects of the event;
- restore the situation to normal;
- inform any person who may be affected.

In addition, only persons with appropriate respiratory protective equipment, protective clothing and other specialised safety equipment may be permitted to enter an affected area.

Confined spaces

3.37 Under the *Confined Spaces Regulations 1997* arrangements must be made for the rescue of anyone who would be trapped, including the provision of appropriate equipment for safe access, first aid and resuscitation equipment.

Compressed air

3.38 The *Work in Compressed Air Regulations 1996* contains provisions similar to those which apply to confined spaces. In addition they require facilities for medical treatment.

Gas supplies

3.39 The *Gas Safety (Management) Regulations 1996* contain various provisions relating to emergencies including provision of a national telephone reporting service for leaks, designation of responsibility for preventing escapes from different parts of the network and a duty on gas conveyors to take steps to avert the danger to any premises. Conveyors of gas are required under these regulations to hold safety cases which, among other things, describe arrangements for gas escapes and investigations, and the appointment of emergency services providers. Gas suppliers, as defined by the *Gas Safety (Installation and Use) Regulations 1998*, have specific responsibility to act upon reports of gas leaks by cutting off the supply or otherwise preventing the escape.

Diving

3.40 Under the *Diving at Work Regulations 1997*, a diving contractor is required to ensure there are sufficient persons available with the competence to deal with foreseeable emergencies and sufficient equipment is available.

Mines and quarries

3.41 The *Quarries Regulations 1999* require there to be means of escape and rescue facilities, written safety instructions for emergency equipment, appropriate training, safety drills and adequate warning and communications for emergencies. *The Escape and Rescue from Mines Regulations 1995* contain broadly equivalent requirements and require the production of emergency plans.

Fire protection

3.42 *The Regulatory Reform (Fire Safety) Order 2005 (RRO)*, which came into force on 1st October 2006 in England and Wales, consolidates previous fire protection legislation. The requirements it places on employers are broadly similar to the requirements which existed under the old regimes, namely the *Fire Precautions Act 1971* and the *Fire Precautions (Workplace) Regulations 1997*, though there are some key differences. These include:

- The removal of the fire certification scheme under which a Fire Authority imposed a Fire Certificate requiring particular fire safety provisions in particular types of premises. Fire Certificates ceased to have effect on 1st October 2006, as did any requirements demanded therein. It is still worth retaining the Certificate however as it will provide good evidence of what the Fire Authority acknowledged as a reasonable standard of precautions.

- The undertaking of a fire safety risk assessment by the newly designated "responsible person" as the main provision by which fire precautions are identified. Such a risk assessment was already required in respect of workplaces; the new order requires it to be undertaken in almost all non-domestic premises (*Articles 6 and 9*).
- The introduction of a statutory duty under which the "responsible person" is required to take positive action to prevent fires occurring, rather than simply mitigating the effects of fire and safeguarding people if a fire is ignited (*Article 8*).
- The introduction of a statutory duty to ensure that equipment and facilities intended for Fire Brigade use and specified in other legislation, such as the Building Regulations, are maintained under a suitable system of maintenance (*Article 13*).

It should be noted that the requirements stipulated under the Building Regulations 2000 remain almost unchanged.

The *RRO* separates the law relating to health and safety from the law relating to fire safety, as is apparent in *Article 47* which disapplies the *HSWA* in relation to fire precautions. The change is, however, largely cosmetic as many of the requirements under health and safety law which applied to general fire safety have simply been reproduced in the *RRO* with little material change being made in terms of what is required.

Similar changes were introduced in Scotland under Part 3 of the *Fire (Scotland) Act 2005*, which also came into force on 1st October 2006. Guidance on compliance with this Act can be found at www.infoscotland.com/firelaw/files/Summary_Guide_Full_doc.pdf

The responsible person and the application of the RRO to certain premises

3.43 The premises to which the *RRO* applies are all non-domestic premises with limited exceptions laid out in *Article 6*. The exceptions include premises such as offshore installations, ships, aircraft, locomotives and rolling stock.

The *RRO* demands that a "responsible person" ensure that the duties set out in the *RRO* are complied with and its practical requirements met. Article 3 of the *RRO* defines the "responsible person" in relation to a workplace as:

- The employer, provided that the workplace is under his control to any extent.
- If the premises are not a workplace under the employer's control, then the responsible person is defined as:
  - the person who has control of the premises (as occupier or otherwise) in connection with the carrying on by him of a trade, business or other undertaking; or

- the owner, where the person in control of the premises does not have control in connection with the carrying on by that person of a trade, business or other undertaking.

The designation of exactly is responsible as "responsible purpose" is one of the most problematic aspects of the *RRO*. It should not be confused with individuals appointed as "competent persons" under *Article 18* of the *RRO* or under the *MHSWR*. Usually the "responsible person" will be the corporate body or other organisation occupying the premises for its business or other activities.

Further confusion may arise because premises may be used by employees of two or more organisations at the same time, for example a bank with outsourced IT personnel working at its offices, a factory with embedded maintenance contract staff or a hotel owner with a management contractor running most of the operations. The *RRO* deals with this by in holding the employers concerned responsible at the same time and requiring them to co-operate and to co-ordinate their compliance arrangements (*Article 22*).

In a scenario such as this it is of no effect and not permissible for the parties to agree (informally or contractually) that only one of them will shoulder all the responsibilities of the "responsible person". This is because a statutory duty cannot be assigned or delegated to another person. They may though agree to undertake the necessary tasks jointly and to share resources and procedures.

To complicate matters further, the fire safety duties are not necessarily restricted to those deemed "responsible persons". *Article 5* provides that *any* person who has, to any extent, control of premises shall have the same fire safety duties, so far as the requirements relate to matters within his control. Furthermore, where a person has, by virtue of any contract or tenancy, an obligation of any extent in relation to (a) the maintenance or repair of any premises, including anything in or on premises or (b) the safety of any premises, that person is treated as being a person who has control of the premises to the extent that these are contractual or other obligations.

The result is a need for the most careful co-operation between all operators, property managers and agents contractors with shared site responsibilities to understand their respective roles in meeting the fire safety duties.

In the remainder of this fire safety section where the duties of responsible persons are described it should assumed that the same duties apply to others who have the requisite degree of control described in *Article 5*.

Under *Article 8* the responsible person must "take such general fire precautions as will ensure, so far as is reasonably practicable, the safety of any of his employees". In respect of relevant persons who are not employees, the responsible person must "take such general fire precautions as may reasonably be required in the circumstances of the case to ensure that the premises are safe". A relevant person is someone who is lawfully on the premises or in the immediate vicinity of the premises, and who could be placed at risk from a fire on the premises. This limb of Article 8 will clearly be of application to sub-contractors in much the same way as *HSWA Sections 3 and 4* apply to non-employees (see page XXX).

Risk assessment

3.44 *Article 9* of the *RRO* requires the responsible person to "make a suitable and sufficient assessment of the risks to which relevant persons are exposed for the purpose of identifying the general fire precautions he needs to take to comply with the requirements and prohibitions imposed on him by or under this Order."

This risk assessment is similar to the assessment that was required under the *Fire Precautions (Workplace) Regulations 1997*. However, as the new assessment is now formally a requirement of the *RRO*, references on documentation should refer to the *RRO*. Although fire certificates have ceased to have legal status they provide a useful starting point when it comes to assessing the building, its means of escape, the presence of risk and devising ways in which to eliminate or reduce them.

The risk assessment must cover the minimisation of fire risk and any risk associated with the use of dangerous substances. The Department for Communities and Local Government has produced an "entry level" leaflet outlining general guidance on compliance with the *RRO* (DCLG, 2006). There are also guidance notes on how to undertake fire safety risk assessment in the following types of premises (Table 3.9):

- offices and shops;
- factories and warehouses;
- sleeping accommodation;
- residential care premises;
- educational premises;
- small and medium places of assembly;
- large places of assembly;
- factories and warehouses;
- theatres and cinemas;
- outdoor events;
- health care premises;
- transport premises and facilities.

These guides are available online from www.communities.gov.uk.

Table 3.9: Fire risk assessments

To outline, there are five recommended steps to carrying out the fire safety risk assessment:

1. Any hazards in the premises need to be identified. These include sources of ignition, sources of fuel and sources of oxygen.
2. People who are particularly at risk from fire then need to be identified. These include those working near fire dangers and those working alone or in isolated areas.
3. The risk then needs to be evaluated and steps taken to remove or reduce the hazard and reduce the risk. The implementation of a safe-smoking policy is one such way by which the risk of fire is reduced.
4. Records should be made of the risks and those at risk. Emergency plans should be drawn up, and employees should be instructed in what to do in case of fire. Information and training should also be given in respect of the risks that exist in the premises.
5. The fire safety risk assessment needs to be kept up to date. It will need to be re-examined if the responsible person considers it no longer to be valid or when there is a significant change to the level of risk present in the premises.

Performance of this risk assessment will allow the responsible person to identify the "preventive and protective measures" which need to be taken in order to comply with the provisions of the RRO.

Principles of prevention

3.45 *Article 10* requires the responsible person to adhere to the principles contained in the *RRO* when implementing preventative and protective measures in respect of fire risk. The same principles are set out in the current *MHSWR*, see 3.18, and broadly speaking require the elimination and minimisation of risk using the development of prevention policies and instruction to employees.

Fire safety arrangements

3.46 *Article 11* of the *RRO* requires the responsible person to make and effect fire safety arrangements. The arrangements apply to the planning, organisation, control, monitoring and review of the protective and preventive measures, and must be appropriate to the size and nature of the undertaking.

The arrangements must be written down where an employer employs more than five people, where a licence under an enactment is in force in respect of the premises, or where an alterations notice requiring a record to be made of the fire safety arrangements is in force in respect of the premises.

The requirement under *Article 11* will usually be covered by existing health and safety procedures which detail fire safety arrangements. Similarly, existing specific fire safety procedures will normally meet the requirement. Health and safety policies should, however, make explicit reference to the fact that the policy has been drawn up in compliance with *RRO*.

Dangerous substances

3.47 *Article 12* stipulates that the responsible person must eliminate or reduce the risk posed to relevant persons by the use of dangerous substances. For the purposes of the *RRO*, dangerous substances are defined as those materials which are classified as explosive, oxidising, extremely flammable, highly flammable or flammable. It also extends to materials which, owing to their chemical properties or use, give rise to a potential risk, and to dust or fibres which could form an explosive atmosphere.

Where possible, less dangerous alternatives should be used, but where it is not reasonably practicable to eliminate completely the risk, the responsible person must apply measures consistent with the risk assessment to ensure that the risk is controlled and the detrimental effects of a fire mitigated.

Article 16 sets out the emergency measures which apply specifically to premises where dangerous substances are present. The responsible person is under a duty to make sure that relevant persons are provided with information on emergency arrangements, that suitable warning systems are in operation, and that, where necessary, escape facilities are provided and maintained.

The provisions relating to dangerous substances are substantially similar to those set out in the *Dangerous Substances and Explosive Atmospheres Regulations 2002*.

Fire fighting and fire detection

3.48 *Article 13* of the *RRO* places a duty upon the responsible person to make sure that premises are, to the extent that it is appropriate to do so, equipped with appropriate fire-fighting equipment, and with fire detection systems and alarms. Whether a measure is deemed "appropriate" will depend on the size and use of the premises, and on the physical and chemical properties of any substance that might be found therein.

The responsible person must, where necessary, elect competent persons to implement measures for fire fighting on the premises, and must also arrange necessary contacts with the emergency services. Non-automatic fire-fighting equipment provided should be easily accessible, simple to use and indicated by signs.

Emergency routes and exits

3.49 The responsible person must ensure that the routes to emergency exits are kept clear at all times. The safety of relevant persons is to be safeguarded by complying with the requirements laid down in *Article 14(2)*, which stipulate the need for the emergency routes and exits to lead as directly as possible to safety, to be of suitable dimension and adequately distributed, and to be illuminated and indicated by signs.

Procedures for serious and imminent danger and for danger areas

3.50 *Article 15* places a duty on the responsible person to establish and effect procedures to be followed in the event of a serious threat to relevant persons. Specifically, the article requires safety drills to be undertaken. Relevant persons exposed to serious and imminent danger are to be informed as to the nature of the hazard and the measures that are to be taken to protect them from it.

Maintenance

3.51 Under Article 17 the responsible person must ensure that the premises and the equipment and facilities required under the Order or under other fire safety legislation are maintained in "an efficient state, in efficient working order and in good repair." *Article 17* then goes on to require the responsible person to liase with other occupiers of the premises, including occupiers of premises to which the *RRO* does not apply, in order to ensure blanket compliance.

In addition to requiring the maintenance of the premises and equipment, *Article 17* also stipulates that a "suitable system of maintenance" be in place. Written procedures and records will therefore be of importance in establishing that such a system has been developed and adopted.

Article 38 contains a specific requirement relating to the maintenance of measures provided for the protection of fire fighters. This includes equipment required under the *Building Regulations* 2000. Again, a suitable system of maintenance must be in place.

Safety assistance

3.52 *Article 18* of the *RRO* requires the responsible person to appoint one or more competent persons to assist him or her in undertaking the preventive and protective measures. Competence for the purpose of Article 18 is defined as having sufficient training and experience or knowledge to be able to assist with the preventive and protective measures. This article makes it acceptable for employers to seek professional advice on fire safety.

Information for employees

3.53 *Article 19* establishes the information which must be passed on to employees. Accordingly, comprehensible and relevant information on the following subjects should be disseminated:

- The risks to employees identified in the risk assessment.
- The preventive and protective measures in place.
- The procedures and measure referred to in *Article 15* (in relation to safety drills, etc.).
- The identities of persons nominated under *Article 13* in respect of the implementation of fire-fighting procedures.
- The risks notified to the responsible person under *Article 22* (see below).

Where the responsible person elects to employ a child he or she must provide that child's parents with information on the risks to the child highlighted in the risk assessment, the preventive and protective measures in place, and the risks notified to the responsible person under *Article 22*.

Article 19 also makes stipulations in respect of information to be supplied to employees concerning dangerous substances. The responsible person must inform employees of the details of the dangerous substance, including its name and the risks it poses in addition to any legislation which applies to the substance. Access to relevant data sheets is also to be afforded to employees.

Under *Article 21* the responsible person has a duty to ensure that employees receive safety training. Such training must be provided when the employee is first employed and should be repeated periodically. It should also be repeated when the employee is exposed to new or increased levels of risk.

Article 23 sets out the general duties of employees when at work. Employees must:

- take reasonable care for their safety and the safety of other relevant persons who may be affected by his acts or omissions at work;

- co-operate with their employers to enable compliance with the Order;
- inform their employers or colleagues with specific responsibility for the safety of their fellow employees of any work situation which represents a serious and immediate danger to safety, and of any matter which could represent a short-coming in the employer's protection arrangements.

Other employers

3.54 *Article 20* of the *RRO* places a requirement on the responsible person to ensure that the employer of other employees working on the premises is provided with information relating to the risks posed to those employees and the preventive and protective measures that have been taken in view of the risks. The responsible person also needs to provide the employees with instructions and information relating to the risks posed to them.

Enforcement, offences and penalties

3.55 The various provisions of the *RRO* are variously enforced by local fire and rescue authorities, the HSE, the fire service maintained by the Secretary of State for Defence, local authorities or fire inspectors. Each of these enforcing authorities has responsibility for a particular type of premises.

Under *Article 27* inspectors may do anything necessary to ensure that the *RRO* is complied with. They have the power to enter premises to ensure that the *RRO's* requirements are being met, and can require the production of information. Fire and rescue authorities can also be granted the same powers if they obtain written authorisation from a fire inspector.

Article 29 gives the enforcing authority the ability to serve an alteration notice on a responsible person. This step will be taken where the authority considers that the premises constitute a serious risk to relevant persons or may constitute such a risk if a change is made to them or their use. The notice will require the responsible person to notify the enforcing authority before any alterations are made. This is a separate requirement from applications for Building Regulations approval. Failure to comply with the alteration notice is an offence.

Enforcement notices can also be served if an enforcing authority is of the opinion that the responsible person has failed to comply with a requirement of the *RRO*. Such a notice will require that steps are taken to remedy the failure within a specified period. Failure to comply with the notice is an offence.

Prohibition notices will be served by an enforcing authority where it considers that the use of the premises involves or will involve a risk to relevant persons "so serious that use of the premises ought to be prohibited or restricted." In assessing whether such a notice is necessary or desirable, particular relevance will be given to anything which affects the ability of relevant persons to escape from the premises in the event of a fire. Failure to comply with any requirement or prohibition imposed under *Articles 8–22* where that failure poses relevant persons at risk of death or serious injury constitutes an offence. This, as with the failures to comply with the alterations or enforcement notices, could be punishable with a fine and/or imprisonment.

Once an alteration, enforcement or prohibition notice has been served the recipient has 21 days in which to appeal to the magistrates court. Upon appeal the court may either affirm or cancel the notice, making such modifications as it sees fit. Appeals against alterations or enforcement notices have the effect of suspending the notice; prohibition notices, however, remain effective. Under *Article 36* the Secretary of State has a power to determine disputes between responsible persons and enforcing authorities relating to the measures to be deployed to remedy any failures under the *RRO*.

Additional fire precautions requirements

3.56 A large number of extra legal controls and potential obligations can arise under miscellaneous legislation which govern specific activities or sectors. These are summarised in Table 3.10.

First aid

3.57 The legal requirements for the provision of first aid arrangements apply to all employers and the self-employed. First aid for these purposes means either treatment of minor injuries which would otherwise not need or not receive medical treatment, or where medical treatment will be needed, treatment for the purposes of preserving life and minimising the consequences of injury or illness until medical help is obtained (*Health and Safety (First-Aid) Regulations 1981, Regulation 2*). The provisions extend to any injury or illness, whether work related or not.

Health and Safety (First-Aid) Regulations 1981

3.58 The *Health and Safety (First-Aid) Regulations 1981* contain the principal requirements for:

(a) appointment for first aid and other staff responsible for helping the injured;

(b) provision of equipment and first aid rooms.

These provisions are described more fully in Chapter 4.

Table 3.10: Legislation governing specific activities and sectors

Sector	*Reference*	*Requirements*
Quarries	*Quarries Regulations 1999*	No use of naked flames in quarries
Gas	*Gas Safety (Installation and Use) Regulations 1998*	Person searching for gas leak must not use of source ignition.
Work Equipment	*Work in Compressed Air Regulations 1996*	Fire-fighting equipment must be available, and no person shall smoke when working.
Construction	*Construction (Design and Management) Regulations 2007*	Carry out steps to prevent injury from construction work explosions, flooding or asphyxiation. Fire detection fighting equipment to be held on site where necessary.
Flammable Substances	*Pipelines Safety Regulations 1996* *Petroleum Spirit (Consolidation) Act 1928*	Requirements for emergency shut down. Licence required for storing petrol, for labelling and safe storage imposes requirements for labelling and safe storage.
Gas	*Gas Safety (Management) Regulations 1996*	Gas escapes. If gas leak causing fire, upstream provider to shut off source, and investigate source of leak.
Offshore Installations	*Of Shore Installations (Prevention of Fire and Explosion and Emergency Response) Regulations 1995*	Duties for protecting installation from fire: risk assessments, preparation for emergencies, response plan, safety equipment, training, evacuation and lifesaving appliances.
Electricity	*Electricity at Work Regulations 1989*	Electrical equipment exposed to potential damage or bad weather must be constructed so as to minimise danger risk.
Docks	*Dangerous Substances in Harbour Area Regulations 1987*	Those handling dangerous substances shall observe safety precautions. Fire-fighting equipment must be available.
Flammable Substances	*Highly Flammable Liquids and liquefied Petroleum Gases Regulations 1972*	Requirements for storage of highly flammable liquids, marking, precautions against spills and escape, no sources of ignition, escape in case of fire, means to fight fire available.
Sports Ground	*Safety of Sports Grounds Act 1975* and the *Fire Safety and Safety of Places of Sport Act 1987*	Requirement for safety certificate for a sports ground.
Entertainment and Leisure	*Cinematography (Safety) Regulations 1955*	Seats to allow free exits, premises to have fire equipment, no smoking allowed and heating appliances to be covered with guards.

(*Continued*)

Table 3.10: (Continued)

Sector	*Reference*	*Requirements*
	Theatres Act 1968	Requirement for premises to be licensed if used for public plays.
	Gaming Act 1968	Gaming premises must be licensed – application to be approved by fire authority. Inspector can enter premises and look for safety breaches.
	Local Government (Miscellaneous Provisions) Act 1982	Places of entertainment must be licensed; restrictions may be imposed for inter alia, securing access for emergency vehicles.
	Private Places of Entertainment (Licensing) Act 1967	Places of private entertainment must (in certain circumstances) be licensed.
Schools	*Education (School Premises) Regulations 1999*	Premises must allow safe escape in case of fire.
Care and Nursing Homes	*Children's Homes Regulations 2001*	In consultation with the fire authority, precautions must be taken against fire.
	Residential Care Homes Regulations 1984	The person registered must consult with the fire authority on fire precautions.
	Nursing Homes and Mental Nursing Homes Regulations 1984	The person registered must take adequate precautions against risk of fire, escape routes, warnings and extinguishing equipment.
	Children's Homes Regulations 2001	
	Residential Family Centres Regulations 2002	
	Private and Voluntary Care (England) Regulations 2001	
Housing	*Building Regulations 2000*	Requirements imposed on building operations; Part B covers smoke detectors, of escape, control of spread of fire.
	Housing Act 1985	Local authorities' powers to require means of escape from fire and other adequate fire precautions.

Other Regulations addressing the provision of first aid

Offshore Installations and Pipeline Works (First-Aid) Regulations 1989

3.59 The usual first aid legislation does not apply offshore. Instead, these Regulations apply requirements which are broadly similar to the 1981 Regulations but which go further in that the employers' duties apply to first aid when employers are not actually at work but off-duty on an installation or vessel. In addition, the duty extends to the provision of appropriate medications, and arrangements for treatment on the direction of a medical practitioner (who may or may not be present). Special exemptions apply for qualified first aid personnel on offshore installations who may administer certain prescription medicines.

Diving at Work Regulations 1997

3.60 These Regulations require diving contractors to make adequate provision for first aid for diving projects. (The 1981 Regulations do not apply where these diving Regulations are applicable.)

Merchant Shipping and Fishing Vessel (Medical Stores) Regulations 1995 and *Merchant Shipping (Training and Certification) Regulation 1997.*

These Regulations implement Directive *92/29/EEC* on minimum requirements for international treaty obligations on mandatory first-aid training and medical services on board vessels. They cross-refer to *Merchant Shipping Notice M 1807* (now *MSN 1726*).

Air Navigation Order 2000

3.61 This order specifies requirements for first aid equipment and a first aid handbook for UK registered aircraft.

Management and Administration of Safety and Health at Mines Regulations 1993

3.62 These Regulations modify the *Health and Safety (First-Aid) Regulations 1981* to place the duties on the mine owner as for an employer. (See also additional signage requirements for mines in the *Electricity at Work Regulations 1989, Regulation 25.*)

Can first aid personnel have personal liabilities?

3.63 Although it is stated in HSE guidance that legal action would be unlikely to be taken against a first aider personally, the fact remains that a person in this role could be sued independent of the employer because liability for negligence is also personal to the individual. The employer's potential vicarious liability for the employer's negligence is not a defence to the individual if sued (or prosecuted under the *HSWA*) (*Lister v Romford Ice and Cold Storage*). If the first aid is given beyond the scope of the employer's duties – for example to a passer-by – the employer may not be liable at all. It is therefore advisable for insurance arrangements for first aid personnel and their limits be clearly established with the employer's insurance brokers.

First aid personnel attract additional statutory duties by virtue of *HSWA, Section 7* and *MHSWR, Regulation 14*. They must co-operate with the employer's system for providing first aid, and they will have a special responsibility to draw the attention of the employers to any shortcomings in the system or the equipment and facilities available which their training would enable them to recognise.

There are no mandatory insurance requirements relating specifically to negligent acts or omissions on the part of the person providing first aid. This would generally be covered by the employer's liability insurance policy when the first aider provision is by a nominated employee because the employer is vicariously liable for employee's negligence, though where a private medical practitioner or nurse is engaged on a service contract basis for these purposes it is possible that claims may be covered by their professional indemnity insurance. Employers' liability insurers' policies can have extensions of insurance indemnity to first aid and medical teams.

Some first aiders may also be insured through voluntary organisations to which they belong for action they take in giving first aid outside the immediate scope of their employment, for example in giving first aid to a member of the public.

Legal compliance checklist

3.64

Health and safety policies and procedures

✓ A written health and safety policy is maintained up to date (see Table 3.6).

✓ Safety management systems are formalised according to POPIMAR model.

✓ Competent persons are appointed, with recourse to external advice when needed.
✓ Risk assessments, safety procedures and training and information are controlled and documented.
✓ Document management systems enable access to relevant legislation, AcoPs and guidance.

Emergency arrangements

✓ Significant hazards are identified.
✓ Risk assessments undertaken and recorded.
✓ Emergency planning has been documented according to all relevant legislation.
✓ Danger areas are operational and effective.
✓ Roles and responsibilities are defined.
✓ External services are identified and emergency contact details are available.
✓ Competent persons have been designated.
✓ Fire risk assessments have been documented.
✓ Emergency routes and exist are kept clear.
✓ Emergency equipment is maintained.
✓ Emergency plans are practised with drills.
✓ Correct signage is employed.
✓ Information on emergency arrangements is appropriately disseminated.

First aid

✓ First aid needs have been assessed.
✓ First aiders or appointed persons are appropriately designated and trained.
✓ First aid facilities equipment are available and correctly maintained.
✓ Staff are adequately informed of first aid arrangements.
✓ First aid records are maintained.

Insurance

✓ Employers liability insurance is correctly maintained.
✓ Historical records of all insurance are available.
✓ Insurance policies taken out by previous undertakings now incorporated into the business are traceable.

References

ABI (1999) *Code of Practice for Tracing Employers' Liability Policies www.dwp.gov.uk/publications/dwp/2004/elci/codedocument.pdf*

Barrett v Ministry of Defence [1995] 1 WLR 1217.

Bernadone v Pall Mall Services Group [2000] IRLR 487.

BSI (1999) *Occupational Health and Safety Management Systems – Specification, OHSAS 18001.*

BSI (2004) *Guide to Occupational Health and Safety Management Systems, BS 8800.*

Capital & Counties plc v Hampshire County Council [1997] QB 1004.

DCLG (2006) *Entry Level Guide – A Short Guide to Making Your Premises Safe from Fire. www. communities.gov.uk*

HSC (2002) *Directors' Responsibilities for Health and Safety* (INDG 343). *www.hse.gov.uk/pubns/indg343.pdf*

HSE (1997) *Successful Health and Safety Management.* HS(G)65. HSE Books.

HSE (1999) *Emergency Planning for Major Accidents: Control of Major Accident Hazards Regulations 1999 (HSG 191).* HSE Books.

HSE (2004) *Management of Health and Safety at Work: Management of Health and Safety at Work Regulations 1999.* L21. HSE Books.

Jebson v Ministry of Defence [2000] ICR 1220.

Kasapis v Laimos [1959] 2 Lloyd's Rep 378.

Kent v Griffiths [2001] QB 36.

Lister v Romford Ice and Cold Storage (1957) AC 55.

Martin v Yorkshire Imperial Metals Ltd [1978] IRLR 440.

McDermid v Nash Dredging and Reclaimation Co Ltd [1987] AC 906.

OLL Ltd v Secretary of State for Transport [1997] 3 All ER 897.

Reid v Rush & Tomkins Group plc [1990] 1 WLR 217.

4 First aid

In this chapter:

The rationale behind first aid requirements

What is first aid?

4.1 In simple terms, first aid is emergency care given immediately to an injured person. The purpose of first aid is to minimise injury and future disability. In serious cases, first aid may be necessary to keep the victim alive.

This definition of first aid shows that it is very broad. First aiders never know when they are going to come across an incident or which skills they are going to be required to use. It could be a colleague who has tripped over and has a cut or injured limb as a result, or someone having a choking fit at the next table in the canteen or a member of the board suffering a heart attack.

Why is it important?

4.2 There are a number of reasons why it is important that employers make first aid provision for their staff, beyond the minimum compliance with legislative requirements described below:

- reducing injury costs,
- business benefits & psycho-social aspects,
- staff development,
- serving the community.

First aid practice is evidence based. There are well-documented examples showing that, in many health emergencies, interventions in the first few minutes, before the emergency services arrive, can make a crucial difference to the outcome of the casualty.

Every year in the UK, there are five and a half million attendances at Accident and Emergency departments for accidents. Of these, there are three million attendances of the types of accident and injury that first aid treatment can benefit.

Many accidents occur in the workplace and first aid skills can help to reduce injury and save lives. For example, evidence shows that:

- immediate cardiopulmonary resuscitation (CPR) can double or triple the chances of survival for a heart attack casualty;
- an unconscious patient with a blocked airway can have only 4 minutes to live – unless a bystander steps forward and does something as simple as tilting the head back to enable them to breathe;
- basic procedures like cooling a burn can reduce the need for skin grafting, lessen injury and promote healing.

Reducing injury costs

4.3 Prompt and effective first aid can help to reduce the cost of injury and consequently the amount of lost working time. Over one million people suffer an injury at work each year. In 2000/2001 7.3 million working days were lost from people taking time off because of these injuries. Work injuries can happen in any business, whatever its size.

Whether this is bandaging a wound so that the employee can return to their workplace or providing treatment to ensure that time off due to injury is reduced, by providing good first aid cover employers can save themselves money reducing the amount of time lost to injury.

Business benefits and psycho-social aspects

4.4 In addition to reduced injury costs businesses derive other benefits from effective first aid cover.

Research shows that first aid training can significantly improve the confidence and motivation of employees. The results of a resilience survey conducted by the British Red Cross demonstrate this (see Table 4.1).

> "We see this all the time. Delegates arrive on Monday morning somewhat reluctantly as many have been 'volunteered' by their employer. By the end of the fourth day of training there is a complete change. There is a great team spirit and everyone really enjoys the experience" (Audrey Edwards, Training Development Manager, British Red Cross).

Table 4.1: British Red Cross resilience survey (August 2005)

After completing this course please indicate what you feel about the following?		
a.	I feel more confident within myself	97%
b.	I feel that I would be able to react to a first aid situation should I come across one	100%
c.	I feel more confident in my daily life knowing I have life saving skills	93%
d.	I will feel more confident in my working life having completed the training	97%
After completing this course please indicate whether you agree with any of the following.		
a.	I believe that I can play a wider role in my community through having first aid skills	93%
b.	I believe that first aid is a vital skill that everyone should know	100%
Please indicate which of the following elements you think have been of most value on your course.		
a.	Learning a life saving skill	100%
b.	Increased confidence levels	90%
c.	Learning a new skill	97%
d.	Having the ability to help others	100%
e.	Having the ability to help my colleagues	100%
f.	Having a skill that colleagues don't have	77%
g.	Personal development	100%

In addition to the employee being trained, other staff also benefit from good first aid provision:

- They feel more relaxed by knowing that if an incident were to occur, they would be able to receive prompt and effective treatment.
- First aiders can help spread good health and safety messages in the workplace.
- First aid training in itself can improve safety behaviour and may contribute to accident prevention, for example, in support of the Health and Safety Executive (HSE) focus on slips, trips and falls, etc.

Staff development

4.5 Training employees in first aid fulfils many functions. It equips them with a life saving skill, increases motivation and can contribute to professional development. The administer first aid qualification forms part of the public service NVQ qualification. The other four compulsory units are:

1. Promote and maintain health, safety and security in the workplace.

CASE STUDY 5

4.6

PA comes to the aid of a colleague

Sarah is a PA and first aider in her workplace. She had been on an HSE approved first aid course and had just completed "refresher training". One lunchtime, Sarah answered a call at her desk to say that a lady called Paula was choking on a chicken bone. Another first aider was also promptly called to assist.

On arriving at the scene Sarah found Paula was distressed and indeed choking. The other first aider went to call an ambulance straight away whilst Sarah began with "back-slaps". After the appropriate amount of these, the chicken bone did not dislodge. Sarah then began administering the abdominal thrusts. On the second attempt the chicken bone released. The ambulance arrived a couple of minutes after this, but once Paula had been checked out, she was sent home to recover for the rest of the day.

Sarah said afterwards that she was extremely glad that all her training stayed with her. There was much panic during this situation and was amazed that she was able to remain calm and confident. She went on to take advanced training in use of defibrillators.

2. Utilise resources to maintain personal effectiveness.
3. Establish, develop and maintain effective working relationships.
4. Maintain personal level of physical fitness for duty.

First aid can also be part of the continual professional development of members of staff.

Serving the community

4.7 By training staff in first aid an organisation not only benefits the individual staff member and itself as a whole, but also its wider community. First aid is a life

CASE STUDY 6

4.8

> ### *Hero saves drowning boy*
>
> Wayne a trained first aider, was out cycling with his daughter in Wales when he noticed a group of distressed children by a pond. Their friend, an eight-year-old boy, had gone swimming and disappeared beneath the water.
>
> The pond, the site of a disused coal mine, was surrounded by "No Swimming" signs. Despite the obvious danger, Wayne grabbed a rubber ring and dived straight in.
>
> Guided by his daughter on the shore, he found the unconscious boy floating just below the surface of the murky water. Grabbing hold of him firmly, he swam back to dry land.
>
> Wayne said: "*The boy wasn't breathing so I gave him immediate mouth-to-mouth resuscitation.*"
>
> *After a short time he started coughing and spluttering, and his eyes started to focus. It was a massive relief.*"
>
> Wayne's actions saved the young boy's life. A woman passer-by called for an ambulance and, after a night's observation in hospital, the boy made a full recovery.
>
> The incident provided a powerful reminder of how valuable first aid skills can be in an emergency situation. Wayne said "*Everything I'd done in training just kicked in. It was like I was on auto-pilot – I just switched off and did what I had to do.*"

saving skill that can also be deployed outside the work place office. First aiders may come across incidents in their daily lives, which their first aid skills equip them to deal with. For example a choking incident in a local restaurant, a heart attack in a shopping centre or a car accident. Their skills may make all the difference to saving someone's life or preventing a more serious injury or illness.

The Health and Safety (First-Aid) Regulations 1981, Approved Code of Practice and Guidance L74 (referred to hereafter as ACoP L74)

4.9 The legal requirements for the provision of first aid arrangements apply to all employers and the self-employed. First aid for these purposes has a specific legal definition which is: either treatment of minor injuries which would otherwise not need or not receive medical treatment, or where medical treatment will be needed, treatment for the purposes of preserving life and minimising the consequences of injury or illness until medical help is obtained (*Health and Safety (First-Aid) Regulations 1981*). The provisions extend to any injury or illness, whether work related or not. However they do not place a legal obligation on employers and the self-employed to make first aid arrangements for others such as contractors or visiting members of the public.

The *Health and Safety (First-Aid) at Work Regulations 1981* are quite short and specify only the barest requirements:

1. *An employer shall provide, or ensure that there are provided, such equipment and facilities as are adequate and appropriate in the circumstances for enabling first aid to be rendered to his employees if they are injured or become ill at work.*
2. *Subject to paragraphs (3) and (4), an employer shall provide, or ensure that there is provided, such number of suitable persons as is adequate and appropriate in the circumstances for rendering first aid to his employees if they are injured or become ill at work; and for this purpose a person shall not be suitable unless he has undergone:*
   (a) *such training and has such qualifications as the Health and Safety Executive may approve for the time being in respect of that case or class of case, and*
   (b) *such additional training, if any, as may be appropriate in the circumstances of that case.*
3. *Where a person provided under paragraph (2) is absent in temporary and exceptional circumstances it shall be sufficient compliance with that paragraph if the employer appoints a person, or ensures that a person is appointed, to take charge of:*

(a) *the situation relating to an injured or ill employee who will need help from a medical practitioner or nurse, and*

(b) *the equipment and facilities provided under paragraph (1) throughout the period of any such absence.*

4. *Where having regard to:*

   (a) *the nature of the undertaking, and*

   (b) *the number of employees at work, and*

   (c) *the location of the establishment,*

   it would be adequate and appropriate if instead of a person for rendering first aid there was a person appointed to take charge as in paragraph (3)(a) and (b), then instead of complying with paragraph (2) the employer may appoint such a person, or ensure that such a person is appointed.

5. *Any first aid room provided pursuant to this regulation shall be easily accessible to stretchers and to any other equipment needed to convey patients to and from the room and be sign-posted, and such sign to comply with regulation 4 of the Health and Safety (Safety Signs and Signals) Regulations 1996 as if it were provided in accordance with that regulation.*

The main factors for consideration in meeting these requirements are outlined in ACoP L74 and comprise:

- The nature of the work.
- Size of the organisation.
- Past history and consequences of accidents.
- The nature and distribution of the work cause.
- The remoteness of the site from Emergency Medical Services.
- The needs to travelling, remote and lone workers.
- Employees working on shared or multi-occupied sites.
- Annual leave and other absences of first aiders.
- Other relevant factors.

ACoP L74 states that the *minimum* first aid provision for each workplace should be:

- a suitably locked first aid container;
- the person appointed to take charge of first aid arrangements;
- information for employees on first aid arrangements.

Table 4.2 contains HSE guidance intended as an aid to a more detailed assessment of first aid requirements.

Table 4.2: Assessment of first aid needs checklist

Aspects to consider	*Impact on first aid provision*
1. What are the risks of injury and ill to health arising from the work as identified in your risk assessment?	If the risks are significant you may need employ first aiders.
2. Are there any specific risks, for example, working with: – hazardous substances; – dangerous tools; – dangerous machinery; – dangerous loads or animals?	You will need to consider: – specific training for first aiders; – extra first aid equipment; – precise siting of first aid equipment; – informing emergency services; – first aid room.
3. Are there parts of your establishment where different levels of risk can be identified (e.g. in a University with research laboratories?)	You will probably need to make different levels of provision in different parts of the establishment.
4. Are large numbers of people employed on site?	You may need to employ first aiders to deal with the higher probability of an accident.
5. What is your record of accidents and cases of ill health? What types are they and where did they happen?	You may need to: – locate your provision in certain areas; – review the contents of the first aid box.
6. Are there inexperienced workers on site, or employees with disabilities or special health problems?	You will need to consider: – special equipment; – local siting of equipment.
7. Are the premises spread out, for example, are there several buildings on the site or multi-floor buildings?	You will need to consider provision in each building or on several floors.
8. Is there shift work or out-of-hours working?	Remember that there needs to be first aid provision at all times people are at work.
9. Is your workplace remote from emergency medical services?	You will need to: – inform local medical services of your location; – consider special arrangements with the emergency services.
10. Do you have employees who travel a lot or work alone?	You will need to: – consider issuing personal first aid kits and training staff in their use; – consider issuing personal communicators to employees
11. Do any of your employees work at sites occupied by other employers?	You will need to make arrangements with the other site occupiers.
12. Do you have any work experience trainees?	Remember that your first aid provision must cover them.

(*Continued*)

Table 4.2: (Continued)

Aspects to consider	*Impact on first aid provision*
13. Do members of the public visit your premises?	You have no legal responsibilities for non-employees, but HSE strongly recommends you include them in your first aid provision.
14. Do you have employees with reading or language difficulties?	You will need to make special arrangements to give them first aid information.

Source: First Aid at Work ACoP (L74), Appendix 1

Offshore, operations, diving and other first aid requirements

Offshore

4.10 There is specific legislation for industries based offshore, as their remote location may mean it is difficult for them to access the health care and expertise which would normally be available. In accordance with the Offshore Installations and Pipeline Works (First-Aid) Regulations 1989, the person in control, for example, the installation operator, needs to take care of all first aid and health and safety arrangements, such as ensuring the required number of offshore medics and first aiders are on site. The regulations also cover any associated equipment and types of drugs that may be needed. Guidance which outlines the minimum requirements is available from the United Kingdom Offshore Operators' Association (UKOOA, 2002).

Any arrangements made also need to cover visitors or contractors who may be present, including those working on certain associated vessels, as may be the case during commissioning or decommissioning. Offshore medics have to be supervised by a registered medical practitioner, who is usually based onshore. Responsibility for the sickbay normally falls to the offshore medic and there are regulations pertaining to the size, siting, layout and facilities which should be available. The sickbay must also be able to provide accommodation and medical support for an ill or injured person for up to forty eight hours.

For more information the UKOOA guidance referred to above should be read in conjunction with the ACoP L74.

Diving

4.11 In accordance with the Diving at Work Regulations 1997, there must be first aid and medical equipment available to participants during a diving project.

The responsibility for meeting these regulations falls to the diving contractor. As part of their professional training and assessment in Great Britain, commercial divers are required to:

- have been taught diving physiology (which includes the function of the nervous and musculoskeletal systems);
- have been taught diving medicine (which includes the function of the nervous and musculoskeletal systems).

In addition to this training and assessment, they are also required to obtain an HSE first aid at work qualification.

Commercial Diving Projects Inland/Inshore

4.12 Not all inland/inshore divers are required to hold a valid HSE first aid certificate. However, the diving contractor does have a responsibility to ensure that there is suitable and sufficient first aid cover at the dive site.

The supervisor and at least one diver in each diving operation should have in-date first aid qualifications. It may transpire that inshore divers have been trained and assessed in other first aid procedures, for example the emergency administration of pure oxygen (which needs to be refreshed every three years).

Commercial Diving Projects Offshore

4.13 According to the HSE, all offshore divers should have an in-date HSE first aid at work qualification and there should be at least one diver medic in every offshore diving project.

Appointing people

Basic minimum legal requirements

4.14 Two kinds of personnel with distinct functions are envisaged under the Regulations: "*first aiders*" and "*appointed persons*".

Strictly speaking a first aider is someone who has undergone an HSE approved training course in the administration of first aid, and who holds a current first aid at work certificate. First aid at work certificates are valid for three years, after which time, employees need to undergo refresher training and re-certification (see 4.20).

Based on the assessment of first aid needs however, there may be additional levels of training beyond the norm required to cover special risks associated with the workplace.

An "appointed person" on the other hand is someone selected to take charge of first aid arrangements when someone is injured or unwell, including calling an ambulance or healthcare professional, and whose role extends to looking after first aid equipment and facilities. An appointed person does not require HSE approved training and they should not administer first aid if they have not been so trained. Short training courses in handling emergency situations are nevertheless recommended for appointed persons.

The effect of the Regulations is that it would be permissible for first aid personnel to comprise "appointed persons" in two situations:

1. where the nature of the business, its location, and the number of employees does not warrant a fully qualified first aider on the premises; or
2. where there is usually a first aider present, based on the assessment of need, that this person is temporarily absent in "exceptional circumstances". For this to apply, the absence must be unforeseen; an absence for annual leave or training courses would not justify reliance on an appointed person instead of a first aider.

Table 4.3 contains the HSE's guidelines on the numbers of first aiders/appointed persons in different situations, although it must be stressed that these are only indicative numbers and based on the assessment of need more extensive cover that this could be required in particular workplaces.

Table 4.3: Guidelines on appointments

Category of risk	*Numbers employed at any location*	*Suggested number of first aid personnel*
Low risk *Shops, offices, libraries*	Fewer than 50	At least one appointed person
	50–100	At least one first aider
	More than 100	One extra first aider for every 100 staff
Medium risk *Light engineering, assembly work, food processing, warehousing*	Fewer than 20	At least one appointed person
	20–100	At least one first aider for every 50 staff
	More than 100	One extra first aider for every 100 staff
High-risk *Most construction, slaughter houses, chemical manufacture, extensive work with dangerous machinery or sharp instruments*	Fewer than 5	At least one appointed person
	5–50	At least one first aider
	More than 50	One extra first aider for every 50 staff

Source: First Aid at Work ACoP Guidance (L74)

What is adequate depends on the type of workplace, the industry sector and not necessarily just the number of staff employed. The following information is intended as a guideline for basis situations:

In a low risk sector for example an office, shop or library the recommendations are as follows:

- One first aider for fifty to one hundred staff, with one additional first aider for every one hundred employed.

- If there are fewer than fifty staff, then the recommendation is for one appointed person.

In a medium risk sector such as assembly work, warehousing, light engineering or food processing the recommendations are as follows:

- One first aider for twenty to seventy staff, with one additional first aider for every one hundred staff after that.

- If there are fewer than twenty staff then the recommendation is for one appointed person.

In a high-risk sector such as construction, chemical manufacture or work with dangerous machinery or sharp instruments, the recommendations are as follows:

- Five to fifty staff – at least one first aider, with one additional first aider for every fifty staff.

- If fewer than five staff then the recommendation is for one appointed person.

Multi-site requirements

4.15 The requirements differ depending on the number of staff working in each site.

If sites are in a low risk sector such as an office or shop then the recommendations are as follows:

- One first aider for fifty to one hundred staff, with one additional first aider for every one hundred employed.

- If fewer than fifty staff, then the recommendation is for one appointed person.

For a medium risk sector such as assembly work, warehousing, light engineering or food processing the recommendations are as follows:

– One first aider for twenty to seventy staff, with one additional first aider for every one hundred staff after that.

– If there are fewer than twenty staff then the recommendation is for one appointed person.

For sites in a high-risk sector such as construction, chemical manufacture or work with dangerous machinery or sharp instruments, the recommendations are as follows:

– Five to fifty staff – at least one first aider, with one additional first aider for every fifty staff.

– If fewer than five staff then the recommendation is for one appointed person.

Self-employed

4.16 Self-employed people need to ensure that they have adequate equipment to provide first aid to themselves while at work. This involves doing a risk assessment of the hazards and risks in the workplace and determining what is an appropriate level of first aid provision.

For example for a low risk activity such as clerical work from home, a simple first aid kit for the home would be sufficient. If the work involves driving long distances then a first aid travel kit should kept in the vehicle.

Where work is carried out at others' premises with other self-employed or employed workers the self-employed person is still legally responsible for his/her own first aid provision. However it will make sense to join forces with the others located on the premises. This means that one employer would take on responsibility for all first aid cover for all workers on the premises. In these circumstances the HSE recommends that a written agreement should document the agreed arrangements.

The implications of having a mobile workforce

4.17 If staff regularly work away from the main site, then adequate first aid provision will have to be made for them at these different locations. Those who are frequently on the road or who travel regularly for business should carry a personal first aid box. If they are working in a remote area there should be a reliable means of summoning help such as a mobile phone. In addition it is prudent to issue lone workers with a single person first aid kit.

Table 4.4: Considerations for selecting first aiders

Who to choose as first aiders?

Some obvious factors will influence your decision on who to choose as your first aider(s). Some of the factors you may want to consider are:

Personality

- Is the employee a good communicator? Will they be able to calm and reassure a victim?
- Sympathetic? Will their manner help to reassure a victim and colleagues?
- Confident? Will they panic when they come across an incident or will they have the confidence to deal with it?
- Reliable? Is their attitude and attendance record good?

Availability

- Is the employee readily available? Can they easily be released to deal with an incident and for periodic training?
- Is the employee often away at meetings or on business? If so they may not be free to attend an incident.
- Is the employee easy to contact in a crisis?

Location

- Does the employee have a fixed work space or are they more mobile around the work place? If it is the latter then they could be more difficult to find in a crisis?
- Do restrictions on access to dangerous areas, personal protective equipment requirements or local competency rules make it sensible to select a first aider only from a pool of employees authorised to work in such locations?

If the organisation is spread out over a large area or has multiple floors, first aiders need to be located at strategic distances to ensure that they can get to an incident quickly. Also, where there are shift workers or out of hours services it will be necessary to ensure that there is cover at all times. Holiday cover should be factored in.

If there is a mobile workforce, ideally each member of staff should have some basic first aid.

Making appointments

4.18 Appointments of first aiders and appointed persons are usually made quite informally as an addition to the individual(s) normal job description, though some businesses which employ medical personnel may have formal employment contracts which are specific to the provision of first aid services. It is possible to outsource the provision of first aid personnel, as long as the employer can ensure their availability and competence.

There are no pre-requisites or particular qualifications for selecting first aid personnel. An employer should be satisfied however as to the person's general reliability, physical and mental robustness, and capability to learn the skills and if necessary obtain a first aid certificate (see the *Management of Health and Safety at Work Regulations 1999 (MHSWR), Regulation 13*). It is also important that the person's normal duties will permit him/her to leave their normal post (Table 4.4).

Having made the appointment, the employee should confirm this in writing. It is not unusual for the employer to make a small increment to an appointed employee's salary in such cases. Should the employer later wish to terminate the appointment the process needs to be handled with care as this could in some cases be grounds for constructive dismissal if it is not dealt with in accordance with the original terms of appointment, or if there is or any procedural irregularity.

Overview of training courses

4.19 Training courses vary according to the requirements of the organisation and are influenced by factors such as the size of the organisation and industry sector (Tables 4.4 and 4.6).

Where the need for first aiders has been identified in the workplace, they must gain a certificate of competence from a training organisation that has been approved by HSE. Courses last for at least twenty four hours of training, usually held over four days or spread over several weeks.

Training qualifications and certificates issued in other countries are not recognised for the purposes of compliance with these UK legal requirements.

There are no specific legal training requirements for appointed persons. However if the need identified is for an appointed person, then it is recommended that the person attends a one-day course for an appointed person as there is a general legal requirement under the *MHSWR* for all employees to have adequate training in their designated tasks.

There are also a variety of tailored courses, which have been designed to meet the needs of specific industries.

The HSE does not run training courses itself but does approve training organisations whose courses have reached the standards required. The First Aid Approval and Monitoring Section (FAAMS) of HSE's Corporate Medical Unit carries out this work.

Currently there are approximately 1200 training providers approved across England, Scotland and Wales. Each organisation is monitored and all are working to the standards required by HSE (Table 4.5).

Table 4.5: The HSE approved first aider course content

First aid at work The HSE approved course Approximately twenty four hours of training
Who is it for? Employees who require a first aid at work certificate to comply with HSE requirements
Course content includes Priorities of first aid Managing an incident Roles/responsibilities of employers and employees for first aid at work Care of the unconscious casualty Cardiopulmonary resuscitation (CPR) Disorders of the airway Dealing with bleeding and shock Causes of unconsciousness Burns and scalds Injuries to bones, muscles and joints Poisoning Recognition and management of major illnesses Recognition of minor illnesses and appropriate action Eye injuries Miscellaneous conditions Record keeping and accident reporting Infection control Communication and delegation in an emergency Manage the transportation of a casualty
Qualification Recognised first aider in the workplace First aid at work certificate valid for three years

In addition, support modules exist for those involved in higher-risk industries, particularly those involved with chemicals, such as resuscitation support course, use of oxygen, entonox, administration of epinephrine, hydrofluoric acid and cyanide.

For a member of staff to be trained as an appointed person, the recommendation is to attend a one-day first aid course for the appointed person (Table 4.6).

Table 4.6: Typical one day appointed person course content

Basic first aid for the appointed person Approximately six hours of training
Who is it for? Those wishing to learn basic first aid Those acting as the appointed person in the workplace Those supporting a qualified first aider Those employed in a low risk working environment with less than fifty people
Course content includes Personal safety, assessing the incident Priorities: DRAB (danger, response, airway, breathing) check and sending for help Unconscious casualty Resuscitation: adults Choking Dealing with blood loss and shock Burns and scalds Fractures Heart attack Diabetes Strokes Seizures Reporting and recording of incidents as required by HSE Contents of HSE approved first aid kit
Qualification Appointed person certificate valid for three years.

Note: This course does not qualify the employee to be an HSE approved first aider nor are the courses monitored by the HSE. However most courses are run by HSE approved trainers and organisations which have their own internal monitoring system

There are other appointed person courses tailored for particular sectors, such as:

- Basic first aid for the care of the elderly person
- Basic first aid for the care of children
- Basic first aid for teachers
- Basic first aid for drivers

Table 4.7: Typical refresher course content

First at work refresher Approximately twelve hours of training
Who is it for? Employees who need to renew a valid first aid at work certificate
Course content includes Priorities of first aid Managing an incident Roles/responsibilities of employers and employees for first aid at work Care of the unconscious casualty Cardiopulmonary resuscitation (CPR) Disorders of the airway Dealing with bleeding and shock Causes of unconsciousness Burns and scalds Injuries to bones, ligaments and joints Poisoning Recognition and management of major illnesses Recognition of minor illnesses and appropriate action Eye injuries Miscellaneous conditions Record keeping and accident reporting Infection control Communication and delegation in an emergency Manage the transportation of the casualty
Qualification Recognised first aider in the workplace First aid certificate that is valid for three years

- Basic first aid for pharmacists
- Emergency aid for sports

There are other shorter courses, which give a basic grounding in the main first aid skills also such as "emergency life support".

Refreshers and re-qualification

4.20 An HSE approved first aid certificate is valid for three years. After this period delegates need to undergo an HSE approved refresher training course. The re-training should be undertaken whilst the certificate is still valid, or within a maximum of one month after the expiry date, otherwise the delegate needs to complete a full first aid at work certificate all over again (Table 4.7).

Some of the more specific first aid training may also require more frequent refreshers. For example, those trained to use a defibrillator (see 4.29) need to update their skills at least every year. This ensures that their skills are up to date, are confident to use the automated external defibrillator (AED) should this be required and are informed of any changes to protocol.

However during the period between courses, delegates who are not called upon to use their skills regularly can suffer a corresponding lack of confidence and skills fade. Training providers have therefore introduced some initiatives to help combat this. There are half-day and full-day courses designed to enhance and update skills and to provide an opportunity to practice them. Delegates recap the main topics and are also updated on any changes to legislation or protocol. This type of course does not renew the first aid at work certificate but is aimed at maintaining competency and confidence.

Facilities and equipment

First aid rooms

4.21 There is no express requirement for the provision of first aid rooms by employers but the provision of such a room will fall within the "facilities" required by *Regulation 3(1)* where the need for it is shown by the assessment. The provisions in Regulation 3(5) for equipping and sign-posting a first aid room were introduced by amendments to the *Health and Safety (First-Aid) Regulations 1981* in 1999. There is no reason in principle why the restroom facility provided for the purposes of Regulation

25 of the *Workplace (Health Safety and Welfare) Regulations 1992* cannot double as a first aid room provided the room is capable of meeting the requirements in containing the facilities recommended by the HSC ACoP L74 and those in *Regulation 3(5)* for stretchers and other means of conveying patients.

What should be kept in the first aid room?

4.22 The room should contain essential first aid facilities and equipment. Typical examples of these are:

- a sink with hot and cold running water;
- drinking water and disposable cups;
- soap and paper towels;
- a store for first aid materials;
- foot-operated refuse containers, lined with disposable yellow clinical waste bags or a container for the safe disposal of clinical waste;
- a couch with waterproof protection, clean pillows and blankets;
- a chair;
- a telephone or other communication equipment;
- a record book for recording incidents where first aid has been given.

Who should have access to the first aid room?

4.23 If possible, the room should be reserved specifically for providing first aid and designated first aiders or appointed persons should be given responsibility for the room. It should be easily accessible to stretchers and be clearly signposted and identified.

Identifying the first aid room

4.24 The location of the room must be easily identified with white lettering or symbols on a green background in accordance with the *Health and Safety (Safety Signs and Signals) Regulations 1996.*

Symbol for use in identifying first-aid rooms. White cross on a green background

What first aid equipment should be provided?

4.25 Once an assessment of first aid needs has been carried out, the findings can be used to decide what first aid equipment should be provided in the workplace. The minimum level of first aid equipment is a suitably stocked first aid box,stored in clean, secure conditions.

There may also be a need for items such as protective equipment where first aiders may have to enter dangerous environments. (Note that assessment of personal protective equipment and special training for such areas may be needed too.) This should be securely stored near the first aid box, in the first aid room or the hazard area, as appropriate. Access to the equipment should be restricted to those trained in its use.

If mains tap water is not readily available for eye irrigation, at least one litre of sterile water or sterile normal saline (0.9 per cent) in sealed, disposable containers should be provided. When the seal has been broken, the container should not be reused. The container should not be used beyond its expiry date.

What should be in a first aid box?

4.26 There is no statutory requirement specifying the contents, or the amounts of materials that should be stored. Indicative lists of items are provided in the HSC ACoP *L74M* and related HSE Guidance leaflets. These comprise very basic equipment – dressings, bandages, sterile wipes and gloves, and an instructions leaflet. This reflects the limited scope of first aid under the regulations, which stops short of any action which could be described as medical treatment. HSC ACoP *L74* also covers the contents of travelling first aid kits for provision to employees working away from the main site, or who travel long distances.

Deciding what you should include should be based on your assessment of the first aid needs of your organisation. The assessment may indicate that additional materials and equipment are required such as scissors, adhesive tape, disposable aprons and individually wrapped moist wipes. They may be put in the first aid container if there is room or stored separately. Requirements of what should be in a first aid kit do differ according to industry. The exact contents should be based on a specific risk assessment (Table 4.8).

It should always be remembered that the employer remains fully responsible for the safety of first aiders and appointed persons when they are undertaking their responsibility, which includes the provision of appropriate personal protection equipment and training in their own protection. It is particularly important that there should always be an adequate supply of disposable gloves, and strict adherence to procedures for protecting against blood-borne pathogens (HSE, 2002). However, first aid treatment should not be delayed due to the absence of gloves. The employer needs to have undertaken a risk assessment of biological hazards for the purposes of the *Control of Substances Hazardous to Health Regulations 2002* which might

Table 4.8: Contents of first aid boxes

Low risk working environment such as shops and offices

- First aid guidance leaflet – containing first aid tips and emergency information
- Twenty individually wrapped sterile adhesive dressings (assorted sizes) – sealed in protective wrappers
- Two sterile eye pads – for treating eye wounds
- Four individually wrapped triangular bandages – to provide support for an arm or hand injury, these are usually made of cloth or strong paper
- Six safety pins – for securing triangular bandages
- Six medium sized (approximately 12 cm by 12 cm) individually wrapped sterile unmediated wound dressings – sealed in protective wrappers
- Two large (approximately 18 cm by 18 cm) sterile individually wrapped unmedicated wound dressings – sealed in protective wrappers
- One pair of disposable gloves – wearing gloves when dressing wounds minimises the risk of a wound becoming infected

Note: Any items in the first aid box that have passed their expiry date should be disposed of safely.

extend to consideration of vaccination where other control measures may not provide adequate assurance.

How often should the contents of first aid boxes be replaced?

4.27 Although there is no specified review timetable, many items, particularly sterile ones, are marked with "best before dates" and the manufacturers will generally not be responsible for their condition or fitness for purpose beyond that time. All items should be checked periodically and replaced by the dates given. In cases where sterile items have no dates, it would be advisable to check with the manufacturers to find out how long they can be kept. For non-sterile items without dates, it is a matter of judgement, based on whether they are fit for purpose.

Storing first aid kits

4.28 Equipment should be stored in a case which is clearly marked with the EU symbol for first aid (a white cross on a green background). The first aid kit should always be prominently displayed on a wall to which there is easy access, such as outside a kitchen or by the door.

Staff need to be informed of the location of first aid boxes. Generally putting up a notice telling people where the first aid kit is located will be sufficient.

Medicines

4.29 The HSC's ACoP L74 states that first aid at work does not include giving tablets or medication to treat illness and such items should not be kept in the first aid box (even aspirin and paracetamol tablets). Although it is not necessarily unlawful to supply medicines, they should not, as a rule, be accessible with the rest of the first aid box contents. (It is possible that some special circumstances would warrant the separate storage of certain medicines where serious risks which can be identified (e.g. a need for anti-venoms), but this in turn requires careful consideration of circumstances in which such medicines might need to be administered.)

The legal difficulty here is that medicines legislation restricts the supply and administration of most medicines to treatment by or on the instructions of a doctor. Narrow exemptions exist (as for example in relation to the administration of adrenaline) for an anaphylactic shock and life threatening cases (see the *Prescription-Only Medicines (Human Use) Order 1997*) but it will generally not be realistic to expect a first aider to diagnose a patient's condition or make a determination about the appropriateness (or the lawfulness) of administrating a particular medicine. The guidance advises that first aiders may nevertheless help persons to administer their own prescribed medication where they are trained to do so and to contact the emergency services as necessary.

There is also no objection by the HSE to paracetamol or aspirin being made available in the workplace or to employers providing vending machines for basic over-the-counter painkillers, provided they are not accessible by the general public.

Automated external defibrillators

4.30 An AED is a small portable device used to treat a person who has collapsed of a heart condition, usually cardiac arrest (often referred to as a heart attack).

The rationale for the use of AEDs is that swift intervention to treat a cardiac arrest is critical to the victim's chances of a successful recovery. Of approximately 150,000 deaths from cardiac arrest each year in the UK, around 50 per cent occur before the victim gets to hospital, and the chances of successful defibrillation are believed to reduce by 7–10 per cent with each minute of delay. (Resuscitation Council, 2005.)

In a cardiac arrest, the heart beats in an uncoordinated way, which means it can no longer effectively pump blood to the brain and other key organs. AEDs are designed to record, to analyse and deliver shocks which interrupt the irregular and uncoordinated activities that occur in cardiac arrests, enabling the heart to restore the normal electrical activity and pumping mechanism.

The AED is attached to the person's chest using an adhesive pad, it then automatically analyses the persons' heart rhythm and advises the operator on whether an electronic shock is required. The shock is then delivered by pressing a button on the device.

Defibrillators are the logical extension to the first aid kit and are very easy to use. Teaching their operation complements routine first aid training – the additional training is minimal, just a few hours. Alternatively courses are available for those who wish to combine a basic knowledge of first aid and use of a defibrillator. An employer should consider the need to have a defibrillator available on the basis of the assessment of first aid needs – based on the profile of the work force, work undertaken, location of site and other factors present.

The Resuscitation Council (UK) has published detailed guidance on the legal status of those who attempt resuscitation – see *www.resus.org.uk/pages/legal.htm*

Keeping records of first aid administration

4.31 There is no specific statutory duty to make or keep records of the first aid measures taken in any incident. The HSE recommends that it is good practice to provide first aiders/appointed persons with a book in which to record incidents that require their attendance. There are various reasons why records of first aid given should be recorded on behalf of the employer in a form which can be accessed centrally:

(a) this will provide a contemporaneous record which is fuller than the basic accident book entry;

(b) in the event that there is a claim that first aid equipment was deficient, or treatment was not provided in time or was performed negligently, to the extent that the original injury or illness has been made worse than it should have been there needs to be evidence of the treatment that was given, and any factors that could have delayed or hindered treatment. (*The Health and Safety (First-Aid at Work) Regulations 1981* do not contain provisions excluding civil liability, so a claimant may base a claim on breach of the regulations.)

There is no statutory defence built into these regulations, and the duty to ensure the provision of adequate facilities and suitably trained staff is an absolute one – it is not subject to the qualification of providing these only so far as is reasonably practicable;

(c) some events which involve first aid thereby give rise to *RIDDOR* reporting obligations; the recording system therefore needs to preserve this information but also to trigger the necessary *RIDDOR* notifications (see page 143);

(d) it provides information with which the person who supervises the first aid personnel will be able to review individuals' experience, give advice, where appropriate, and review their capabilities and training needs where necessary (see *MHSWR Regulation 13(3)*);

(e) The data obtained will supplement other information available for monitoring accident and occupational health trends.

This record can be retained with the accident book records but should not be written into the accident book itself which is a separate statutory record (see page 138).

See First Aid Treatment Record Form on next page.

Informing employees of arrangements

4.32

> "An employer shall inform his employees of the arrangements that have been made in connection with the provision of first aid, including the location of the equipment, facilities and personnel" (Health and Safety (First-Aid) Regulations 1981, Regulation 4).

The information which should be relayed in any employee induction programme and the arrangements also need to be documented as part of the duty to establish written safety arrangements for the purposes of the *MHSWR* (see page 72).

It is generally the case that employers should go further and display first aid information on the premises. In large organisations, each floor should contain information on who the first aiders are. In smaller organisations, the information should be displayed in an area where all staff are likely to see it, such as the kitchen area. If there is an intranet, then the information can also be displayed here.

Special consideration is needed in the assessment of the provision of information to the visually impaired and those with language difficulties.

FIRST AID TREATMENT RECORD FORM

Name of Injured Person		
Accident/Illness		
Date: **How Injury Occurred:**	**Time:**	**Location:**
Injury/Illness		
Nature of injury/illness: **Treatment given:** **Did the injured person:**	**go home** ☐ **go to hospital** ☐ **go back to work** ☐ **other (state)** ☐	
Person Making Entry		
Name:		**Date:**
Signature:		
Please send this form to the Occupational Health Dept		

First aid for the care of the public

4.33 *The Health and Safety (First-Aid) Regulations 1981* do not oblige employers to provide first aid for visitors and members of the public. However, many organisations provide a service for others, for example places of entertainment and shops, and HSE strongly recommends that employers include the public and others on their premises when making their assessment of first aid needs.

There is an element of uncertainty about the wider obligations of employers and when first aid provision might be an element of their duty of care. *Section 3* of the *HSWA* (see page 54) requires them to take reasonably practicable steps to ensure the safety of employees. When an organisation interacts closely with a member of the public and there are significant risks – such as with large crowds or certain activities like sports, which are prone to injuries or vulnerable categories such as children are involved – it becomes strongly arguable that there is a duty to make at least basic arrangements for first aid. A common law duty of care could also arise in these situations.

An evaluation of the *Health and Safety (First Aid) Regulations 1981* carried out for the HSE found that 67 per cent of respondents considered non-employees/members of the public when assessing their first aid needs (HSE, 2003). Employers in workplaces where there is a large public presence, for example airports, shopping centres and places of entertainment, generally made first aid provision for them. This was seen as being of mutual benefit to both employers and the public. Indeed, such organisations may feel they have a "moral obligation" to cater for the first aid needs of the public and in turn this might help create a positive public image. There was also a perception that failure to deal with injury or illness to members of the public could have a negative economic impact. In smaller businesses where resources are limited, first aid provision for non-employees may not be feasible.

In 2003 the HSE issued a Discussion Document that included a question about whether employers should continue to cater for the first aid needs of the public on a voluntary basis or whether this provision should be made compulsory. This recognised that there would be difficulties in determining precisely who should have responsibilities for providing public first aid, and how widely "the public" should be defined. The majority of respondents were in favour of maintaining the current position. In analysing the views of stakeholders, HSE concluded that there was already a good voluntary response, especially in sectors dealing with large numbers of the public. In addition, moving to a compulsory regime would require a change to primary legislation and it would place an unreasonable burden on small businesses. First aiders were concerned about litigation to the extent that some would reconsider volunteering to undertake their duties if they were obliged by law to provide first aid to the public. The Health and Safety Commission agreed with HSE's recommendation to continue with a voluntary approach.

Specific guidance on first aid for the public

4.34 In addition to its general guidance on first aid at work, HSE also produces more specific guidance, which incorporates first aid, in relation to specific activities/sectors where there might be a large public presence. This includes information and advice for the organisers of large events, and for those in charge of swimming pools, fairgrounds and amusement parks and care homes. This guidance, which is periodically reviewed, encourages employers to include non-employees in their first aid provision.

Below is a summary of these guidance documents and what they cover.

The event safety guide, a guide to health, safety and welfare at music and similar events (HSG195)

4.35 This document outlines that employers are normally only responsible for providing first aid for their employees. However, at events, such as pop concerts, it is the organiser's responsibility to ensure that there is adequate first aid provision, in order to satisfy the requirements of the Health and Safety at Work Act 1974 and associated regulations.

It covers everything from entertainment, sustenance and equipment to medical, ambulance and first aid management.

Managing health and safety in swimming pools (HSG179)

4.36 Operators are required to carry out a suitable and sufficient risk assessment of their operations and to identify necessary control measures. A suitable and sufficient risk assessment would have to take account of the whole user population of the swimming pool and that drowning can occur very quickly indeed.

This guide covers both the ideal physical environment of a pool and steps which can be taken to increase a pool's safety if physical changes are not possible, such as increased supervision, barriers and signage. It also states that additional life guards may not in themselves provide adequate safeguards.

Fairgrounds and amusement parks (HSE, 1997b)

4.37 This HSE guidance develops the good practice concerned with the overall safety management of attractions contained in the previous Code of Safe

Practice with increased emphasis on risk assessment, management of safety and inspection. It points out that, although not a requirement of the 1981 Regulations, provision of first aid for the public would be a common law duty, and part of the site emergency planning arrangements.

Health and safety in care homes (HSE, 2001)

4.38 This document outlines how owners and managers of care homes, as well as employees and safety representatives, can understand and meet their duties under health and safety legislation. The main risks found in care homes are

Table 4.9: First aid for child carers

First aid for child carers Duration: approximately twelve hours
Who is it for? Members of the National Child Minders Association (NCMA) and Pre-School Learning Alliance (PLA). Those requiring a detailed knowledge of first aid for young children Those wishing to become an appointed person
Course content includes Accidents – how and what type occur Unconscious casualty: adult/child/infant Resuscitation: adults/children/infants Contents of the first aid kit Dealing with blood loss and shock Burns and scalds Causes of unconsciousness Electric shock Foreign objects Recognition and management of fractures Administration of medication Recognition of infectious diseases and minor illnesses Recording and reporting – NCMA/PLA and HSE
Qualification First aid for child carers certificate valid for three years

Note: This is not an HSE approved course

covered in detail and guidance given on what should be done to safeguard both workers and service users. It is recommended that first aid provision extends to service users and visitors.

Childcare and education services

4.39 It is an Ofsted requirement that all those involved in the care of children have a recognised first aid qualification. There is a good Department for Education and Employment good practice guide for schools on first aid provision (DfEE, undated) (Tables 4.9 and 4.10).

Table 4.10: Basic first aid for teachers

Basic first aid for teachers Duration: approximately eight hours
Who is it for? Teachers and staff wishing to learn basic first aid for adults and children within a school environment Those wishing to become an appointed person
Course content includes Personal safety, assessing the incident Priorities: DRAB (danger, response, airway, breathing) check and sending for help Unconscious casualty: adult/child Resuscitation: adults/children Choking: adults/children Dealing with blood loss and shock Burns and scalds Emergency treatment for diabetes and asthma Faints Contents of HSE approved first aid kit Reporting and recording of incidents
Qualification Basic first aid for teachers' certificate Basic first aid for teachers' display certificate Record of achievement

Note: This is not an HSE approved course

4.40

A large office

This company's premises are three floors. Fifteen first aiders are appointed to cover a workforce of 550 people. Their certified training and refresher courses are supplemented by "first aider lunches" where procedures and training needs are discussed and there are opportunities to practise. At these meetings the part-time occupational health nurse also presents a specific topic for discussion (e.g. asthma or treating eye injuries).

First aiders positions are advertised internally. The company pays first aiders an extra £250 pa on top of their normal salaries (which is taxable). Training is carried out off-site and in normal working hours.

Two small converted offices serve as first aid rooms and these are equipped according to the HSE guidelines. Additional items though include instant one-use ice packs used to reduce inflammation and bruising (slips and trips being the most frequent risk in this work environment). Painkillers are available for staff to purchase from vending machines in the toilets.

Given the age profile of the staff the company also has an AED, which a number of first aiders have been trained to use.

The company makes use of technology to improve its first aid provision. First aid policies, procedures and contacts are available on the company's intranet and so the information can be readily updated. First aid help is summoned by dialling one number which then rings simultaneously at the desks of all first aiders. The first one to pick up the call then acts. If assistance is required then first aiders have extension numbers of others listed on the intranet and in the internal telephone directory.

First aid treatment form pads are available which are filled in and then scanned and linked to HR records.

The company also has a small satellite office across town which is to service a particular customer. This is staffed full-time by a receptionist and a rotating staff of up to six people from the head office. Based on the low risk environment, and the low numbers at the location at any one time, the company designated the receptionist the "appointed person" for the satellite office and the health and safety manager produced a written task list for maintaining the first aid boxes and for calling for assistance in an emergency and gave the receptionist a one-to-one briefing on this which was recorded for training purposes. The restroom was designated as a first aid room.

CASE STUDY 8

4.41

A retail chain

This well-known stores group prides itself on its customer service, and care of the public featured heavily in its assessment of its first aid needs.

Health and safety manuals describe the company's policies and organisation of first aid which closely follow the HSE ACoP. It classifies the shops and offices as low risk, and the warehousing operations as medium risk.

First aiders are usually volunteers and are selected by the occupational health team using a broad range of criteria based around aptitude and reliability. Employees who are known to be HIV positive or carry other blood-borne diseases are not accepted as first aiders.

As well as the usual certified refresher courses the company maintains awareness and skill levels of first aiders using various additional training methods. Videos are shown twice a year and first aiders attend at least one hour of training every six months. This will include practice of techniques including CPR and care of the unconscious casualty.

The company standards for building design specify the size and location of medical rooms near to sales areas. Levels of sound-proofing are specified to provide confidentiality and a calm environment.

First aid equipment is based on HSE guidelines and includes portable first aid bags for use in emergencies. Because of the likelihood of first aiders treating the public careful attention has been paid to their insurance position, and this is clearly described in the manuals for their reassurance.

The manuals also contain requirements for record keeping, and guidelines on appropriate discussions that can take place with customers (e.g. where it may not be possible to verify that there has been a genuine accident).

CASE STUDY 9

4.42

A manufacturing site

This facility comprises several process plants, with around 650 people working on the site in three shifts. It is assessed as a medium risk working environment. At any one time the company has thirty to thirty five first-aiders (more than double the number suggested by the HSE guidelines). Training takes place in their own time and to reflect this the company pays a salary increment of £700 pa in compensation.

First aiders are contacted in emergencies directly or by a group pager. They are supported by the full-time occupational health professional and administrator who provides an additional resource for emergency treatment. As part of its routine reviews of risk assessments the company has concluded that while its first aid provision was adequate, it could be improved still further by "zoning" its first aiders so that cover was evenly spread across the large site, and it is now implementing more sophisticated planning arrangements for selecting employees in the desired locations.

First aid boxes are located all around the large site. There is a central occupational health suite accessible to first aiders twenty four hours a day which serves as the first aid centre. It is fully equipped with treatment rooms and defibrillators.

Contractors are permanently on site for facilities maintenance. They are required to make their own trained first aider provisions, but the site facilities are available to them as well.

Policy review and the future of first aid course structures

4.43 In 2006/2007 the HSE is concluding a lengthy consultation exercise with both employers and training providers about possible changes to the requirements and contents of first aid training. The research carried out identified a number of key aspects for further consideration:

- Small organisations find it difficult to release employees to attend initial four-day first aid at work courses.
- Many training providers and first aiders consider the current three-year period between the initial four-day course and subsequent refresher training is too long as skill decay will occur.
- The role of the appointed person is considered a valid one, particularly for small, low risk organisations. However, in many cases there is confusion between the role of appointed persons and that of first aiders. Some organisations are using appointed persons as first aiders after they have completed only a one-day first aid course.
- Some organisations have sent employees on a one-day course in first aid and use such personnel as "basic first aiders". This approach, while not complying with the regulations, may be considered more proportionate to the needs of smaller organisations where only low risk activities are carried out.

Overall, the research showed general support for shorter first aid courses, more frequent refresher training and "basic first aiders" trained in emergency first aid.

Further support for more frequent refresher training has come from an HSE review of the scientific and medical literature on the issue of skills and knowledge retention. The findings of this review indicated that some individuals cannot adequately perform basic life support within two months of first aid training and after 3 years the results are generally very poor. This reinforces the view that first aiders in the workplace should undertake refresher training more frequently than every 3 years.

Findings from the research and consultation described above indicate that the current first aid at work training regime has a number of problems that need to be addressed. This is crucial if first aid training is to meet the needs of modern business. Any new training arrangements have to take account of the changes that have occurred in industry since the *Health and Safety (First-Aid) Regulations 1981* first came into force. There has been a shift away from manufacturing to service industries and there are fewer large firms and far more small ones. Over 90 per cent of the 3.5 million businesses employ fewer than ten people, but nearly half the workforce is employed in large organisations.

Introducing a one-day emergency first aid course

4.44 The evidence presented indicates that there is sufficient justification for introducing a one-day (six contact hours) course in emergency first aid and such a change has received support from stakeholders. It would give employers a choice when assessing their first aid provision. If a first aider is needed, a suitable employee could be sent on either a one-day emergency first aid course or a full first aid at work course depending on the findings of the needs assessment. In this way, it will be easier for employers to ensure that first aid provision is proportionate to the needs of their workplaces.

It is anticipated that the one-day course would be particularly applicable to workplaces with a small number of employees and low hazards and that this will increase the provision of first aid in such workplaces where currently there may be no one with a first aid qualification. To ensure consistency of training standards, HSE is considering approval for the one-day course. This has received support from first aid at work training providers.

Introducing a shorter first aid at work course

4.45 The introduction of a first aid at work course lasting eighteen contact hours that can be delivered over three days has also received support from stakeholders. However, some first aid at work training providers have reservations about this change on the basis that it will not provide sufficient time for training and final assessment, but this assumes no change to the syllabus.

Reducing the number of contact hours for the first aid at work course is important for several reasons. Firstly, shorter, more clearly focused first aid at work courses should help first aiders by reducing the amount of non-essential detail they are required to assimilate. This will leave them to concentrate on the key skills and knowledge that will provide the tools to deliver high-quality first aid in the workplace. Secondly, it may also encourage employers to send employees on first aid at work courses where they have previously been reluctant to do so because of concerns over releasing employees to attend the current four-day course. Finally, a shorter course will also help offset the additional time first aiders will need to spend away from work on annual refresher training courses under the new regime.

Introducing annual refresher training

4.46 Annual refresher training has received widespread support, mainly in recognition of the evidence of decline in first aid skills and knowledge following completion of training courses and the lack of opportunities for practice. The exact frequency of refresher training that would maximise its potential benefits is difficult to define. It is influenced not just by the evidence from scientific studies but also by the practicalities of its implementation.

In research, 18 per cent of those surveyed stated that annual refresher training was provided for first aiders, even though it is not a statutory requirement. In addition, the research also found that most companies (generally medium–large) would sanction more frequent training provided it did not amount to more than one day per year.

Under these proposed changes, employers will benefit from less time out of the office for staff but with improved skills retention and confidence levels (Figure 4.1).

Figure 4.1: Proposed new scheme of first aider courses

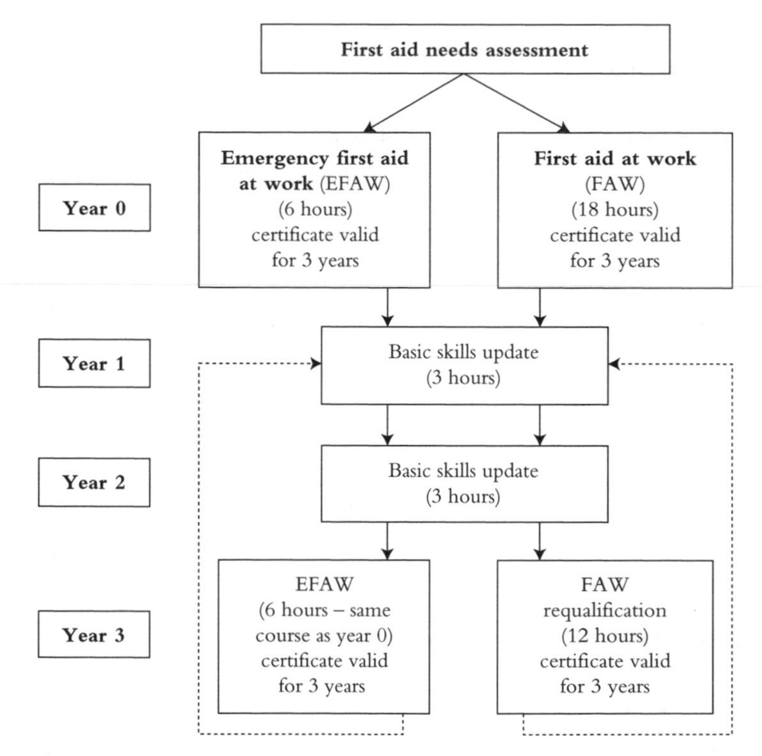

Source: HSE, 2006

The role of appointed persons

4.47 Since it has been decided that the *Health and Safety (First-Aid) Regulations 1981* will not be changed the role of the appointed person will remain unchanged. The HSE's review nevertheless found that that there was some confusion over the role and responsibilities of the appointed person. The appointed person will continue to be the minimum requirement for small, low hazard organisations. As is the case now, formal first aid training will not be mandatory to carry out this role.

However to clarify the role of appointed persons there will be a clearer description of the separation from the role of first aiders in the workplace.

The time period for introducing the new changes is still under consultation but is expected to be in 2008. As part of this process it is likely too that the HSE's functions in approving and monitoring first aid training organisations will be transferred to an industry body representing all parts of the first aid training industry.

Sources of further information on first aid at work

4.48

1. HSE first aid web page at *www.hse.gov.uk/firstaid*
2. British Red Cross *www.redcross.org.uk*
3. St John's Ambulance *www.sja.org.uk*
4. St Andrew's Ambulance Association *www.firstaid.org.uk*

Legal compliance checklist

4.49

- ✓ First aid needs have been assessed.
- ✓ First aiders or appointed persons are appropriately designated and trained.
- ✓ Any required risk assessments have been carried out.
- ✓ First aid facilities and equipment are available and correctly maintained.
- ✓ Staff are adequately informed of first aid arrangements.
- ✓ First aid records are maintained.
- ✓ Provision has been made for the care of the public (if appropriate).

References

HSE (1997a) *First aid at work. The Health and Safety (First-Aid) Regulations 1981. Approved Code of Practice and Guidance L74*. HSE Books.

HSE (1997b) *Fairgrounds and amusement parks: guidance on safe practice (HSG175)*. HSE Books.

HSE (1999a) *The event safety guide. A guide to health, safety and welfare at music and similar events (HSG195)*. HSE Books.

HSE (1999b) *Managing Health and Safety in Swimming Pools (HSG179)*. HSE Books.

HSE (2000) *Health care and first aid on offshore installations and pipeline works. Offshore Installations and Pipeline Works (First-Aid) Regulations 1989. Approved Code of Practice and Guidance L123*. HSE Books.

HSE (2001) *Health and Safety in Care Homes (HSG 220)*. HSE Books.

HSE (2003) *Evaluation of the Health and Safety (First-Aid) Regulations 1981 and the approved code of practice and guidance*. HSE Research Report 069. http://www.hse.gov.uk/research/rrpdf/rr069.pdf.

HSE (2005) *Blood-borne Viruses in the Workplace (IND342)*. www.hse.gov.uk/pubns/indg342.pdf.

HSE (2006) *Update on the review of the Health and Safety (First-Aid) Regulations 1981*. www.hse.gov.uk/firstaid/review/june06.htm.

Resuscitation Council (2005) *Resuscitation Guidelines 2005*, Resuscitation Council (UK). www.resus.org.uk.

UKOOA (2002) *First Aid and Medical Equipment on Offshore Installations (Reference EHS12)*. UKOOA Publications.

DfEE (undated) *Guidance on first aid for schools*. www.teachernet.gov.uk/_doc/4421/GFAS.pdf.

5 Legal aspects of reporting and investigating accidents

In this chapter:

Introduction

5.1 All accidents involving any injury need to be recorded. Depending on the type of injury, severity and consequences an accident may need to be reported to the enforcing authorities. Some incidents which do not result in any injury must also be reported.

Care is needed to ensure that each of the different formalities are understood and complied with in an organisation. It can also be essential to preserve potentially important evidence for a number of reasons. It is likely to be needed for the organisation's own investigation of an accident and to avoid its recurrence (see Chapter 9). Contemporaneous evidence can also be influential in determining liability in any subsequent legal proceedings. Perhaps most pressingly, there can be criminal penalties for not reporting and retaining records of accidents.

Information on individual accidents and their causes is also needed for compliance with wider *Health and Safety at Work Act 1974* (*HSWA*) duties to ensure safety and related requirements of the *Management of Health and Safety at Work Regulations 1999* (*MHSWR*) to review risk assessments to ensure that they are suitable and sufficient and that they remain valid.

In the absence of an express statutory duty to undertake formal accident investigations – and the government's decision in 2002 not to introduce such a duty in British health and safety legislation – the duty to review of risk assessments has attracted more scrutiny by enforcing authorities and health and safety professionals. Clearly this type of information can be sensitive and potentially damaging to the interests of some of those involved. Later in the chapter the issues of confidentiality are considered, in particular the issue of confidentiality which can arise when litigation is anticipated or has been commenced. The chapter concludes with a compliance checklist.

Obligations on workers to report accidents and dangerous situations

5.2 Although the most significant responsibilities for accident recording and reporting are naturally placed on employers, it is useful to begin with an overview of the rather less known obligations the law imposes on employees.

There are two different types of statutory reporting requirements which apply directly to employees. The first relates only to accidents, and exists not within the framework of health and safety law, but instead under social security law. This requirement has become the basis of the familiar Accident Book in which entries of injuries are routinely recorded. The second set of requirements stems more recently from EU health and safety directives and are found in the *MHSWR*.

Notification of accidents by employees

5.3 All employees need to be made aware that there are notification requirements imposed on them by the *Social Security (Claims and Payments) Regulations 1979*. (This legislation is highly complex and has been subject to much revision.) An accident for these purposes is one "arising out of and in connection with the course of [an] employed earner's employment".

Although not actually bound in with the separate scheme of health and safety legislation, these reporting requirements nevertheless complement those which employers have under the *Reporting of Injuries, Diseases and Dangerous Occurrences Regulations 1995 (RIDDOR)*, and there is some overlap in the record-keeping requirements for both sets of provisions (see 5.8).

Every employed earner who suffers personal injury by an accident for the purposes of the Regulations is required to give either *oral or written* notice to the employer. There is considerable flexibility about the ways of doing this. The notification can be given directly to an individual employer, to any foreman or other official under whose supervision the person works at the time of the accident, or to any person designated for the purpose by the employer. It can be done by the accident victim or by any other person on his behalf. However, the alternative method permitted by the Regulations – and the one which has probably become the norm through custom and practice for most organisations – is by way of a written entry as soon as practicable after the happening of an accident of the appropriate particulars in a *book kept specifically for these purposes under the Regulations* (i.e. the Accident Book).

The familiar old workplace Accident Book (Form BI 510) was replaced in 2003 by a new HSE-approved version with the aim of making it compliant with the *Data Protection Act 1998*. It now contains tear-out pages which, after completion, are meant to be detached and kept separately in a secure location. (Use of the old version of the book was allowed for a transitional period only until 31 December 2003 and unused copies should be disposed of.) The up-to-date version (still referenced BI 510) is available from HSE Books. As well as sections for completing details of accidents the publication includes pages of instructions to employees and employers about their responsibilities.

Relevant accidents

5.4 An accident to which the employees' notification requirement applies is strictly speaking one in respect of which one of the following types of social security benefits may be payable: Disablement benefit, reduced earnings allowance, retirement allowance and industrial death benefit.

It is totally impractical for individuals (or their employers for that matter) to apply anything other than a broad-brush approach to these technical provisions. In practice workers must be advised to report any injury, however minor, which is work related.

There are some provisions in the legislation which clarify the need for notification of an accident where there might otherwise be a question whether it should be counted as occurring in the course of normal employment:

- Where the person was at the time of the accident contravening any regulatory requirements or instructions given by the employer (or acting without instructions) this will still be deemed to be in the course of employment as long as the activity concerned was being undertaken in connection with the business.
- An accident occurring while a person is travelling to and from a place of work as a passenger in a vehicle operated by the employer or by someone else the

employer has made the arrangements with will be covered (accidents on public transport are not included).

- If the accident occurs while a person is taking steps on an actual or supposed emergency, it should still be notified.
- Accidents caused by another's skylarking or other misconduct, or by the behaviour of animals, or being struck by a falling object or by lightning will generally be covered.

However, it is important to appreciate that these obligations arise in relation to an accident as an *event* and not the *injury* or *condition* as such. This is clear from the 1979 Regulations, and this distinction is supported by case law, in particular in the area of aviation law where the term is important in limiting liability under the Warsaw Convention, where it had been determined that *accident* must mean a cause of injury and does not constitute the injury itself. An accident has been construed as meaning a sudden, unusual or unexpected event.

This means that an employee who has a condition, for example repetitive strain injury (RSI), which might have been aggravated by work on an occasion is not required to notify it and is not entitled to insist on it being recorded as an accident. An accident must also be distinguished from a disease such as is more explicitly the case under *RIDDOR* (where there are separate disease provisions).

What accident details should be notified?

5.5 The *appropriate particulars* which have to be given by or for the accident victim are specified by the Regulations as follows:

- Full name, address and occupation of injured person.
- Date and time of accident.
- Place where accident happened; cause and nature of injury.
- Name, address and occupation of the person giving the notice, if other than the injured person.

Providing the right information is straightforward if one simply fills in the necessary entries in the Accident Book (BI 510) (HSE, 2003).

See Accident Book Record Form on opposite page.

Are there penalties for non-compliance?

5.6 No penalty is laid down in the 1979 Regulations for any failure by an employee to meet these notification requirements. By incorporation of the required

1 About the person who had the accident

Name

Address

Postcode

Occupation

2 About you, the person filling in this record

▼ If you did not have the accident write your address and occupation.

Name

Address

Postcode

Occupation

3 About the accident *Continue on the back of this form if you need to*

▼ Say when it happened. Date / / Time

▼ Say where it happened. State which room or place.

▼ Say how the accident happened. Give the cause if you can.

▼ If the person who had the accident suffered an injury, say what it was.

▼ Please sign the record and date it.

Signature Date / /

4 For the employer only

▼ Complete this box if the accident is reportable under the Reporting of Injuries, Diseases and Dangerous Occurrences Regulations 1995 (RIDDOR).

How was it reported?

Date reported / / Signature

procedures in the employer's staff handbook or other similar procedures observing the rules can be made a term of employment and persistent failure to co-operate in providing accident information could be a disciplinary matter.

Nor do the regulations impose any specific sanction for misuse of an Accident Book, for example by the employer or employee altering a recorded account, any defacing or removal of an entry, or the insertion of information which is inappropriate, inaccurate or is not strictly speaking required to be reported.

Occasionally individuals may seek to use the book to record additional information or their opinions on safety in the workplace. In order to take steps to avoid the inclusion of any inappropriate details in an Accident Book, an employer would have to fall back on the general terms, expressed or implied, of the employment contract which might justify appropriate disciplinary action for misuse of the Accident Book. However, any employee complaint about an accident should be properly investigated.

Employees' duties to highlight dangers

5.7 All employees have a general duty under *HSWA, Section* 7, to take reasonable care for their own safety and that of others who could be affected by their acts or omissions at work. This implies alerting others to a danger if it would be reasonable to recognise the risk and to warn others about it. When this duty can arise and what it requires is clarified and narrowed in scope by the *MHSWR, Regulation 14* which now requires any employee to inform his employer of:

(a) any work situation which a person with [that] employee's training and instruction would reasonably consider represented a serious and immediate danger to health and safety;

(b) any matter which a person with [that] employee's training and instruction would reasonably consider represented a shortcoming in the employer's protection arrangements for health and safety.

Essentially the same duties are imposed by fire safety legislation in relation to informing the employer about fire risks.

The duty will not arise though if the danger or shortcoming in question is not one affecting the particular individual personally, and it does not relate to his own work activities. Nor is it necessary to report further a matter which has already been reported by another employee.

For practical purposes therefore an employee who is injured, or who is involved in a "near miss" or other dangerous occurrence or failure in a safety system, or who just subjectively perceives the threat of serious danger ought to make the employer aware of the event. It is not necessary for the employee to discern the cause of the danger, the nature of the shortcomings or to offer any reason why the precautions were inadequate. The purpose is simply to put the employer on notice; and the onus always remains with the employer to comprehend the work-related risks and to take all reasonably practicable steps to ensure the employees' health and safety.

Duties on employers and others to notify and report accidents

RIDDOR

5.8 The *Reporting of Injuries, Diseases and Dangerous Occurrences Regulations 1995 (RIDDOR)* require employers and other responsible persons who have control over employees and work premises to *notify* and *report* to the relevant enforcing authority the following events occurring at work:

- Reportable accidents causing injuries, fatal and non-fatal.
- Reportable occupational diseases.
- Reportable dangerous occurrences, irrespective of whether injury results.

The reasons for these requirements are twofold. First, to enable inspectors of the enforcing authority to undertake investigations and deploy their statutory powers of inspection within a short time of any occurrence while the evidence is still available (if necessary by seizing items or documents) (see page 176). Second, the data collected forms an essential part of the HSE's monitoring of accident trends across sectors and the working population as a whole. Under-reporting of occurrences by duty holders has been a longstanding concern, particularly in some industries, but as the legislation provides no incentive to duty holders to report themselves in the form of immunity from prosecution or a reduction in sentencing on conviction – the threat that self-reporting will lead to prosecution will always deter some from complying with *RIDDOR*. Nevertheless, it is an offence under health and safety legislation to fail to observe the notification and reporting requirements and prosecutions are regularly brought where inspectors only learn of serious occurrences later through other channels.

RIDDOR is complex and requires consideration of many statutory definitions of terms and schedules which describe events which trigger the duties to inform the authorities. In practice it is always necessary to check the regulations carefully because of significant variations in the reporting obligations, and because in certain circumstances the duties can apply not only in the case of incidents involving employees, but also to visitors, customers and members of the public killed or injured by work activities. The basic requirements are outlined in Figure 5.1.

Figure 5.1: Basic requirements of RIDDOR

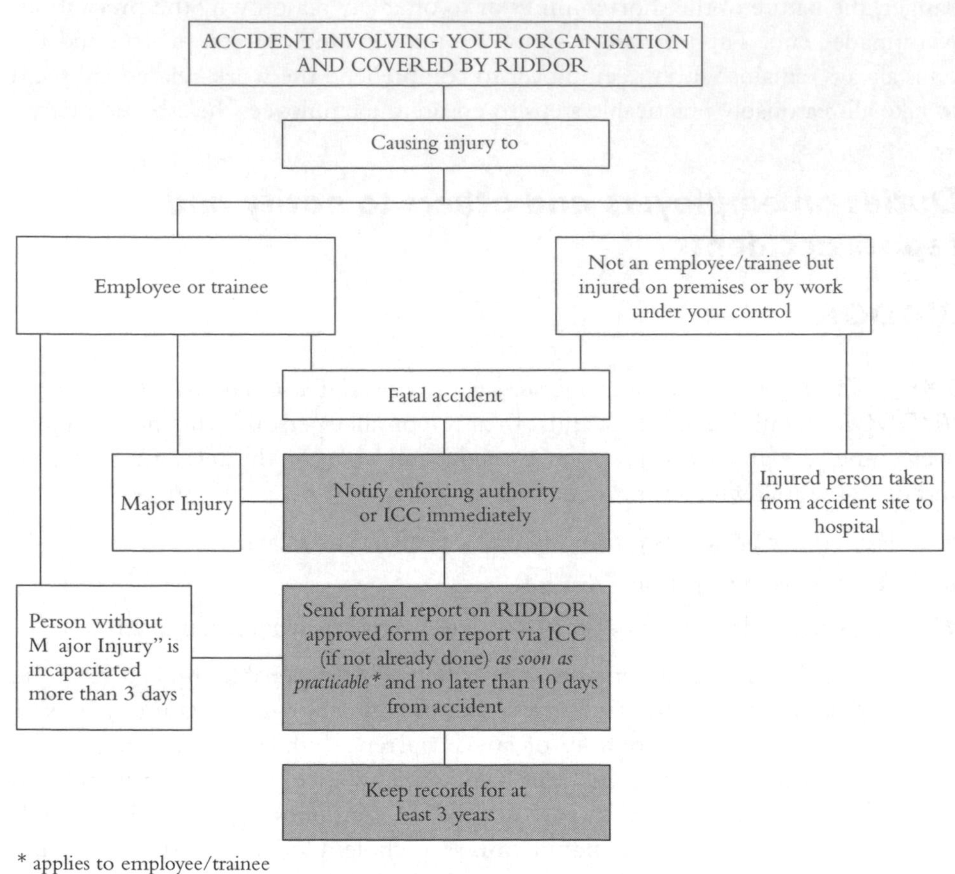

Source: Tolley, 2006

Deaths and major injuries

5.9 Reportable injuries consist of deaths and the following types of "major injuries" arising out of or in connection with work:

1. Any fracture, other than to fingers, thumb or toes.
2. Any amputation.
3. Dislocation of shoulder, hip, knee or spine.
4. Loss of sight (whether temporary or permanent).
5. A chemical or hot metal burn to the eye or any penetrating injury to the eye.
6. Any injury resulting from an electrical shock or electrical burn (including any electrical burn caused by arcing or arcing products) leading to unconsciousness or requiring resuscitation or admittance to hospital for more than 24 hours.

7. Any other injury
   (a) leading to hypothermia, heat-induced illness or unconsciousness;
   (b) requiring resuscitation;
   (c) requiring admittance to hospital for more than twenty four hours.
8. Loss of consciousness caused by asphyxia or by exposure to a harmful substance or biological agent.
9. Either of the following conditions which result from the absorption of any substance by inhalation, ingestion or through the skin:
   (a) acute illness requiring medical treatment; or
   (b) loss of consciousness resulting from the absorption of any substance by inhalation, ingestion or through the skin.
10. Acute illness which requires medical treatment where there is reason to believe that this resulted from exposure to a biological agent or its toxins or infected material.

An accident for these purposes has an extended meaning which includes acts of non-consensual violence done to a person at work. It also includes suicides on railways, tramways and similar transport systems.

Over-3-day injuries

5.10 This category of reportable injury only applies to accidents to persons at work. (As above, this includes assaults.) It arises where such a person is incapacitated for more than three consecutive days from his or her normal contractual work days (excluding the day of the accident itself but including weekends and other days which would not have been working days) by an injury resulting from an accident at work. If the injury was in any case a major injury, it becomes excluded from this category.

Special rules for road accidents

5.11 Road accident deaths and injuries are only notifiable and reportable if the death or injury is caused by or connected with:

(a) exposure to any substance conveyed by road;

(b) loading or unloading vehicles;

(c) construction, demolition, alteration or maintenance activities on alongside roads; or

(d) an accident involving a train.

Dangerous occurrences

5.12 Assessing whether there is a duty to notify and report a dangerous occurrence involves examining *RIDDOR, Schedule 2, Part 1* which lists twenty one categories for serious events of general application which involve the following:

- Lifting machinery
- Pressure systems
- Freight containers
- Overhead electrical lines
- Electrical short circuit
- Explosives
- Biological agents
- Malfunction of radiation generators, etc.
- Breathing apparatus
- Diving operations
- Collapse of scaffolding
- Train collisions
- Walls
- Pipelines or pipeline works
- Fairground equipment
- Carriage of dangerous substances
- Collapse of a building or structure*
- Explosion or fire
- Escape of flammable substances
- Escape of substances

* Different provisions apply offshore

It is necessary to refer to the text of Schedule 2 of *RIDDOR* under each of the headings above to see if the type of occurrence is reportable. See Appendix I to this chapter.

In addition, there are twenty seven listed categories relating to occurrences in mines or quarries, twenty four for transport systems and eleven for occurrences offshore.

An additional category of occurrences in the *RIDDOR* is gas incidents. For these purposes they are:

- Incidents involving flammable gas in fixed pipe distribution systems or refillable LPG containers. Where notice of a death or major injury is received by a gas conveyor (or in the case of LPG containers, a filler, importer or supplier (other than by a retail supplier)) the notification and reporting duties will arise.
- Where an employer or self-employed person who is an approved gas fitter has sufficient information to decide that a gas fitting, flue or ventilation is, or is likely to cause death or major injury because of leakage or inadequate combustion, he is required to report that fact to the HSE on a prescribed form within 14 days.

Diseases

5.13 Regulation 5 of *RIDDOR* creates two separate classes of diseases for these purposes. The first is in relation to all workplaces and employment relationships, and comprises a list of forty seven occupational diseases. These diseases are reportable after they have occurred only if the persons work involves them (i.e. their current job) in one or more of the activities listed against that disease in Schedule 3 of *RIDDOR*. The second class of diseases comprises a list of twenty five diseases which (in addition to those in the first class) are reportable if they occur in offshore workplaces.

These reporting requirements are outlined in Figure 5.2. See also Appendix II to this chapter for the relevant disease and the workplaces' activities in relation to which these are reportable.

Figure 5.2: Occupational disease identification within an organisation

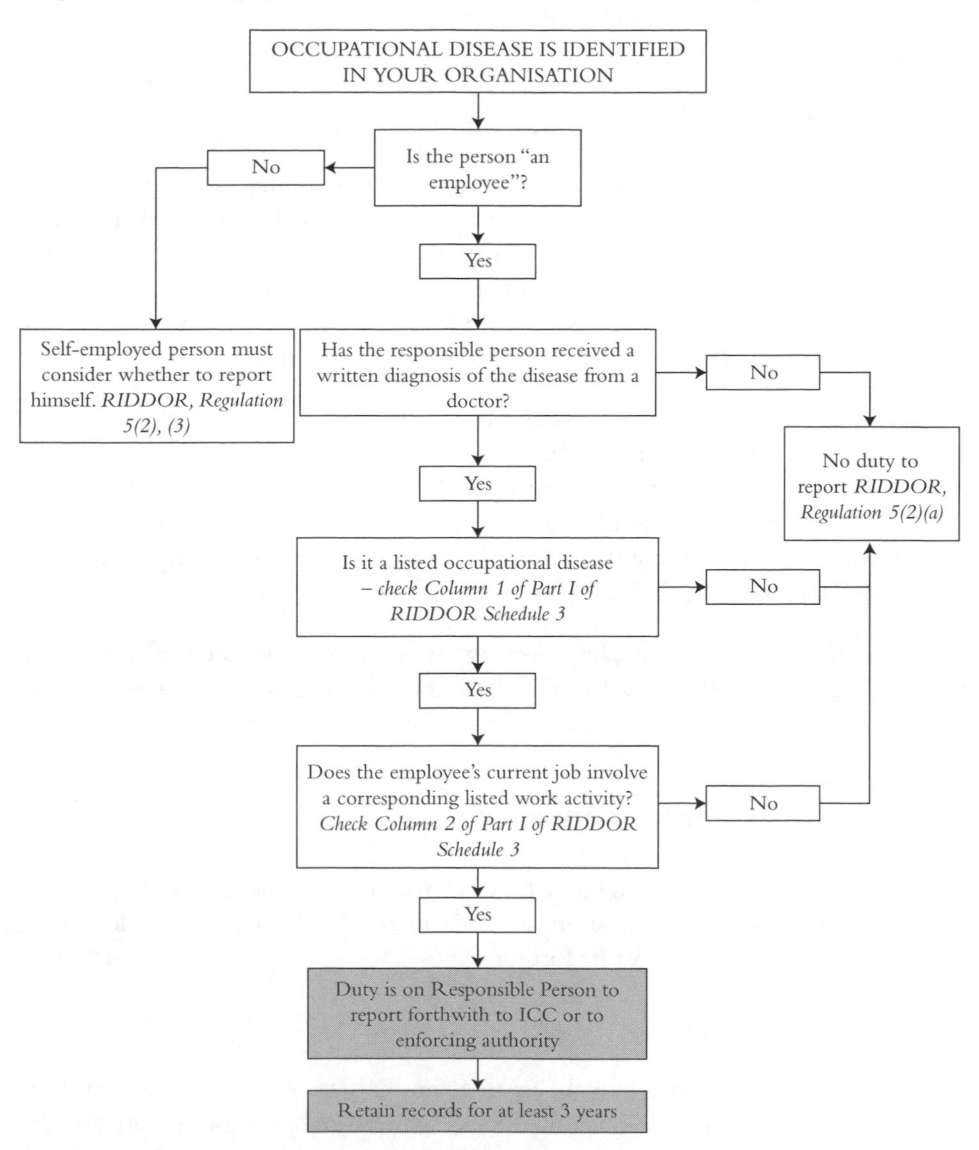

Most of the listed diseases are ones which, if diagnosed, would be quite obviously serious and would probably come to the attention of senior persons in the line management. Caution is needed though because some apparently less serious conditions are included, for example cramp of the hand or forearm due to prolonged typing or writing, or traumatic inflammation of the tendons.

This reporting duty arises only when the responsible person has received information in writing from a registered medical practitioner diagnosing one of the reportable diseases. (In the case of the self-employed, the duty is triggered regardless of whether or not the information is given to him in writing.)

Responsible persons

5.14 The person usually responsible for reporting injury-causing accidents, deaths or diseases will be the employer of the victim; or, if the victim is not an employee, the person having control of the work premises or activity may be the responsible person. In some situations the position is modified and there are different designated responsible persons. The position is summarised more fully in the HSE's guidance on *RIDDOR* (HSE, 1999) as shown in Table 5.1.

In situations where the responsible person is difficult to identify because of the shared control of a site or operations, or because the employment status of the victim is hard to discern, the true responsible person risks breaching *RIDDOR* if appropriate action is not taken. Consequently, co-ordination arrangements may need to be discussed and recorded to ensure that *RIDDOR* notification and reporting occurs correctly (see *MHSWR, Regulation 11*).

Strictly speaking there is no obligation on someone who is not the responsible person to undertake *RIDDOR* notification and reporting where the responsible person as defined by the regulations fails to do so for any reason.

Duty to notify and/or report events

5.15 The regulations deal slightly differently with obligations to make the authorities aware of incidents (notification) and the formality of supplying a document summarising the circumstances (reporting).

Duty to notify the event

5.16 Responsible persons (see above) must notify the relevant enforcing authorities (either the HSE or the local authority, whichever has responsibility for the

Table 5.1: Identifying "responsible persons"

Reportable event (under RIDDOR 1995)		*Responsible person*
1. Special cases		
All reportable events in mines		The mine manager
All reportable events in quarries or in closed mine or quarry tips		The owner
All reportable events at offshore installations, except cases of disease reportable under Regulation 5		The owner, in respect of a mobile installation, or the operator in respect of a fixed installation (under these Regulations the responsibility extends to reporting incidents at sub-sea installations, except tied back wells and adjacent pipeline)
All reportable events at diving operations, except cases of disease reportable under *Regulation 5*		The diving contractor
2. Injuries and disease		
Death, major injury, over-3-day injury, or case of disease (including cases of disease connected with diving operations and work at an offshore installation):	Of an employee at work	That person's employer
	Of a self-employed person at work in premises under the control of someone else	The person in control of the premises: • at the time of the event; • in connection with their carrying on any trade, business or undertaking
Major injury, over-3-day injury or case of disease:	Of a self-employed person at work in premises under their control	The self-employed person or someone acting on their behalf
Death, or injury requiring removal to a hospital for treatment (or major injury occurring at a hospital):	Of a person who is not at work (but is affected by the work of someone else), for example a member of the public, a student, a resident of a nursing home	The person in control of the premises where, or in connection with the work going on at which, the accident causing the injury happened: • at the time of the event;

(*Continued*)

Table 5.1: (Continued)

Reportable event (under RIDDOR 1995)	*Responsible person*
	• in connection with their carrying on any trade, business or undertaking
3. Dangerous occurrences	
One of the dangerous occurrences listed in *Schedule 2* to the Regulations, except:	The person in control of the premises where, or in connection with the work going on at which, the dangerous occurrence happened:
• where they occur at workplaces covered by part 1 of this table (i.e. at mines, quarries, closed mine or quarry tips, offshore installations or connected with diving operations); or	• at the time the dangerous occurrence happened
• those covered below (which are *paragraphs 13, 14(a) to (f), 16* and *17* of *Schedule 2, Part 1*)	• in connection with their carrying on any trade, business or undertaking
A dangerous occurrence at a well (see *paragraph 13 of Schedule 2, Part 1*)	The concession owner (the person having the right to exploit or explore mineral resources and store and recover gas in any area, if the well is used or is to be used to exercise that right) or the person appointed by the concession owner to organise or supervise any operation carried out by the well
A dangerous occurrence at a pipeline (see *paragraph 14(a) to (f) of Schedule 2, Part 1*), but not a dangerous occurrence connected with pipeline works (*paragraph 14(g) of Schedule 2, Part 1*)	The owner of the pipeline
A dangerous occurrence involving a dangerous substance being conveyed by road (*see paragraphs 16 and 17 of Schedule 2, Part 1*)	The operator of the vehicle

Source: HSE (1999)

premises by virtue of the *Health and Safety (Enforcing Authority) Regulations 1998*) by the quickest means practicable (normally by telephone) of the following:

(a) Death as a result of an accident arising out of, or in connection with, work.

(b) Any major injury to a person at work as a result of an accident arising out of, or in connection with, work.

(c) Can injury suffered by a person not at work (e.g. a member of the public or a customer) as a result of an accident arising out of, or in connection with, work, where that person is taken from the accident site to a hospital for treatment.

(d) A major injury suffered by a person not at work, as a result of an accident arising out of, or in connection with, work at a hospital.

(e) A dangerous occurrence.

(f) Gas incidents.

Duty to submit a report

5.17 A responsible person must also provide a written report to the enforcing authority of events causing death or major injury, dangerous occurrences and details associated with workers' occupational diseases. In cases of death, major injury and accidents leading to hospitalisation, the duty to report extends to visitors, bystanders and other non-employees. (The timescales in which this must be done are set out in Table 5.2.)

There is no requirement to notify or report the injury or death of a patient undergoing treatment in a hospital or a doctor's or dentist's surgery or a member of the armed forces of the Crown.

Responsible persons can now choose between dealing directly with the local office of the relevant enforcing authority, or reports can be made by telephone or electronically to the national Incident Contact Centre (ICC).

Direct notification and reporting to the enforcement authority

5.18 If this method is used it will be necessary to deal with the obligations to give notice by the quickest practicable means by telephone, or failing that to fax, to the local HSE office or local authority (whichever is appropriate under the *Health and Safety (Enforcing Authority) Regulations 1998*) (see page 232). (It is advisable to make a file note of the call in case there is any question about timing or content of the notification.)

Table 5.2: Reporting timetables

Events	*Timescale for report*
Death resulting from accident (*RIDDOR*)	Within ten days
Major injury	Within ten days
Injury to person not at work taken straight to hospital	Within ten days
Person not at work suffers major injury as a result of accident arising out of or in connection with work at a hospital	Within ten days
The death of an employee if it occurs within a year following a reportable injury (whether or not reported under (a) above)	As soon as employer is aware
Incapacitation for work of a person at work for more than 3 consecutive days as a result of an injury caused by an accident at work	As soon as practicable and in any event within ten days
A dangerous occurrence	Within ten days
Reportable diseases relating to persons at work, and also specifically those suffered by workers on offshore installations provided that, in both cases: • the responsible person has received a written statement by a doctor diagnosing the specified disease, in the case of an employee; or • a self-employed person has been informed by a registered medical practitioner that he is suffering from a specified disease	Immediately the medical evidence is obtained (in the case of an employee, written evidence)
Gas incidents	Within fourteen days of knowledge of incident

The duty to submit the statutory report form is discharged by sending either Form 2508 (accidents and dangerous occurrences) or Form 2508A (reportable diseases) (see Appendix II to this chapter) to the same HSE or relevant local council office. The regulations do not require proof of posting or use of registered or recorded delivery, although this is advisable as occasionally documents go astray and inspectors do not receive the *RIDDOR* form. A copy of the *RIDDOR* form should also be retained.

Incident Contact Centre

5.19 Since 2001 it has been possible to deal with notification and reporting either over the Internet using *RIDDOR* website www.riddor.gov.uk or by telephone on

Health and Safety at Work etc Act 1974 ?

The Reporting of Injuries, Diseases and Dangerous Occurrences Regulations 1995

Click here for report guidance

HSE Health & Safety Executive

Report of an injury or dangerous occurrence

Filling in this form

This form must be filled in by an employer or other responsible person.

Part A

About you

1 What is your full name?

2 What is your job title?

3 What is your telephone number?

About your organisation

4 What is the name of your organisation?

5 What is its address and postcode?

6 What type of work does the organisation do?

Part B

About the incident

1 On what date did the incident happen?

2 At what time did the incident happen?
(Please use the 24-hour clock eg 0600)

3 Did the incident happen at the above address?

Yes ☐ Go to question 4

No ☐ Where did the incident happen?

☐ elsewhere in your organisation – give the name, address and postcode

☐ at someone else's premises – give the name, address and postcode

☐ in a public place – give details of where it happened

If you do not know the postcode, what is the name of the local authority?

4 In which department, or where on the premises, did the incident happen?

F2508 (05.00)

Part C

About the injured person

If you are reporting a dangerous occurrence, go to Part F. If more than one person was injured in the same incident, please attach the details asked for in Part C and Part D for each injured person.

1 What is their full name?

2 What is their home address and postcode?

3 What is their home phone number?

4 How old are they?

5 Are they

☐ male?

☐ female?

6 What is their job title?

7 Was the injured person (tick only one box)

☐ one of your employees?

☐ on a training scheme? Give details:

☐ on work experience?

☐ employed by someone else? Give details of the employer:

☐ self-employed and at work?

☐ a member of the public?

Part D

About the injury

1 What was the injury? (eg fracture, laceration)

2 What part of the body was injured?

Next Page

3 Was the injury (tick the one box that applies)

- [ ] a fatality?
- [ ] a major injury or condition? (see accompanying notes)
- [ ] an injury to an employee or self-employed person which prevented them doing their normal work for more than 3 days?
- [ ] an injury to a member of the public which meant they had to be taken from the scene of the accident to a hospital for treatment?

4 Did the injured person (tick all the boxes that apply)

- [ ] become unconscious?
- [ ] need resuscitation?
- [ ] remain in hospital for more than 24 hours?
- [ ] none of the above.

Part E

About the kind of accident

Please tick the one box that best describes what happened, then go to Part G.

- [ ] Contact with moving machinery or material being machined
- [ ] Hit by a moving, flying or falling object
- [ ] Hit by a moving vehicle
- [ ] Hit something fixed or stationary

- [ ] Injured while handling, lifting or carrying
- [ ] Slipped, tripped or fell on the same level
- [ ] Fell from a height

How high was the fall?

[ ] metres

- [ ] Trapped by something collapsing

- [ ] Drowned or asphyxiated
- [ ] Exposed to, or in contact with, a harmful substance
- [ ] Exposed to fire
- [ ] Exposed to an explosion

- [ ] Contact with electricity or an electrical discharge
- [ ] Injured by an animal
- [ ] Physically assaulted by a person

- [ ] Another kind of accident (describe it in Part G)

Part F

Dangerous occurrences

Enter the number of the dangerous occurrence you are reporting. (The numbers are given in the Regulations and in the notes which accompany this form)

Part G

Describing what happened

Give as much detail as you can. For instance

- the name of any substance involved
- the name and type of any machine involved
- the events that led to the incident
- the part played by any people.

If it was a personal injury, give details of what the person was doing. Describe any action that has since been taken to prevent a similar incident. Use a separate piece of paper if you need to.

Part H

Your signature

Signature

Date

If returning by post/fax, please ensure this form is signed, alternatively, if returning by E-Mail, please type your name in the signature box

Where to send the form

Incident Contact Centre, Caerphilly Business Centre, Caerphilly Business Park, Caerphilly, CF83 3GG.
or email to riddor@natbrit.com or fax to 0845 300 99 24

Continue

For official use

Client number | Location number | Event number | INV | REP | Y | N

HSE Health & Safety Executive

Health and Safety at Work etc Act 1974
The Reporting of Injuries, Diseases and Dangerous Occurrences Regulations 1995

?

Click here for report guidance

Report of a case of disease

Filling in this form

This form must be filled in by an employer or other responsible person.

Part A

About you

1 What is your full name?

2 What is your job title?

3 What is your telephone number?

About your organisation

4 What is the name of your organisation?

5 What is its address and postcode?

6 Does the affected person usually work at this address?

Yes ☐ Go to question 7

No ☐ Where do they normally work?

7 What type of work does the organisation do?

Part B

About the affected person

1 What is their full name?

2 What is their date of birth?

3 What is their job title?

4 Are they

☐ male?

☐ female?

5 Is the affected person (tick one box)

☐ one of your employees?

☐ on a training scheme? Give details:

☐ on work experience?

☐ employed by someone else? Give details:

☐ other? Give details:

F2508A (05.00)

Next Page

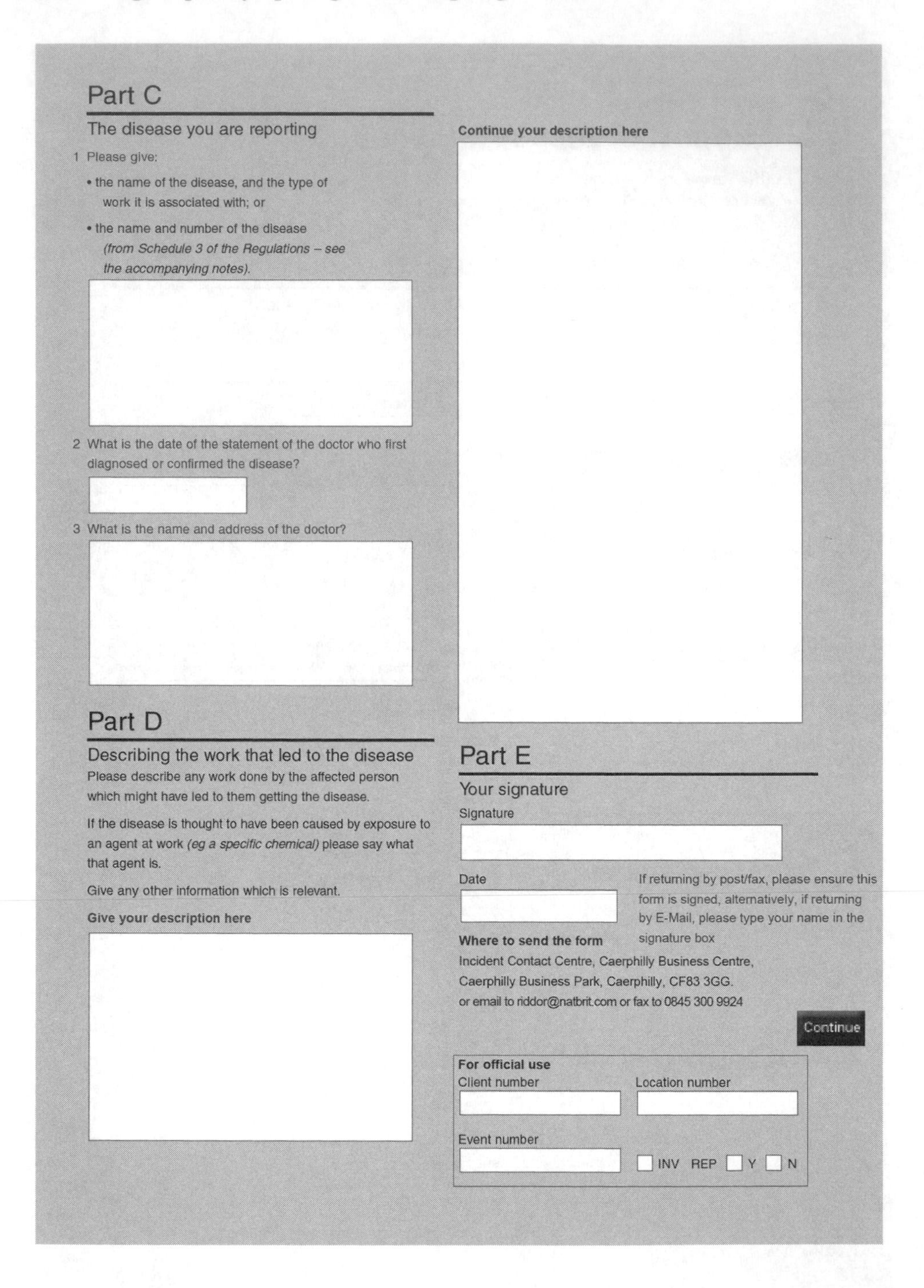

Part C

The disease you are reporting

1 Please give:

- the name of the disease, and the type of work it is associated with; or
- the name and number of the disease *(from Schedule 3 of the Regulations – see the accompanying notes).*

2 What is the date of the statement of the doctor who first diagnosed or confirmed the disease?

3 What is the name and address of the doctor?

Part D

Describing the work that led to the disease

Please describe any work done by the affected person which might have led to them getting the disease.

If the disease is thought to have been caused by exposure to an agent at work *(eg a specific chemical)* please say what that agent is.

Give any other information which is relevant.

Give your description here

Continue your description here

Part E

Your signature

Signature

Date

If returning by post/fax, please ensure this form is signed, alternatively, if returning by E-Mail, please type your name in the signature box

Where to send the form

Incident Contact Centre, Caerphilly Business Centre, Caerphilly Business Park, Caerphilly, CF83 3GG.

or email to riddor@natbrit.com or fax to 0845 300 9924

Continue

For official use

Client number

Location number

Event number

INV REP Y N

0845 300 9923. This eliminates the need to work out if the local HSE office or other local authority is the correct enforcing authority for reporting purposes (HSE, 2001a). The ICC will then log the details and pass the report on to them. Reporting to the local HSE office or local enforcement authority by phone and on the current statutory forms is still an option and this information will be forwarded in any event to thc ICC.

Incidents may also be reported to the ICC by fax on 0845 300 9923.

Alternatively a paper form can be downloaded and sent by post to:

Incident Contact Centre
Caerphilly Business Park
Caerphilly
CF83 3GG

RIDDOR record-keeping requirements

5.20 Records of injury-causing accidents, dangerous occurrences and specified diseases must be kept by responsible persons for at least 3 years, *RIDDOR, Regulation* 7. Again, the rules vary considerably depending on what type of event has to be recorded.

In the case of injuries and dangerous occurrences, such records must contain:

1. Date and time of the accident or dangerous occurrence.
2. In the event of an accident suffered by a person at work, the following particulars of that person:
   (a) full name;
   (b) occupation;
   (c) nature of the injury.
3. In the event of an accident suffered by a person not at work, the following particulars of that person (unless they are not known and it is not reasonably practicable to ascertain them):
   (a) full name;
   (b) status (e.g. "passenger", "customer", "visitor" or "bystander");
   (c) nature of injury.
4. Place where the accident or dangerous occurrence happened.

5. A brief description of the circumstances in which the accident or dangerous occurrence happened.
6. The date on which the event was first reported to the relevant enforcing authority.
7. The method by which the event was reported.

In the case of specified diseases, such records must contain:

1. Date of diagnosis of the disease.
2. Name of the person affected.
3. Occupation of the person affected.
4. Name or nature of the disease.
5. The date on which the disease was first reported to the relevant enforcing authority.
6. The method by which disease was reported.

The record-keeping duty is satisfactorily complied with by maintaining a file of copies of duly completed RIDDOR report forms F2508 and F2508A. Computerised records of the same information, as long as they are printable, will also be sufficient to comply with these record-keeping requirements.

If the ICC route has been used to file a report it will be allocated a unique incident number on the ICC computer. Taking advantage of these arrangements, the employer or other reporting person will not necessarily have a copy of a statutory *RIDDOR* form (though one can be printed immediately prior to completing the form on the ICC website); so the ICC will send out confirmation copies of the report which should be checked for accuracy and retained.

In addition, it should be noted that the statutory Accident Book form reminds the employer to complete a box confirming details of the accident it is one which is reportable under *RIDDOR* (see 5.5).

Defence in proceedings for breach of RIDDOR

5.21 It is a defence under the Regulations for a person to prove that he was not aware of the event requiring him to notify or send a report to the enforcing authority that he had taken all reasonable steps to have all such events brought to his notice *RIDDOR, Regulation 11.*

In practice this defence is difficult to prove, as someone in an employer's organisation will generally be aware of an accident or occurrence falling within the definitions.

If no one knew the employer can be criticised for having ineffective monitoring and management control.

Other reporting requirements

Notification of accidents to insurers

5.22 Contracts of insurance are entered into subject to various conditions and other legal obligations which require the insured to disclose material matters that could affect the risks which the insurer takes on. Most insurance contracts will be entered into – or renewed – based on information the insured organisation has provided and vouched for circumstances which have occurred and which might give rise to claims. If the information turns out to be incorrect or incomplete the insured may lose the entitlement to cover under the policy.

It is common that while the insurance cover is in place there is a continuing obligation on the insured to give notice, as soon as possible, of the occurrence of any event which might later give rise to a claim under the policy. Notification for these purposes is usually made via the insurance broker of the insured. It is normal also for there to be policy wording requiring the insured also to send to the insurer, on receipt, any notice of a claim or court proceedings that might be served and requirements that no admission of liability may be made without the insurer's consent. The reason for these terms is that the insurer will normally have control over the conduct of any claim, and is entitled to assess its position and determine whether to dispute the claim, and if so, how.

The issue of whether an insurer can take advantage of the late notification of an accident or other occurrence to avoid liability under a policy can be a complex one involving interpretation of the wording on the condition and the insurer's conduct after the matter in question has eventually been notified. This issue cannot arise with employer's liability policies because such conditions are not permitted to prevent an insurer meeting a legitimate claim brought by an employee (see page 65). (Potentially the insurer could still seek to reclaim the amount paid back from the employer if the policy wording permits.) Cover under other forms of insurance (of which public liability insurance and professional indemnity insurance are particularly important for these purposes) will be at risk though if accidents or diseases are not promptly notified.

It is therefore necessary to have reporting systems which capture not just *RIDDOR* reportable accidents and diseases, but also more minor incidents, shorter absences through work-related injury, diseases not covered in the schedules to *RIDDOR* and injuries to persons for whom there may be a liability but for whom the insured is not necessarily the responsible person.

Internal notification of accidents, etc.

5.23 Internal notification of accidents and ill health or near misses is not a specific requirement in any health and safety legislation. It merits attention as an issue however because if the safety management system does not require information on accidents, etc. to reach the right individuals in an organisation there is a significant risk that:

- a number of other statutory duties will fail to be met;
- accidents will not be checked to see what degree of investigation is required;
- directors may be left unaware of critical information;
- injured employees, their representations and colleagues will regard the organisation as indifferent to accidents;
- a poor safety culture will emerge.

So, internal reporting needs to be addressed as one aspect of compliance with the duty under *MHSWR, Regulation 5* to have appropriate written safety arrangements. An outline is provided in Figure 5.3.

In particular, the scheme of the arrangements must cover the processes by which information is retained and transmitted so that:

- emergency action is initiated and duly completed;
- *RIDDOR* reporting is correctly handled;
- Accident Book reports are reviewed;
- risk assessments and assessments such as first aid needs are reviewed in light of events;
- other internal investigations are undertaken in accordance with the safety policy;
- appropriate actions are considered by line managers;
- directors are able to receive accurate and comprehensive information on safety performance and respond appropriately;
- accurate accident data is maintained;
- insurance policy conditions are observed.

Figure 5.3: Accident/near miss investigations report form

Name: .... Occupation: ....

Supervisor/Manager: ....

Date: .... Time........................ Location: ....

Type: *Injury* *Hazard* *Fire* *Damage*

Details of above (e.g.: nature of injury)

....

....

Medical treatment received:

First aider *Occupational health* *Other:* ....

First aid treatment form completed: *Yes* *No*

Accident book entry has been checked: *Yes* *No*

What task was injured person doing at the time:

....

....

Description of accident/How did accident occur:

....

....

Names of any witnesses:

....

....

Further investigation/other action:

....

....

Signature: .... Date: ....

For circulation to:

....

....

....

....

Safety representatives' information and inspection

5.24 Where the employer has recognised a trade union for the workforce there are likely to be safety representatives appointed on their behalf who have a number of rights and functions which can overlap with the employer's duty record and evaluate accident information (see 5.30). Safety representatives are entitled under the *Safety Representatives and Safety Committees Regulations 1977* to carry out inspections of the workplace, and more specifically, they have a role in relation to events to which *RIDDOR* applies. The relevant provisions are as follows:

1. Where there has been a notifiable accident or dangerous occurrence in a workplace or a notifiable disease has been contracted there and:
   (a) it is safe for an inspection to be carried out;
   (b) the interests of employees in the group or groups which safety representatives are appointed to represent might be involved those safety representatives may carry out an inspection of the part of the workplace concerned and so far as is necessary for the purpose of determining the cause they may inspect any other part of the workplace; where it is reasonably practicable to do so they shall notify the employer or his representative of their intention to carry out the inspection.
2. The employer shall provide such facilities and assistance as the safety representatives may reasonably require (including facilities for independent investigation

by them and private discussion with the employees) for the purpose of carrying out an inspection ... but nothing ... shall preclude the employer or his representative from being present in the workplace during the inspection.

Evaluation and investigation of accidents

Investigations and confidentiality

5.25 In 2001 the HSC issued a consultative document on proposals for a new duty to undertake formal investigations into accidents, diseases and dangerous occurrences. This suggested the amendment of the *MHSWR* to introduce a duty on employers to take reasonable steps (steps which are practicable and proportionate to the scale of the incident) to investigate accidents, diseases and occurrences which fall within the reporting requirements of *RIDDOR*. In the event the government announced after the consultation period that these proposals would not be adopted, and instead the HSE developed and published a new guide on the investigation of accidents that cause injuries and ill health in the workplace (see Chapter 9).

One reason why these proposals were not advanced was that it was felt that the process of accident investigation was likely to be encouraged more by education and encouragement than by a legal duty. However, it is also the case that health and safety legislation already indirectly requires all employers to evaluate accidents, diseases and dangerous occurrences (whether reportable under *RIDDOR* or not).

Liability considerations have been identified as a significant barrier to improving the quality of accident investigation (HSE, 2001b). There can be fundamental conflicts between the objectives of, on the one hand, fully exploring the underlying causes of accidents in a blame-free setting and, on the other hand, seeking to protect an organisation and individuals within it from potentially unfair criticism, disciplinary action, criminal proceedings or punitive claims. These conflicts have to be resolved pragmatically when they arise on a case-by-case basis, bearing in mind that the rules governing civil litigation combined with the duty to re-appraise risk assessments and employee safety representatives' rights, require a degree of openness and disclosure on the part of employers. Those within the organisation who record information and undertake any investigations must be alert to the different objectives of liability investigations and all the purposes to which reports (and the information obtained to produce them) might later be put.

Legal investigations can arise in a number of settings. These include:

- routine enquiries insurers or their loss-adjusters might make to obtain factual information about an accident before deciding whether to accept liability;

- inquiries undertaken in parallel with an HSE investigation where legal advice is needed to advise on formal interviews and requests for documents;.
- the collation of evidence with which to defend criminal proceedings against an organisation or an individual;
- inquests, fatal accident inquiries or other formal investigations into a major accident;
- internal disciplinary or Employment Tribunal proceedings.

In each case a range of other parties may have a strong interest in the outcome of these investigations, and might find it highly advantageous to gain an insight into the evidence they generate and the conclusions reached by the advisors. Information of this nature is usually however protected from disclosure by legal professional privilege.

The concept of legal professional privilege has developed in common law for reasons of public policy: it exists to enable everyone to obtain advice or assistance in litigation in a relationship of confidentiality without fear that the person might incriminate or prejudice him or herself by doing so. As it was put in a case in the House of Lords:

> "The principle that runs through all [the authorities] is that a man must be able to consult his lawyer in confidence, since otherwise he might hold back half the truth. The client must be sure that what he tells his lawyer in confidence will never be revealed without his consent. Legal professional privilege is thus much more than an ordinary rule of evidence, limited in its application to the facts of a particular case. It is a fundamental condition on which the administration of justice as a whole rests" (*R v Derby Magistrates Court ex p B*).

The courts have been careful not to extend this exceptional form of confidentiality. The same protection is not afforded to anyone else's involvement in an accident investigation for non-legal purposes (such as health and safety professionals or engineering experts), even though there are arguably powerful public policy reasons why effective accident investigation should also not be jeopardised by the reluctance of witnesses or investigators to address issues because liability consequences could follow.

The position at present is that questions whether documents, communications, statements and other materials that relate to specific accidents might have to be disclosed to health and safety inspectors or in the case of proceedings are to be determined by what is called the *dominant purpose* test. The need to determine a document's purpose to see if it is privileged stems from the case of *Waugh v British Railways Board*. This involved an internal inquiry report into a train collision which had led to a fatality. This was in a form expressed as being prepared for the Board's solicitor, in anticipation of litigation, and for the solicitor to advise. Although this had the usual trappings of being privileged, the House of Lords ruled that in reality it was not. The report in fact served a dual purpose: one was obtaining legal advice in relation to a potential claim but the other, which was held to be of equal rank or weight, was for operational safety reasons (i.e. to understand what had gone wrong). Thus the legal purpose of this report was not *dominant* and the report was fully disclosable.

Where an accident investigation process involves liability issues it will therefore be necessary to make an earlier decision about what purpose is intended. This purpose must not change, and if the relevant documents are not kept confidential (e.g. they are distributed as learning tools to managers) privilege will be lost. As legal correspondence and documents created for privileged purposes should be kept in separate files, it is good practice to mark these files "confidential and privileged".

So, involving legal advisors in the process will not, of itself, render it privileged. In practice, accident investigations carried out along the lines described in Chapter 9 and the statements, notes and drafts associated with them will rarely qualify for privilege, and inevitably legal enquiries might need to proceed separately and in parallel if those involved are concerned about not benefiting from or potentially waiving privilege. Even so, if the employer – or a witness or other participant – wishes to seek legal advice on the accident investigation process or, for example, the proposed findings or report wording, privilege could apply to that communication and the documents those involved produce in order to obtain the advice.

It will always be advisable to seek specific legal advice on whether any such communication will be privileged before the communication is made.

Employer's duty to re-evaluate risk assessments

5.26 An accident investigation is in effect needed to comply with the *MHSWR, Regulation 3(3)* which provides that all risk assessments need to be reviewed, if

(a) there is reason to suspect they are no longer valid; or
(b) there has been a significant change in the matters to which they relate.

Where, as a result of any such review, changes to an assessment are required, the employer or self-employed person is required to make them. This is in effect an elaboration of the general duty in the *HSWA, Section 2* to ensure the health and safety of all employees and the *HSWA, Section 3* duty to not expose other persons to risks. The risk re-evaluation duty concerns not just risks to employees, but also non-employees: visitors, contractors, the public, etc.

It is often the case that following an investigation of an incident re-evaluation of the risk assessment will identify errors or incompleteness in the earlier processes or conclusions, and that in identifying this the employer will generate information which can be used in evidence against it in criminal or civil proceedings. The duty to review, and if necessary improve the risk assessment – and record the new version – is an absolute one. There is no privilege against self-incrimination in this context, nor can it usually be claimed that legal professional privilege applies. For this reason the re-evaluation of risk assessments and consequent improvements in safety measures are the commonest evidence used in claims against employers and prosecutions for breaches of health and safety legislation. Nearly always this is an inevitable and unavoidable risk for employers.

Where the incident in question involves exposure to a hazardous agent, particular consideration will be needed of subsequent health surveillance needs. Whereas an original risk assessment may not have deemed health surveillance to be necessary, once an exposure to an individual has occurred, the risk assessment could change and there may be an identifiable disease or adverse health condition related to the exposure which will require ongoing monitoring.

Health surveillance: retention of health records and confidentiality

5.27 Where an incident has occurred and the review of the risk assessment or other action identifies a potential disease or other adverse condition could arise it will be necessary to consider the duty to undertake health surveillance under the *MHSWR* (or other regulations which apply more specifically to the type of exposure). The *MHSWR* provides that: "*Every employer shall ensure that his employees are provided with such health surveillance as is appropriate having regard to the risks to their health and safety which are identified by the assessment.*"

It will usually be necessary to obtain a medical or occupational health advisor's opinion on the nature of the exposure and the likelihood of an identifiable condition developing. The duty does not extend however to the provision of testing, treatment or rehabilitation services at the employer's cost. It is possible that such costs – if incurred by an employee – could be recoverable as damages in a civil claim if it can be established that the exposure was caused by the employer's negligence or breach of a relevant statutory duty.

The results of such health surveillance would in most organisations form part of a larger collection of routine health monitoring and personnel records, and as such it will come within the scope of the *Data Protection Act 1998*.

There are two types of data which are caught under the legislation: personal data and sensitive personal data. Personal data includes any information that can be used to identify a living individual such as names and addresses. Sensitive personal data includes data relating to an individual's political opinions, racial or ethnic origins, religious beliefs and, importantly in relation to the field of health and safety, any details about their health.

By storing this data or passing it to various third parties such as relevant authorities, employers are deemed to be controlling and processing the data. In order to be allowed to control an individual's data an employer would normally be required to notify the Information Commissioner's office and obtain a registration certificate.

Injury records will normally comprise *sensitive personal data* for these purposes and principles of good practice for data handling specified under the Act must be observed. Notice of this type processing data will generally be required to be given to the Data Protection Commissioner.

Access to the data needs to be strictly controlled. There is a substantial amount of guidance available on maintaining confidentiality in health records, the uses to which it can be put in occupational health management. This can be found on the website of the Information Commissioner referred to above.

There are eight principles set out in the *Data Protection Act*:

1. Personal data shall be processed fairly and lawfully.
2. Personal data shall be obtained only for one or more specific and lawful purposes and shall not be processed in any manner incompatible with that purpose or purposes.
3. Personal data shall be adequate, relevant and not excessive in relation to the purpose or purposes for which they are processed.
4. Personal data shall be accurate and, where necessary, kept up to date.
5. Personal data processed for any purpose or purposes shall not be kept for longer than is necessary for that purpose or those purposes.
6. Personal data shall be processed in accordance with the rights of the data subject under this Act.

7. Appropriate tactical and organisational measures shall be taken against unauthorised or unlawful processing of personal data and against accidental loss or destruction of, or damage to, personal data.
8. Personal data shall not be transferred to a country or territory outside the European Economic Area, unless that country or territory ensures an adequate level of protection of the rights and freedoms of the data subjects in relation to the processing of personal data.

Guidance from the Information Commissioner on this issue (see www.ico.gov.uk) states that where possible information relating to sickness and injury should be kept separate from absence and accident records (ICO, 2005). This then reduces the possibility of this information being made public or being misused. This data should only be disclosed, other than where there is a legal obligation to do so or it is necessary in connection with legal proceedings or obtain legal advice when the worker has given explicit consent.

The new Accident Book produced by the HSE (2003) should be used to record incidents with its removable pages it allows the records to be torn out and stored in a separate secure place and so be compliant with the *Data Protection Act*.

In order to obtain or renew insurance it may be necessary to release sensitive personal data to the insurance company. In this situation, the employer will need to comply with one of the conditions of Schedule 3 to the Act as there are no relevant exemptions. One method would be to obtain the explicit consent of each individual whose data is to be released. However, the more practical approach is simply to anonymise the data. It would therefore no longer fall into the category of sensitive personal data and the restrictions of the legislation could be avoided.

Monitoring of employees' health falls within various other requirements in health and safety regulations. This does not mean there has to be duplication of records where more than one set of regulations apply – one record will suffice for both purposes – but it does mean that particular attention must be given to different requirements, particularly those which relate to the duration of the retention period.

The requirement of the *MHSWR* above implies the retention of a documentary health record, but the duration is only covered by a comment in the *MHSWR ACoP* (HSE, 2004) which states that it should be maintained throughout an employee's employment unless the risk to which the worker is exposed and associated health effects are rare and short term.

The retention period of health records under other regulations are more specific (see Figure 5.4).

Figure 5.4: Retention period of health records

Regulation	*Where surveillance required*	*Record retention*
Control of Substances Hazardous to Health Regulations 2002 Regulation 11	Where an identifiable disease or adverse health effect will result from exposure (Also, special provision for exposure to certain substances).	At least 40 years from date of last entry
Control of Asbestos at Work Regulations 2006 Regulations 3(2) and 22	Exposure of any employee to asbestos unless it is minimal and below the control limits	At least 40 years from date of last entry
Ionising Radiation Regulations 1999 Regulation 24	Classified persons (likely to receive dose above threshold given) and persons over-exposed or subject to exposure restrictions	Until the person to whom record relates is 75 years old but in any event 50 years from date of last entry
Work in Compressed Air Regulations 1996 Regulation 10	Employees engaged on work in compressed air	At least 40 years from date of last entry

In most cases the regulations specify that if an employer ceases trading the HSE should be informed and all the health records made available to the HSE.

Rights of access and disclosure of accident records and related information

5.28 This is a complex subject where the common law has a number of statutes and regulations that interact, sometimes in a way that has not been fully co-ordinated.

As a general rule the employer is not under a duty to disclose any information relating to accidents or ill health, but this is merely the starting point. The *RIDDOR* is clearly an important inroad into confidentiality since it requires the reporting of the basic circumstances of an occurrence. Disclosure to insurers of information on the other hand will normally be treated as being in confidence and subject to legal privilege if it is for the purposes of a particular accident which could rise to a claim for compensation.

Individuals' rights

5.29 The rights of individuals to obtain information are restricted to two areas:

1. Documents relevant to legal proceedings which are in contemplation (*pre-action disclosure*) or which have begun (*standard disclosure* and *specific disclosure*). See page 267.
2. Documents which comprise the individual's own health records. The rules are most complex in the area of health records, where it will generally be necessary to seek specialist advice on the interaction of the *Data Protection Act 1998*, the *Access to Medical Records Act 1988* and the *Access to Health Records Act 1990*, and the various subject access provisions for health records which exist in the Regulations described in Figure 5.4.

   It will usually be possible for an employee to obtain a copy of his or her own records through these provisions, although care may be needed to prevent the inadvertent disclosure of information in the records that might be confidential in respect of some other person or which would cause the individual serious physical or mental harm if it were disclosed. (See in particular the *Data Protection (Subject Access Modification) (Health) Order 2000*.)

Safety representatives and consultation

5.30 *The Safety Representatives and Safety Committees Regulations 1977* entitle trade unions recognised by an employer to appoint safety representatives. Once appointed they have various rights, among which are the rights to investigate employee complaints and, to investigate the cause of accidents, and the right to receive facilities and assistance where an accident or other occurrence is one which is reportable under the *RIDDOR*, *Regulation 7(1)* provides that:

> "Safety representatives shall for the performance of their functions … if they have given the employer reasonable notice, be entitled to inspect and take copies of any document relevant to the workplace or to the employees the safety representatives represent which the employer is required to keep by virtue of any relevant statutory provision within the meaning of Section 53(1) of the 1974 Act except a document consisting of or relating to any health record of an identifiable individual."

As these rights are restricted to documents required only by statutory requirements, it will only apply to items such as the original and any revised risk assessment, *RIDDOR* records and background information such as instructions and written safety procedures under the *MHSWR*. Supplement disclosure rights are however given by *Regulation 7(2)* which provides that:

> "An employer shall make available to safety representatives the information within the employer's knowledge, necessary to enable them to fulfill their functions ..."

The scope of this provision is potentially very wide and the boundaries are unclear. It could include information on the results of any accident investigation, or for example information relating to relevant advice or warnings the employer might have received in the past which are relevant to the issue. However, this provision does not require an employer to produce or allow inspection of any document (or part of a document) which does not relate to health, safety or welfare. In addition, certain restrictions are placed on these safety representatives' rights of access. Exempt from disclosure by an employer are the following:

(a) Any information the disclosure of which would be against the interests of national security.
(b) Any information which he could not disclose without contravening a prohibition imposed by or under an enactment.
(c) Any information relating specifically to an individual, unless he has consented to its being disclosed.
(d) Any information the disclosure of which would, for reasons other than its effect on health, safety or welfare at work, cause substantial injury to the employer's undertaking or, where the information was supplied to him by some other person, to the undertaking of that other person.
(e) Any information obtained by the employer for the purpose of bringing, prosecuting or defending any legal proceedings.

Where there is no recognised trade union the *Health and Safety (Consultation with Employees) Regulations 1996* apply instead. If consultation is with elected employee representatives their rights include access to *RIDDOR* records and other information necessary for their functions. There are similar exempt categories of documents which an employer need not disclose to those described above under the *Safety Representatives and Safety Committee Regulations 1977*.

Separate regulations apply offshore: see the *Offshore Installations (Safety Representatives and Safety Committees) Regulations 1989*. Inspectors of the HSE or local authorities with enforcement responsibility under the *Health and Safety (Enforcing Authority)*

Regulations 1998 have extensive powers to require disclosure of documents and information which include, but are not restricted to, documents created for the purposes of statutory requirements. Any information they might request for the purposes of carrying out their functions will be disclosable, except documents which are subject to legal professional privilege (see 5.25).

Disclosure in civil proceedings

5.31 The onus will be on an employer to preserve all documentation that could be relevant to any claim which is notified in England and Wales. Once legal proceedings in a civil case are contemplated, virtually all protection of confidentiality in documentation an employer has (other than privileged material) is likely to be lost unless there is an early admission of liability on the employer's part which renders it unnecessary to disclose the documents. The reason for this lies in the Woolf reforms to the civil justice system which introduced the feature of the pre-action protocol for personal injury claims. The effect of the protocol is to bring forward the stage at which a defendant to an action is required to reveal documents relevant to the case so that each side is forced to show its hand, take a realistic view of its prospects of success and seriously consider reaching a settlement without proceedings having to be commenced. Documents revealed under the process of the pre-action protocol (and disclosure rules once proceedings commence) will normally be protected as claimants and their representatives may not disseminate them further unless they are used later in court. Under the civil procedure rules there is an undertaking given by the recipient not to use the documents or disclose them to another person except for the purposes of the claim.

Legal compliance checklist

5.32

Reporting arrangements

- ✓ All internal and external reporting requirements are identified in the organisation's procedures manuals.
- ✓ Employees are adequately informed about their roles and responsibilities for reporting and recording incidents and hazards.
- ✓ Accidents, disease and dangerous occurrences are monitored.
- ✓ Individuals' responsibilities are clear for notifying and reporting *RIDDOR* events.

- ✓ *RIDDOR* reporting procedures and guidelines are available.
- ✓ Insurance reporting requirements are in place and notifications are recorded.

Accident and health records

- ✓ Accident Books are available and reports are monitored.
- ✓ *RIDDOR* records are retained for requisite periods.
- ✓ Health surveillance records are maintained for requisite periods.
- ✓ *Data Protection Act* procedures are correctly applied and audited.
- ✓ Management controls are in place for the release of confidential accident/ill health information and evidence.

Accident investigation

- ✓ Risk assessments are reviewed after all incidents.
- ✓ Arrangements are in place for provision of information to safety representatives.
- ✓ Documents which may be legally privileged are controlled.
- ✓ Legal advice is available on the taking of statements and other investigations.

References

HSE (1999) *A Guide to the Reporting of Injuries, Diseases and Dangerous Occurrences Regulations 1995* (*L73*).

HSE (2001a) *Proposals for a New Duty to Investigate Accidents, Dangerous Occurrences and Diseases CD169*. www.hse.gov.uk/consult/condocs/cd169.pdf

HSE (2001b) *Accident Investigation – The Drivers, Methods and Outcomes – HSE Contract Research Report 344/2001*, HSE Books. www.hse.gov.uk/RESEARCH/crr_pdf/2001/crr01344.pdf

HSE (2003) *Accident Book*, HSE Books.

HSE (2004) *Management of Health and Safety at Work: Management of Health and Safety at Work Regulations 1999* (L21), HSE Books.

ICO (2005) *The Employment Practices Code*, Information Commissioner's Office. www.ico.gov.uk

R v Derby Magistrates Court ex p B [1996] 1AC 487.

Tolley (2006) *Tolley's Health and Safety at Work Handbook 2006*, Annual Editions, Lexis Nexis, Butterworths.

Waugh v British Railways Board [1980] AC 521.

6 Formal accident investigations

In this chapter:

Introduction

6.1 After an accident has occurred, a number of official investigations can be undertaken at the discretion of many different agencies. These may include the police, the fire authorities, the Environment Agency and a number of more specialised bodies, such as the Maritime and Coastguard Agency, the Marine Accident Investigation Branch, the Air Accident Investigation Branch and, when established, the Rail Accident Investigation Branch.

The principal bodies responsible for investigations of accidents are, however, the Health and Safety Executive (HSE) and local authorities where they are the relevant enforcing authority for the purposes of the *Health and Safety at Work etc Act 1974* (*HSWA*). The HSE has entered into various concordats with other agencies to determine the extent of their respective roles and responsibilities.

In fatal accident cases and major disasters, yet more authorities will become involved. Later in this chapter, the functions of coroners' inquests, Scottish fatal accident inquiries and Health and Safety Commission (HSC) sponsored full public inquiries are explained. However, the investigatory role of the HSE and local authority enforcement officers is the one encountered routinely when accidents occur and the powers which are available to them will be considered first.

Health and safety inspectors' investigatory powers

6.2 These powers all stem principally from the *HSWA, Section 20.* Inspectors have no legal authority to go beyond the limits set by these sections, but they will generally seek, and receive, co-operation on a voluntary basis when undertaking inquiries and investigations. It is rare for there to be any significant dispute about their authority to obtain information relating to an accident when at the scene or in interviewing people immediately afterwards.

Knowingly or recklessly providing false information, obstructing or failing to co-operate with the instruction of an inspector using the powers is an offence under *HSWA, Section 33(1).* An inspector must however produce proof of his appointment if requested to do so. Once an investigation progresses to the stage where possible criminal offences are suspected these statutory powers of inspectors are exercised with more circumspection and formality because of the potential for evidence obtained unfairly or with other irregularity being rendered inadmissible in subsequent criminal proceedings. In England and Wales, this issue is put into sharp focus by the procedural requirements of the *Police and Criminal Evidence Act 1984* (*PACE*) and its associated Codes of Practice.

Entry onto premises

6.3 An inspector does not require a search warrant or other prior authority to enter and inspect premises. An inspector can at any reasonable time (or, in a situation which in his opinion is or may be dangerous, at any time) enter any premises which he has reason to believe it is necessary for him to enter for the purposes of his enquiries. It is very rare, but he may also take with him a police constable if he has reasonable cause to apprehend any serious obstruction in the execution of his duty *HSWA*, and he can also attend with any other person duly authorised by the inspector's enforcing authority, for example a specialist engineer.

Examination and search

6.4 Once on premises an inspector can make any examinations and investigations necessary for the purpose of his enquiries. In particular any article or substance found in any premises which appears to him to have caused or to be likely to cause danger to health or safety can be dismantled or subjected to any process or test (but not so as to damage or destroy it unless this is in the circumstances necessary for the purpose of the enquiries).

Two conditions are placed on this power being exercised. First, anything done with the items must normally be done in the presence of the person present and responsible for the premises if that person requests it. Second, the inspector must consult such persons as appear to be appropriate to ascertain what dangers, if any, may be involved in doing what is proposed.

Control over premises and their contents

6.5 An inspector can direct that premises or anything in them, shall be left untouched for so long as is reasonably necessary for the purpose of any examination or investigation. A written "Notice to Leave Undisturbed" is usually issued in such cases.

Where an inspector has reasonable cause to believe that any item is a cause of imminent danger of serious personal injury he can seize it, or have it destroyed or made safe.

Recording of information

6.6 These powers supplement those of investigation and examination, and enable an inspector to take such measurements and photographs or make other recordings as he considers necessary. They can also take samples of any articles or substances found in the premises, and of the atmosphere in or in the vicinity.

Preservation of evidence

6.7 When inspectors encounter articles or substances which appears to be (or have been) a cause of danger they can take and detain it for so long as is necessary for all or any of the following purposes:

- to examine it or have it dismantled or tested;
- to ensure that it is not tampered with before the examination of it is completed;
- to ensure that it is available for use as evidence in any proceedings for a health and safety offence or any proceedings relating to enforcement notice.

Inspectors in these circumstances are nevertheless subject to rules of the *Police and Criminal Evidence Act 1984* which means that aspects of *Code of Practice B: Searching of Premises and Seizure of Property* can apply. In particular, there should normally be a written explanation of the legal basis upon which the search and seizure is undertaken and the rights of the occupier.

Production and copying of documents

6.8 This is a very commonly used power. An inspector can demand to see or receive copies of any documents which are required to be kept by virtue of any of the relevant statutory provisions. The term "documents" is generally accepted to mean any form of paper or electronic records and communications. Some inspectors have a practice of issuing a formal notice to produce document, but it is equally valid (and more common) for the request to be made verbally or by letter.

This is wide-ranging which would include for example risk assessments, *Reporting of Injuries, Diseases and Dangerous Occurrences Regulations 1995* (*RIDDOR*) records, written safety procedures and maintenance logs.

More importantly, inspectors can insist on being shown *any* other documents which it is necessary for them to see for the purposes of any investigations. It has been argued that this power does not extend as far as providing an inspector with access to an accident investigation report, on the basis that it is not necessary to have this for an inspector's *own* investigation pursuant to his other powers.

The only category of documents explicitly protected from being seen by an inspector is those which are subject to legal privilege (see pages 165–166). There is no power under the *HSWA* for an inspector to compel the production by any person of a document of which legal professional privilege, and it be wrong to take these without agreement being given by an officer of the organisation.

Questioning witnesses and other persons

6.9 An inspector's most effective power in practice is that of being able to compel anyone to answer questions. The *HSWA, Section 20* expresses this as a power:

> "… to require any person whom he has reasonable cause to believe to be able to give any information relevant to any examination or investigation … above to answer (in the absence of persons other than a person nominated by him to be present and any persons whom the inspector may allow to be present) such questions as the inspector thinks fit to ask and to sign a declaration of the truth of his answers".

This power is considered in more detail in connection with interviews by inspectors (see 6.11–6.12).

Ancillary powers

6.10 The ways in which inspectors carry out questioning and gather information from witnesses are not as well defined as most of the other powers in legislation. Depending on the purpose of the exercise different practices will apply. Inspectors may need the information for enquiries to ascertain simply what has happened, as the enforcing authorities are required to do, to determine whether future investigation is called for or for use as evidence in any proceedings for possible health and safety offences. It must always be remembered that enforcing authorities have this dual fact-finding and enforcement role, and (apart from in Scotland) decide on and conduct their own prosecutions.

In the immediate aftermath of an accident an inspector will need to establish the basic facts from witnesses and those responsible for the workplace. The powers of questioning under *HSWA, Section 20* were designed to be used at this stage. Once enquiries have progressed a stage further to the point where a health and safety or other criminal offence is suspected formal interviews will be conducted under the rules of *PACE*. More recently HSE inspectors have begun to request so-called "voluntary interviews" outside their *HSWA, Section 20* powers.

Section 20 interviews

6.11 A person questioned by an inspector using his powers of investigation is required to answer them fully and truthfully. These provisions enable inspectors to interview witnesses, employees and senior managers at will. There is no limit on the extent or period of the questioning and it is permissible to put questions the answers to which might, or in fact do, incriminate the witness, his or her colleagues, his or her employer or anyone else. Safeguards are provided in the legislation against self-incrimination. No answers given by a person would be admissible in evidence against that person (or his or her spouse). However any answers – once given – could subsequently be exploited to obtain evidence of the same facts from another source.

Persons being interviewed in these conditions are permitted to have a person of their choice in attendance. This could be a friend, trade union representative or any other person as desired, including a solicitor. No one else (e.g. the employer or a trade union) has a right to participate in or observe the questioning unless the inspector and the individual concerned agree. Law Society guidelines introduced in 2006 for solicitors for considering conflicts of interest make it difficult for the employers in-house or other solicitors to accompany and assist staff at the interview (see *www.lawsociety.org.uk*).

The practice varies on the taking of statements at this stage. An inspector may just wish to make his own note of the answers, but more often a signed handwritten version of the answers will be prepared at the time by the inspector and given to the interviewee to sign in a prepared witness statement form. This statement can subsequently be used in later proceedings.

PACE interviews

6.12 It can be hard to identify the point at which the rules of the *PACE* come into play for the purposes of proper questioning. Inspectors are trained in the relevant requirements and appreciate that evidence obtained in contravention of *PACE* requirements may be excluded by a court in any subsequent legal proceedings so when in doubt they commonly employ *PACE* procedures in order to ensure that interviewees are treated fairly.

The rules for questioning under *PACE* are contained in *PACE Code of Practice C*, which apply to inspectors by virtue. In fact not all of these rules apply because they are written primarily from the perspective of police station interviews, and one very significant limit placed on inspectors is they cannot arrest or detain anyone for questioning in custody.

An interview for the purposes of *PACE* is described as "*the questioning of a person regarding their involvement or suspected involvement in a criminal offence or offences which … must be carried out under caution*". The Code goes on to provide that whenever a person is interviewed they must be informed of the nature of the offences. The purpose of a caution is twofold: first, to make it clear that at this stage of the questioning there is no requirement for the person to answer any questions and no penalty for not doing so; and second to warn the person that if the right not to answer questions is exercised or, answers are given which are unsatisfactory, this can be drawn to the attention of a court in any subsequent proceedings. What are referred to as "adverse inferences" might be drawn about the motive for not answering or being unable to do so. This means the magistrates or jury as the case may be (see page 247) might fairly conclude that the interviewee would have spoken up at any early stage if he had a genuine defence. The rules about declining to answer questions under *PACE* and adverse inference are highly complex, which is one reason why it is always sensible to take legal advice about interviews.

The recommended wording for these cautions is:

> "You do not have to say anything. But it may harm your defence if you do not mention when questioned to something which you later rely on in court. Anything you do say may be given in evidence".

A person to be questioned under *PACE* conditions will have been advised in advance of the right to have legal representation present. This right should always be exercised, and free legal aid may be available to individuals in these circumstances.

It is not permissible for an interviewer to use oppressive tactics to elicit answers. This could include unduly prolonging the interview, or saying to the interviewee that the lack of co-operation would make it more likely that inspector would lay charges for an offence. (There is nothing wrong if an inspector gives such information however in response to a direct question as to the consequences of a person not answering questions.)

Interviews conducted under the *PACE* conditions will be tape recorded. The questioning may be wide-ranging going beyond the immediate circumstances of any accident into the background such as the adequacy of safety management systems, risk assessments and so on. There is no obligation on an inspector to advise that certain questions might incriminate the individual or anyone else. This is another reason why legal advice will normally be needed during an interview.

Where the inspectors are investigating a possible offence by a company or other corporation (such as a local authority) it is still necessary to interview an individual. The normal practice is for a director or other senior manager to be formally nominated as the organisation's representative. This person's answers are treated as the evidence of the company.

PACE Code of Practice C states that questioning should cease once:

(a) the inspector is satisfied that all questions have been put which are relevant to obtaining accurate and reliable information about the offence (including allowing the person an opportunity to give an explanation);

(b) has taken into account any other available evidence;

(c) the inspector reasonably believes there is sufficient evidence to provide a realistic prospect of conviction for that offence if the person was prosecuted.

It is unusual at the conclusion of interviews for inspectors to ask if there is anything the interviewee wishes to add. This is an opportunity to put over the interviewee's (or more usually his or her organisation's) point of view about the incident in question, any explanation or excuse that might have some bearing on the inspector's decision whether to bring a prosecution (see page 234).

At the end of the interview the interviewee may apply for a copy of the tape recording.

Some inspectors, rather than conducting a *PACE* interview, choose (or offer as an alternative) to put questions in writing and request a signed statement from the person under suspicion or, in the case of a corporation, an executive authorised to answer on behalf of the company. In these circumstances the usual caution will still be given by letter when putting the questions, and there is no obligation to answer any of them or to provide a written statement.

The practice has been growing of individual interviews or company representatives or employers offering written responses to questions whilst declining to attend *PACE* interviews. This is permissible, and no adverse inference should be made against the company in illegal proceedings for adopting this approach. Sometimes this is helpful where the accident or incident being investigated involved complex technical evidence, or occurred a long time ago and a single interview would struggle to provide meaningful answers under questions.

There are two last provisions which are designed to support the other powers. First, an inspector can "require any person to afford him such facilities and assistance with respect to any matters or things within that person's control or in relation to which that person has responsibilities as are necessary to enable the inspector to exercise any of the powers conferred on him ..." This enables, for example, an inspector to ask for locked areas to be opened for examination or enquiries to be made concerning the whereabouts of records he wishes to see *HSWA, Section 20(2)(1)*. Second, there is a curious provision that an inspector can exercise "any other power which is necessary" for the purposes of "carrying into effect any of the relevant statutory provisions". This provision appears to give unlimited opportunities to take any step an inspector may care to choose. It does not however override the limits placed on the other powers which cover almost every conceivable action an inspector could take, nor does it operate to exclude the rules of the *PACE* and its Codes of Practice.

Enforcement notices

6.13 In addition to their power to investigate for breaches of the health and safety legislation, inspectors also have powers to serve enforcement notices. Non-compliance with a notice is an offence punishable with a fine which is not subject to any limit, or up to two years imprisonment or both. It is common for an enforcement notice to be imposed in the aftermath of an accident to prevent any recurrence pending further investigations. Often too, the measures taken to comply with such notices demonstrate the kinds of precautions which could (and, it will be argued, should) have been in place already.

Details of enforcement notices issued by the HSE are publicly available on its notices database website at www.hse.gov.uk/enforce, which contains the recipients' names and address, together with the nature of the risk which the notices have addressed.

Prohibition notices

6.14 A prohibition notice contains a direction to stop the work activities specified in the notice and is used where an inspector considers that work activities involve or will involve a risk of serious personal injury. The notice may have immediate effect if the risk of injury is imminent. Otherwise it may allow a period in which to take remedial action.

Prohibition notices (see page 184) have anticipatory effect and there is no need for the inspector to believe or prove that any statutory provision has actually been – or will be – breached. The temporary or even indefinite voluntary suspension of the work activity, or assurances not to engage in them in the future, will not prevent the service of a valid prohibition notice (*Railtrack plc v Smallwood*).

Improvement notices

6.15 Improvement notices (see page 185) can be served if an inspector is of the opinion that a person is contravening one or more statutory provisions and that the contravention will continue or be repeated. Improvement notices allow a specific period of time in which the offending matter may be put right, which must not be less than 21 days (giving time to lodge an appeal with an employment tribunal). Improvement notices may be served even if the breach of the health and safety regulations is merely a technical one and there is no risk of personal injury.

Appeals against enforcement notices

6.16 Where an improvement notice or prohibition notice is served, the person on whom it is served may appeal to an Employment Tribunal within twenty one days from the date of service. When an appeal is lodged against an improvement notice, the bringing of the appeal automatically suspends the operation of the notice. If the tribunal affirms the notice, it may do so with modifications, for example, it may extend the time within which improvements required by an improvement notice must be carried out.

Health and Safety at Work etc Act 1974, sections 21, 22, 23 and 24

Prohibition Notice

Serial Number ABC/2006/01

	Name	Newco Ltd
	Address	London Road, Bradfield, Berks
	Trading as	xxx
Inspector's full name	I,	Mandy Buff-Sussex
Inspector's official Designation	One of her Majesty's Inspectors of Health and Safety being an inspector appointed by an instrument in writing made pursuant to Section 19 of the said Act and entitled to issue this Notice.	
Official address	of	1 Priestly House, Priestly Road, Basingstoke, Hants
	Telephone Number	01256 404000

hereby give you notice that I am of the opinion that the following activities namely:

Operation of the Striker Power Press serial number 00987 functions without guard supplied by the manufacturer or a guard affording the same protection

under your control at

Location of premises — at — the above premises

Or place of activity — involve, or will involve, a risk of serious personal injury and that the matters which give rise to the said risks are:

the potential for serious injury to operators' hands and arms

and the said matters involve contravention of the following statutory provisions:

Section 2(1) of the Health and Safety at Work Act 1974 Regulation 3 of the Provision and Use of Work Equipment Regulations 1998

because

the said guard has been removed because it was broken and without it being in place and functioning correctly operators may come into contact with dangerous moving parts of the equipment

and I hereby direct that the said activities shall not be carried on by you or under your control immediately/~~after~~* unless the said contraventions and matters have been remedied.

Signature *M Buff-Sussex* Date 13/03/2007

Health and Safety at Work etc Act 1974, sections 21, 22, 23 and 24

Improvement Notice

Serial Number
001001

To:	Name	Newco Ltd
	Address	London Road, Bradfield, Berks
	Trading as	xxx
Inspector's full name	I,	Mandy Buff-Sussex
Inspector's official Designation	One of her Majesty's Inspectors of Health and Safety being an inspector appointed by an instrument in writing made pursuant to Section 19 of the said Act and entitled to issue this Notice.	
Official address	of	1 Priestly House, Priestly Road, Basingstoke, Hants
Telephone Number	01256 404000	
Location of premises Or place of activity	at	the above premises
	You, as	employer

have contravened in circumstances that make it likely that the contravention will continue or be repeated. the following statutory provisions:

Regulation 3, Management of Health and Safety at Work Regulations 1999 Regulation 11, Provisions and Use of Work Equipment Regulations 1998 Section 2(1), Health and Safety at Work etc Act 1974

The reasons for my said opinion are:

The risk assessments undertaken of the office chair production lines are not suitable and sufficient in that they do not address risks of nips and entrapment by machine parts and the control measures are inadequate

Date for compliance

And I hereby require you to remedy the said contraventions or, as the case may be, the matters occasioning them, by **13 June 2007**

And I direct that the measures specified in the Schedule which forms part of this Notice shall be taken to remedy the said contraventions or matters

Signature *M Buff-Sussex* Date *13/03/2007*

An Improvement Notice is also be served on

of

Related to the matters contained in this Notice

Environment and Safety Information Act 1988

This is not a relevant notice for the Environment and Safety Information Act 1988
This page only with form the register entry

Signature *M Buff-Sussex* Date *13/03/2007*

Overlapping police investigations in fatal accident cases

6.17 Fatal accidents present special difficulties for investigating authorities, who may have overlapping powers of investigation. Considerable improvements have been made in coordination between the various agencies, but there can still be tensions between them in dealing with the accident scene and in carrying out subsequent enquiries.

The manner in which the police, the HSE and other enforcing authorities and the Crown Prosecution Service (CPS) jointly conduct investigations is covered by a document entitled *Work-Related Deaths: A Protocol for Liaison* (HSE, 2003). This accords the initial responsibility for securing the area to the police and indicates that they should:

- identify, secure, preserve and take control of the scene and any other relevant place;
- supervise and record all activity;
- inform a senior supervisory officer;
- enquire whether the employer or other responsible person in control of the premises or activity has informed the HSE, the local authority or other enforcing authority and agree arrangements for controlling the scene, for considering access to others and for other local handling procedures to ensure the safety of the public.

The protocol continues that any HSE or other inspector arriving first should preserve the scene until the police arrive. This approach of police primacy is, however, not always effective where specialist inspectors or investigators (e.g. appointed by employer or insurers), may need immediate access to check that there are no continuing dangers that might require urgent action or to record vital evidence. The protocol is not entirely conducive to the objective of conducting the most open thorough and blame-free accident investigations.

Which agency takes the lead role in interviewing witnesses and decision-making depends on whether there is thought to be sufficient evidence of a homicide offence – in practice, this would generally be the crime of manslaughter. The police will control the investigation, taking advice from the CPS, if manslaughter is suspected. If it is not, or once it has been investigated and ruled out, the HSE will take over the investigation and consider any health and safety offences for which charges might be brought. The protocol encourages liaison with the CPS, whose advice must be sought before any charge or manslaughter by an individual or of corporate manslaughter is laid.

Powers of the HSC to direct investigations and inquiries

6.18 The powers of the HSC derive from the *HSWA, Section 14* and comprise two separate processes for detailed investigations of accidents (and any other matters the HSC considers necessary), special reports and formal inquiries.

Special reports

6.19 The HSC can issue a direction to the HSE or authorise another person to investigate and make a special report. This power is used to investigate matters such as train crashes (e.g. the Potters Bar rail accident). The reports are generally published and available on the HSE website.

CASE STUDY 10

6.20

Piper Alpha inquiry, 1989

On 6 July 1998, a series of fires and explosions occurred on the *Piper Alpha* installation in the North Sea, causing the loss of the platform and claiming the lives of 165 people onboard. Two members of the crew of the fast rescue craft *Sandhaven* also lost their lives while attempting to rescue survivors. It was the worst disaster in the history of offshore operations.

Lord Cullen, a Scottish judge, was appointed by the Secretary of State for Energy to conduct a public inquiry under the *Mineral Workings (Offshore Installations) Act 1971* and the *Offshore Installations (Public Inquiries) Regulations 1974* ("the 1974 Regulations"). These Regulations have since been repealed (ironically, as a result of one of the Cullen Inquiry's recommendations) and any similar inquiry in the future would be held under *HSWA*. These are not identical to but are broadly similar in most key respects.

Extent of inquiry

Lord Cullen's remit was:

1. to report on the circumstances of the accident and its cause;

2. to make any observations and recommendations he saw fit with a view to the preservation of life and the avoidance of similar accidents in the future.

The inquiry was therefore intended to be a wide-ranging one. As a result, the inquiry sought to explore the impact of an inter-relationship between the whole range of factors which bore on the accident and its escalation. Also considered were issues such as the nature and extent of the induction given to new arrivals on the platform, the performance of the Department of Energy in its capacity as safety regulator in the year prior to the disaster and the physical layout of the platform. The inquiry was also very much concerned with the way in which safety issues were supervised and managed on the installation and more generally within its operators, Occidental. Broad-ranging as it was, however, the inquiry had its limits. Lord Cullen noted at the outset that it was not "a roving excursion into every aspect of safety at work in the North Sea". Only evidence which had a tenable connection to the events which occurred was permitted to be heard.

Preliminary hearings

Two preliminary hearings held before the inquiry proper commenced. At the first, the court determined who was to be a party to the inquiry and timetabling issues and set out its rules of procedure. Parties who intended to criticise others were ordered to notify their intention to the inquiry's secretariat, so that advance notice of those criticisms could be given. The first preliminary hearing was also noteworthy in that the Lord Advocate (head of the Scottish prosecution system) took the opportunity to announce that the evidence of any witness appearing at the inquiry would not be used against him or any other person in respect of whose acting he may be held criminally liable in any subsequent criminal prosecution. No one was ever charged of a criminal offence arising out of *Piper Alpha.*

The second preliminary hearing was held to assess how far the parties had got in the process of providing documents voluntarily. Progress was found to be satisfactory and the inquiry ultimately commenced without having to use its compulsory powers to obtain documentary evidence. During the inquiry itself, compulsory recovery was sought in respect of only one document – the interim report of the Occidental board of inquiry, which Occidental refused to voluntarily provide. The document was ordered to be produced initially to the Counsel to the Inquiry only, and was never led in evidence, as Counsel to the Inquiry concluded it contained no information which would materially add to what was already available.

Assistance to the inquiry

Three assessors were chosen for their technical expertise and industry knowledge (they comprised a professor of plant engineering, the former managing director of a major oil company and a director of a major petrochemical company). In addition, a firm of consulting engineers was retained to assist in obtaining and preparing lines of technical evidence, and to enable the inquiry to develop lines of cross-examination of technical evidence led by the various parties.

The parties to the inquiry were legally represented, with those who had a community of interest (e.g. the representatives of the deceased and the survivors; the majority of the contractors present on the installation) being ordered to be represented collectively. A team of experienced members of the Scottish bar acted as Counsel for the Inquiry.

Inquiry

Part 1 of the Inquiry was concerned with how and why the disaster happened. Evidence was heard from 196 witnesses, including eyewitnesses such as survivors and those involved in responding to the emergency, as well as witnesses speaking to broader issues such as the systems of management and supervision within the various companies involved on the installation, expert witnesses and witnesses from the Department of Energy and the HSE. Parties were permitted to provide written submissions, briefly highlighted orally.

The inquiry was hampered somewhat by a lack of physical evidence (much of the plant and equipment associated with the accident was never recovered from the seabed), and also by the fact that many of the key witnesses to the accident did not survive it. These difficulties were overcome to an extent by the creation and testing of a number of potential scenarios by the consulting engineers who had been appointed to assist the inquiry. Ultimately, the inquiry was able to ascertain the cause of the explosion (the ignition of a gas leak which had been caused by night-shift personnel running a pump which they did not know had been removed from service for maintenance and which was therefore not leak-tight) and to piece together the series of events which caused the accident to escalate so tragically.

Part 2 of the inquiry was concerned primarily with evidence which would permit Lord Cullen to make observations and recommendations. This part of the inquiry heard evidence from 64 witnesses, including witnesses employed by various operators within the United Kingdom Norwegian Continental Shelves, witnesses from trade unions, independent experts and witnesses with knowledge of emergency equipment and training and response procedures. At the end of the inquiry, an application was heard from the Trade Union Group, who had represented survivors and the bereaved, for a recommendation that their costs should be paid from central funds. Lord Cullen recommended that 40% of their costs should be borne by the public purse.

Report of the inquiry and its effects

Lord Cullen's report ran to some 400 pages plus appendices and concluded with 106 recommendations. It was critical of the way in which the operators had managed their installation and the way in which offshore safety had been regulated. The regulatory function had historically been split between the HSE and the Department of Energy. Lord Cullen perceived a need for a single regulator and ultimately this occurred, with the HSE assuming the functions of the Department of Energy.

All the report's recommendations have been implemented, making root and branch changes to health and safety practice offshore. The report advocated a shift away from the detailed and prescriptive regulations which existed under the old regime. These have been replaced by a system of goal-setting regulations, which state the overall objectives to be met, rather than the precise manner in which those objectives are to be secured. At the heart of the new system is the formal safety assessment which operators are required to carry out for each offshore installation, and in which the operator must identify and assess all hazards which could occur throughout the entire life cycle of the installation, in order to demonstrate its safety. Through such measures, the Cullen report sought to leave behind the box-ticking mentality encouraged by the previous system's reliance upon set rules and to encourage companies and individuals to take responsibility for offshore safety.

The wholesale adoption of the recommendations is a strong testimony to the power of report. The fact that, since the report, the number of offshore accidents leading to loss of life has dramatically reduced is a yet stronger one.

Source: Cullen (1990)

Formal inquiries

6.21 The HSC can, with the agreement of the Secretary of State, direct that a formal inquiry should be held. Usually, such inquiries have been conducted by a judge or another senior lawyer as chairman. One or more technical assessors may be appointed as well. An example of when the process has been used was the Ladbroke Grove Rail Accident Inquiry. These inquiries can involve very large volumes of documentation, witnesses and legal teams and last many months.

Coroners' inquests (England and Wales)

6.22 The office of coroner is an ancient one which previously had more functions and powers than it does today. The coroner's court's principal role in modern times is to establish certain basic, limited facts about violent or unnatural deaths, sudden deaths where the cause is unknown and deaths occurring in prison. All work-related fatalities fall within this remit, though in exceptional circumstances they all sometimes are the subject of a public inquiry instead.

Coroners may be lawyers or legally qualified medical practitioners; in practice most are solicitors or barristers. They hold office under the Crown, but unlike judges they are appointed by the local authority for the relevant area. Deputy and assistant deputy coroners are appointed to provide cover when coroners are unavailable.

Role of coroner's court

6.23 For victim's families, witnesses and other participants the process can be confusing and frustrating. The rights of representatives of interested persons to put questions and explore evidence are inquest limited and do not extend to calling witnesses. They can only question witnesses on matters relevant to the limited remit of the inquest. Legal aid is not generally available for legal representation. The findings of the inquest are very narrowly confirmed.

Coroners' inquests are commonly misunderstood and wrongly taken to be a public investigation into the causes of an accident, or the identity of someone who is responsible for the death. Sometimes even the lawyers representing interested persons at inquests do not fully appreciate how limited the role of the inquest is.

The role of the coroner's court is narrowly delineated by the *Coroners Rules 1984*, which provides the following.

The proceedings and evidence at inquest shall be directed solely to ascertaining the following matters; namely:

(a) who the deceased was;

(b) how, when and where the deceased came by his death;

(c) the particulars … required by the Registration Acts.

The question of how to interpret the phrase "how, when and where the deceased came by his death" has given rise to difficulty. In the case *R v Walthamstow Coroner, ex p Rubenstein* it was described as follows:

> " 'How' means by what means. It does not mean in what circumstances. They are not concerned as to whether he died in the cleanest, the most distressing or the most appropriate of circumstances or surroundings. This is not part of their function. They are concerned with those matters which caused death or led to it in part".

Neither the coroner nor the jury are allowed to express any opinion on any other matters. The effect of these rules is to render largely irrelevant to the coroner the detailed processes of investigating the underlying causes of accidents discussed elsewhere in this book.

The usual adversarial legal context of parties to proceedings each arguing their respective cases is also irrelevant, as the coroner's role is purely inquisitorial: through his own inquiries and questioning of witnesses whom he chooses to call, the coroner's job is to establish just the facts needed to enable himself or the jury to answer the questions set out. The rules specifically bars there being any verdict which would "appear to determine any question of (a) criminal liability on the part of a named person, or civil liability".

Although the verdict of the court has no legal consequence in determining liability, the proceedings nevertheless provide an important opportunity for relations to hear exactly how the deceased died, and for other interested persons to have an opportunity to hear key witnesses and assess the prospects of their evidence would support any claim.

Investigations by and on behalf of the coroner

6.24 A work-related fatality will be reported to the coroner by the police as a matter of routine. Initially inquiries will be made by coroner's officer who will be the court's main point of contact with the family, witnesses to the accident, the employer or other organisations involved, the pathologist and the relevant enforcing authority's inspectors. Any decision whether to require a post-mortem examination of the deceased is however made by the coroner, who will notify the family and other persons with a sufficient interest (such as the HSE or other enforcing authority) of the arrangements and their rights to attend or to have a representative present.

The post-mortem report is addressed to the coroner, but any person who is, in the coroner's opinion, a properly interested person can apply for a copy. As well as the deceased's family a copy would normally be released if requested by the employer, a trade union representative of the deceased, a life insurer for the deceased or an HSE or other inspector.

Where it is decided by the coroner that an inquest will be held the coroner's officer will contract witnesses and take written statements from them. There is no power however to compel a witness to give a written statement if the witness declines to co-operate for any reason. It is possible for interested persons (see 6.27) below, to submit statements or other documents to the coroner which they consider will assist in the inquest and the coroner has a discretion to take this evidence into account if no other person objects.

The coroner has discretion over the release of evidence to interested persons before an inquest opens. In addition to post-mortem report there may be requests that he release statements from witnesses, and while there is no entitlement on the part of anyone to early disclosure coroners may in some cases agree to release copies or to discuss the evidence they have obtained with interested persons.

Inquest proceedings: some common questions

How long will it take?

6.25 In many cases, an inquest will be formally opened within a few days of the death and then immediately adjourned. The purpose of this is so that a funeral can take place without delay. The coroner will at that point issue an order allowing for the release of the deceased's body and enabling death certificate formalities to be dealt with.

After this timescales vary considerably. Depending on the complexity of the investigations and the coroner's workload it can take several months or sometimes up to a year for a full hearing to take place.

However, other legal proceedings may cause delay. The coroner's inquest cannot be allowed to prejudice any significant criminal proceedings that might be taken in relation to a death, and there are various safeguards for this in the rules:

- Requests by police for adjournments – the police or DPP can request an adjournment where a person may be charged with murder, manslaughter, causing death by dangerous driving or certain other serious offences. The coroner then adjourns the inquest for twenty eight days or more.
- A magistrates' court dealing with an accused charge of murder, manslaughter, death by dangerous driving or one of the other relevant offences will inform the coroner who will adjourn a hearing until the conclusion of the criminal proceedings.
- If during the course of an inquest it appears to the coroner from the evidence that the death is likely to be due to manslaughter or one of the other serious offences, and that a person might be charged, then the coroner is required to adjourn for fourteen days and to notify the DPP.

There is no provision for adjournments of inquests pending any health and safety prosecutions in relation to the circumstances surrounding a death. Inquests are commonly concluded before charges for health and safety are laid against an employer, but this is encouraged by *Work-Related Deaths: a Protocol for Liaison* which states that the HSE or other enforcing authority will await the inquest verdict, unless to wait would prejudice the criminal case.

Inquest hearings themselves can vary in duration. Often a workplace death will be dealt with by the court in less than one day. Some inquests however can last for up to a week or more where the evidence is complex.

Will a jury hear the evidence and decide the verdict?

6.26 Yes, workplace deaths are one of the categories of cases for which a coroner must sit with a jury. A jury is also required where a death occurs in prison or public custody, or when a death occurs "in circumstances, the continuance or possible recurrence of which is prejudicial to the health or safety of the public, or any section of the public".

Who can take part in an inquest?

6.27 Anyone who can satisfy the coroner that he is within the categories of "interested persons" under the *Coroners Rules* is entitled to attend and examine witnesses at an inquest. These categories are:

(a) a parent, child, spouse and any personal representative of the deceased;
(b) any beneficiary under a policy of insurance issued on the life of the deceased;
(c) the insurers who issued such a policy of insurance;
(d) any person whose act or omission or that of his agent, or servant may in the opinion of the coroner have caused, or contributed to, the death of the deceased;
(e) any person appointed by a trade union to which the deceased at the time of his death belonged, if the death of the deceased may have been caused by an injury received in the course of his employment or by an industrial disease;
(f) an inspector appointed by, or a representative of, an enforcing authority, or any person appointed by a government department to attend the inquest;
(g) the chief officer of police;
(h) any other person who, in the opinion of the coroner, is a properly interested person.

In practice it is usually only the deceased's family and the employer or other business involved in the activity in the course which a person died who will be represented at an inquest into a workplace death, either by a solicitor or barrister or other authorised advocate.

Can written statements be used as evidence?

6.28 The coroner's court has more relaxed rules of evidence than other courts, and within broad limits coroners can choose what evidence to rely on as long as it is made available in open court. Procedures permit the statements of witnesses to be used without them having to attend and give evidence in person.

These rules are commonly used to avoid family members and colleagues having to give harrowing evidence personally, and to admit uncontroversial evidence such as for example the nature of the work and the workplace.

Photographic evidence of the scene of the death and the deceased is often available to coroners in cases of workplace deaths where they show the deceased they generally will be shown only to the jury, although anyone who is an interested person would be entitled to apply to inspect them.

Do witnesses have to attend?

6.29 Yes, if the coroner requires this. There are powers to compel reluctant witnesses to attend if need be, by court summonses. However, witnesses who are outside England and Wales cannot be forced to attend as witness summonses are not effective outside this jurisdiction.

How do witnesses give evidence?

6.30 The coroner will take the witness on oath through their prepared statements and questions them on any points which he feels require clarification. This includes the pathologist who undertook post-mortem and any other professional or expert witnesses. There is no bar to witnesses giving the coroner evidence which is hearsay – that is evidence of matters which they did not see or hear personally, but which others have told them about.

Do people giving evidence risk incriminating themselves?

6.31 Interested persons or their representatives are entitled to question witnesses, but the coroner is required to prevent any questioning which in his opinion is not relevant or which is otherwise not proper. Given the circumscribed function of the inquest and the prohibition on determining liability issues at above, this means that attempts to question for example a manager of the employer or an HSE inspector about matters concerning compliance with regulations or the outcome of any wide-ranging accident investigation might be ruled out.

Witnesses are not required to answer any questions from the coroner or other representatives which would tend to incriminate them. The coroner is obliged to inform a witness of this right not to answer when it appears that answering a question could self-incriminate the witness. The protection afforded here is limited to answers that would incriminate only in criminal, not civil, proceedings.

A witness will usually have great difficulty in determining on the spot whether an answer would actually be self-incriminating, and he or she may not be in a position to seek legal advice on the issue. The good sense of the coroner usually protects the witness.

A witness is not entitled to decline to answer questions which might incriminate another person or, for example, his or her employer.

What verdicts can be given?

6.32 The verdict consists of the findings of fact in relation to the questions in *Coroners Rules*, that is who the deceased was, how, when and where the deceased came by his death and the particulars required for the registration of the death.

The verdict is recorded in a document called an inquisition. The coroner sums up the evidence for the jury and directs them on the law and the possible verdicts.

The verdict cannot be framed in a way which would appear to determine civil liability, or criminal liability on the part of a named person, and there is a substantial body of case law on the appropriateness of wording (such as "lack of care" or "aggravated by neglect") which has sometimes been used to elaborate on the circumstances preceding an accident. A verdict incorporating a finding of neglect is now only appropriate in cases where a person is dependent on another and basic care, sustenance, shelter or medical treatment is not provided in circumstances where there would be a duty to provide it.

The jury (and the coroner) are not required to use any particular terms in reaching their verdict. Any simple suitable formula can be used to describe how the deceased came by his or her death. An acceptable example would be "the deceased was killed when he was struck by a reversing lorry". In practice, however, a coroner will normally direct the jury as to the appropriateness of one of the "suggested verdicts" contained on the inquisition form, which for a death at work could be as follows.

Death as a result of accident or misadventure: This is a very common verdict for workplace deaths. The terms "accident" and "misadventure" are now viewed as legally synonymous and nothing meaningful turns on the choice of one or other. This verdict is broad enough to cover any unintentional or unexpected event bringing about a death, but it should not be used in cases where manslaughter or one of the other serious offences is believed to have caused the death.

Death from industrial disease: This verdict is heavily dependent on medical and pathology evidence, but can involve any disease causally related to the death and it is not necessary for the disease to be prescribed formally as an industrial disease (e.g. for social security benefits purposes). Normally, the nature of the disease would be stated on the inquisition form.

Unlawful killing: For such a verdict to be reached, the coroner would need to give the jury directions as to the conduct which could constitute manslaughter by an individual or company.

A jury is permitted to reach the conclusion that a death was caused by unlawful killing, despite the prohibition on findings of criminal liability. This is because the finding of criminal liability does not relate to any *named* person. As a consequence, there can be no reference made on the inquisition of the person whose unlawful act or omission is thought to have been involved. Such a verdict may be uncontroversial if manslaughter proceedings have already taken place. However, where no criminal investigation has yet been undertaken, or the CPS have concluded that there is insufficient evidence to charge any individual or organisation, a conflict can arise.

An unlawful killing verdict does not require the CPS to open (or re-open) a criminal investigation or to charge anyone, but it may in practice give the CPS no alternative but to review the criminal case. It can also provide support if relatives or other representatives seek a judicial review of any decision not to bring a manslaughter prosecution, although the High Court will not lightly overturn a decision made by the CPS and will generally only do this where an erroneous view of the law of manslaughter has been used in the course of evaluating the criminal case.

Death in a disaster: This verdict was created in 1999 for rare cases where an inquest has taken place after a public inquiry has already taken place into a major disaster. It is now exceptional for there to be an inquest in such cases.

Open verdicts: This is appropriate only where there is insufficient factual evidence to determine the means whereby death occurred. It is meant to be used in cases where there is real doubt in the jury's mind as to how the death came about. In workplace cases, it will usually be clear that another verdict is appropriate, unless there is evidence pointing to an unrelated cause, such as natural causes or suicide, which cannot be ruled out.

If a jury cannot agree on a verdict, the coroner may accept the verdict of a majority, provided the number in the minority is not more than two. If such a majority is not attained, the jury has to be discharged and another inquest will be held with a new jury.

Can the coroner or jury require anything to be done to prevent similar accidents in future?

6.33 No, but the *Coroners Rules* permit the coroner (although not the jury) to make recommendations where, as a result of the inquest, he believes that action should be taken to prevent a recurrence of similar fatalities. This will be announced at the inquest. The recommendations are addressed to the person or authority who may have power to take such action, such as a minister or the HSE. These representations do not though have any legally enforceable effect, nor do they appear in the inquisition. An example occurred at an inquest in Grantham in 2002, when a recommendation was given to the HSE to investigate the storage and movement of pallets after a forklift accident.

Will there be media coverage of the inquest?

6.34 Inquests are held in public, unless the coroner considers this would be harmful to national security. There may be because this is a public hearing and journalists may attend. The media will report the evidence and verdict of the case if of local or national interest.

There is no appeal against the verdict of a coroner's court in the strict sense. However, in limited circumstances, and with the permission of the Attorney General, a coroner can be ordered to hold an inquest, or a fresh inquest where one has already taken place. Reasons for this would include procedural irregularities, failure to consider evidence sufficiently or the emergence of important new facts about the death. An alternative procedure is to seek a judicial review which, as well as challenging the coroner's correct application of the law or procedure at the inquest, could cover situations where the verdict is alleged to be irrational or biased, or where the inquest is alleged to have contravened rights of interested persons under the *Human Rights Act 1998.*

A successful application was made in the *Marchioness* disaster case to replace the coroner with a new one when the court of appeal agreed with relatives that remarks he had made in public indicated that there was a real possibility that he had unconsciously allowed himself to be influenced against them and their arguments for a resumed inquest some year after the accident.

Can the evidence of witnesses be used again in later proceedings?

6.35 A substantial amount of information about a fatal accident emerges in the course of an inquest and parties to any subsequent proceedings (criminal or civil) will have an interest in examining it in more detail.

Coroners are required to take full notes of the inquest hearing and, as well as making these notes, there will often be a taped recording from which a transcript will be produced if an interested party requests one afterwards. In addition, coroners will supply on request a copy of the inquisition and any documents and exhibits produced in evidence on payment of the prescribed fee, including the post-mortem report. Alternatively, these papers can be inspected free of charge.

Any subsequent proceedings are entirely separate and require the parties to them to adduce evidence afresh, even from witnesses who have already given evidence in full at the inquest. A transcript or the coroner's notes of that evidence will not be admissible as evidence in its own right, unless the parties to the later proceedings reach agreement to allow it to be used in evidence or, in civil proceedings, the court permits it to be admitted as hearsay evidence.

In spite of these restrictions, inquest evidence can still be deployed in support of subsequent proceedings in a number of ways. It provides a basis for interviewing the witnesses (if they agree to this) in order to obtain clarification or expansion of evidence given previously which may assist in determining the liability issues. There is also the opportunity for a party to civil proceedings to tailor allegations and pleadings precisely to matters which can be identified from the evidence as being breaches of a duty of care or of statutory duties. A witness giving evidence in later proceedings can also adopt a prior statement in the coroner's court as his or her evidence in the trial of the subsequent proceedings. Prior evidence can also be used in cross-examination of witnesses where they were called by another party, in order to draw attention to any inconsistencies between evidence given at the inquest and at a trial.

Fatal accident inquiries (Scotland)

6.36 In Scotland, the coroners' courts do not exist. Instead the sheriff court will conduct fatal accident inquiry (FAI) in respect of fatal accidents, deaths of persons in legal custody, sudden, suspicious and unexplained deaths and deaths occurring in circumstances giving rise to serious public concern. *The Fatal Accidents and Sudden Deaths (Scotland) Inquiries Act 1976* govern the holding of an FAI. The FAI process is usually longer and comprises a fuller enquiry than a coroner's inquest on account of the wider remit of determining the cause or causes of death. There are only about fifty–sixty FAIs in Scotland each year. Sheriffs are legally qualified judges. The sheriff does not sit with a jury. The FAI is held in public. Once all the evidence has been presented the sheriff will issue his judgement, which is known as a "determination".

For workplace deaths (and deaths in custody) it is mandatory for the sheriff to hold an FAI unless there have already been criminal proceedings concluded against a person in respect of the death or the accident from which the death resulted and the Lord Advocate is satisfied that the circumstances of the death have been sufficiently established in the course of such proceedings. For example, the mandatory requirement could be waived if there had been a criminal trial (including a health and safety prosecution) where evidence of the circumstances of the death had been heard by the court. This will obviate the need for the same evidence having to be heard twice and it lessens the risk of adding to the distress of bereaved families or other persons who witnessed the fatality. Alternatively, the Lord Advocate has frequently waived the requirement for an FAI in circumstances where the accused has pleaded guilty and sufficient details of the death have been included in the prosecution's narrative, which is presented to the court at a sentencing hearing. It must, however, always be borne in mind that the power to waive the mandatory requirement is discretionary. The sheriff court has the power to award expenses against the

Crown in circumstances where the decision by the Crown to hold an FAI after completion of a criminal jury trial was vexatious (*Smith (Fatal Accident Inquiry)*).

Where the death occurred in circumstances that would not lead to a mandatory FAI, there may be an FAI if the Lord Advocate considers it expedient in the public interest. However, under these circumstances, the death must be sudden, suspicious or unexplained or it must have occurred in circumstances such as to give rise to serious public concern. The Lord Advocate has a wide discretion in deciding whether to recommend an FAI. In deciding whether to exercise his discretion, the Lord Advocate will consider several factors such as the wishes of the relatives of the deceased or whether the facts of the death have been fully reiterated in criminal proceedings.

Powers of the sheriff court and the Procurator Fiscal to require evidence and witnesses

6.37 If a death occurs which may be the subject of an FAI, the Procurator Fiscal (i.e. a state employed public prosecutor for the sheriff court district with which the circumstances of the death are most closely connected) must investigate the circumstances of the death and apply to the sheriff to hold an inquiry. Typically it will be in response to a police report or from the HSE or from some other agency such as a hospital authority.

Obtaining evidence of witnesses

6.38 In carrying out his investigations, the Procurator Fiscal is entitled to ask witnesses for "precognitions". This is a statement or record of the evidence a person may be expected to give if called as a witness in proceedings. Witnesses are given a citation, which is a formal request from a party who requires the witness to attend a court hearing. The precognition is taken down in writing but will not be signed (unless it is taken on oath – see below). Generally speaking, precognitions are not admissible in evidence.

A copy of the precognition will not be given to the witness. Furthermore, the witness is not entitled to be accompanied by a representative. Occasionally a witness may default and fail to appear or refuse to give information. If a person fails to attend without reasonable excuse or refuses to give information within his knowledge regarding any matter relevant to the investigation, the Procurator Fiscal can apply to the sheriff for an order requiring the witness to attend to give a precognition or to give such information as required by the order. If the witness still refuses to attend or to give the information required by the order, he commits an offence.

Precognition on oath

6.39 The Procurator Fiscal has a common law power to have a witness brought before the sheriff for precognition on oath. For instance, a witness may have defaulted in attending the Procurator Fiscal. Under these circumstances, the Procurator Fiscal would apply to the sheriff under his powers. If ordered to do so, the witness appears before the sheriff and will be placed on oath or affirmation. The Procurator Fiscal then questions the witness. A shorthand writer will produce a verbatim record of the entire proceedings. The witness will not normally be allowed to have his solicitor present at the hearing. The court then fixes a date for the witness to appear before the sheriff again. At the continued hearing, the witness will be asked to read out the transcript, sign it and acknowledge its accuracy.

Decision on whether to hold an inquiry

6.40 At the conclusion of his investigations, the Procurator Fiscal will have amassed a body of evidence taken from precognitions and other sources (e.g. documents and evidence such as equipment or substances from the scene of the fatality). He must then assess whether an inquiry is necessary. Should the Procurator Fiscal decide that an FAI is necessary he will apply to the sheriff in whose sheriffdom the death has the closest connection. Once the application is made, the sheriff must make an order for an FAI. He cannot refuse to make an order unless the application has been improperly made.

Holding the inquiry

6.41 The FAI must take place as soon as is reasonably practicable. This depends on the circumstances of the case. For instance, if the FAI is likely to be long and complicated, all interested parties will need to be given a proper opportunity to carry out their own investigations. In making an order for an FAI, the sheriff is required to undertake the following:

- Fix a time and a place for holding the inquiry. The FAI will normally be held at a sheriff court house, but this is not essential. It may be more comfortable and convenient to hold the FAI at some other venue, especially if there is likely to be a large number of witnesses and documents. In the case of the Lockerbie disaster, the FAI was held at a hospital in Dumfries.
- Instruct witnesses and persons having documents in their possession to attend the FAI at the instance of the Procurator Fiscal or any other person who may be entitled by virtue of the 1976 Act (e.g. the deceased's next of kin or the deceased's employer).

Once the sheriff has granted an order to hold the FAI into a work-related death, the Procurator Fiscal must inform the following of the details:

- spouse or nearest known relative of the deceased;
- inform the deceased's employer;
- the HSC;
- in the case of a death in legal custody, any minister or government department or other authority in whose custody the deceased was at the time of his death;
- in the case of an offshore industry fatality the Secretary of State for Employment;
- in the case where it would be competent for a minister or government department under any statute other than to hold a public inquiry into the death, the relevant minister.

There may also be several other parties who wish to be present or represented at the inquiry. The 1976 Act provides that any person with an interest in the FAI and who is entitled to receive intimation of the FAI is entitled to appear and adduce evidence either in person or to be represented by a solicitor or an advocate or any other person (with permission of the sheriff).

The Procurator Fiscal must give twenty one days notice in at least two newspapers circulating in the sheriff court district where the FAI will be held.

Role of the sheriff

6.42 The sheriff is the arbiter of the proceedings before the FAI. He is the judge and he has a statutory duty to produce a written determination after the conclusion of the evidence and the submissions. He has no role to play in the investigation or in the presentation of the evidence. He will, however, question the witnesses and generally manage the conduct of the proceedings. In discharging his duty, the sheriff may either at his own instance, or at the request of the Procurator Fiscal or of any party entitled to appear at the inquiry, summon any person having special knowledge and being willing to do so, to assist him by acting as an assessor at the inquiry.

Pre-hearing reviews

6.43 It is becoming increasingly common in more complex FAIs for the sheriff to hold a pre-hearing review. This is generally about two weeks before the opening of the FAI. The pre-hearing review enables the parties to meet with the sheriff and the Procurator Fiscal to ensure that the parties will be fully prepared for the FAI and to enable the sheriff to make any orders that he considers necessary to ensure the expeditious progress of the FAI. The sheriff may make such orders on his own motion or on the motion of the Procurator Fiscal or another party to achieve that objective.

Evidence

6.44 A "production" is an article produced as evidence in court. This could be a document such as an accident report, a company safety management procedure, photographs of the site of the accident, the pathologist's report of the deceased's autopsy, computer records, time sheets, etc. Productions form an important part of evidence at the FAI and will be lodged by an interested party such as the Procurator Fiscal or the deceased's employer. Witnesses may be asked to refer to a production as part of their examination (e.g. because they are the authors of the documents).

Each witness will be required to swear an oath or to affirm (if he is of suitably mature years to understand the significance of the oath or affirmation). Witnesses are allowed to refresh their memories from notes made by them at the time of the incident in respect of which they are being examined. The Procurator Fiscal will normally present his evidence first and will be the first to call witnesses. The first witness is likely to be the husband or wife or nearest relative of the deceased. They would normally be asked to identify the deceased and to confirm his personal details such as date of birth, address and occupation. Alternatively, the Procurator Fiscal will begin the formal evidence (e.g. the pathologist's report on the autopsy). In many instances the formal evidence is agreed and it would be usual to lodge a sworn affidavit or another signed statement instead of the person giving evidence personally.

Evidence of the facts of the death will almost certainly be adduced from witnesses who saw the accident or who can testify directly as to the facts and circumstances of the death. The Procurator Fiscal would expect to present evidence on the basis of the precognition taken from the witness. After he has finished his examination, the witness will normally remain in the witness box and will be subject to cross-examination by another interested party or his representative. If any other party or representative is calling a witness to give evidence, then his witness will also be subject to cross-examination. Once a witness has given his evidence, he is not permitted to discuss what has happened to him in court with the other witnesses.

The rules permit written statements to be admitted in evidence and to be used in lieu of oral evidence. The written statement, if admitted, will have the same evidential status as if the witness had given his evidence orally. The written statement must have been sworn or affirmed by the deponent before a notary public, a commissioner for oaths or a justice of the peace or before a commissioner appointed by the sheriff for that purpose. The witness must sign the statement. A written statement will not automatically be admitted in evidence. A written statement may only be admitted if all persons who are appearing or are represented at the inquiry agree, and the sheriff considers that it will not result in unfairness to any person appearing or represented at the inquiry.

In practice, written statements will tend to be admitted in circumstances where the evidence is formal in nature (e.g. the pathologist's report) or in circumstances where it is unlikely that there will be any dispute over the evidence.

If a witness or haver gives evidence at the inquiry, this fact will not be a bar to subsequent criminal proceedings being taken against him. However, the witness has the right to claim privilege against self-incrimination and is not required to answer any question tending to show that he is guilty of any crime or offence.

Submissions

6.45 On the conclusion of the evidence, the parties will be invited to make submissions to the sheriff. This is not a mandatory requirement although, in practice, parties do present submissions in the expectation that this will assist the sheriff in making his determination.

The inquiry will generally be open to the public unless a person under the age of 17 years is in anyway involved in which case the sheriff may limit or impose reporting and broadcasting restrictions of the proceedings.

Transcripts

6.46 Generally speaking, the evidence will be recorded by a shorthand writer. Any person with an interest in the FAI can request a transcript. For example, a potential in a civil action would normally request a transcript of the evidence in order to prepare for the future civil action. The Procurator Fiscal may require a transcript if he is contemplating commencing criminal proceedings against an interested party.

Verdicts

6.47 The sheriff must make a determination setting out the following circumstances of the death so far as they have been established to his satisfaction:

(a) where and when the death and any accident resulting in the death took place;

(b) the cause or causes of such death and any accident resulting in such death;

(c) the reasonable precautions, if any, whereby the death and any accident resulting in the death might have been avoided;

(d) the defects, if any, which contributed to the death or any accident resulting in the death;

(e) any other facts which are relevant to the circumstances of the death.

In forming his conclusions, the sheriff is entitled to be satisfied that any of the circumstances referred to above have been established by evidence notwithstanding that the evidence has not been corroborated.

The purpose is meant to be purely fact-finding. It is not to apportion blame, or criminal or civil liability, only to determine the facts above. Any person may obtain a copy of the determination from the sheriff clerk on payment of the prescribed fee. Copies of determinations are also posted on the Scottish Courts Administration website *www.scotcourts.gov.uk*

The legal implications of a determination

6.48 The sheriff's determination may not be admitted in evidence or founded upon in any judicial proceedings, of whatever nature, arising out of the death or out of any accident from which the death resulted. Notwithstanding these restrictions, a determination may have a significant impact on other proceedings.

Facing health and safety investigations: checklist

6.49

Preparation for inspection

- ✓ Has the immediate danger been dealt with? Is the site safe for inspectors to visit?
- ✓ Has the key evidence been preserved?
- ✓ Have senior managers been informed of the visit?
- ✓ Has the manager been appointed as the liaison person? Is he or she briefed to deal with the inspector's questions?
- ✓ Are the key documents available: health and safety procedures, risk assessments, Accident Book and RIDDOR report?

At inspection

- ✓ Ascertain what the inspector wishes to see and offer full co-operation.
- ✓ Documents requested should be provided to the inspector. (If in doubt about confidential documents refer to the legal advisor to check whether they are covered by legal privilege.)

- ✓ Maintain a log of documents given to inspectors.
- ✓ Ensure that samples are kept of items removed by inspector where this is practicable.
- ✓ Deal promptly with enforcement notices or other actions requested by the inspector.

During interviews

- ✓ Check if interviewees wish to have a colleague, union representative, lawyer or other person to accompany them to interviews (as appropriate to type of interview).
- ✓ Where the right to silence exists it should normally be exercised until the company's lawyer is present.
- ✓ Give factual answers which you can vouch for – not opinions, beliefs or second-hand reports.
- ✓ Avoid speculation or unchecked details: arrange to check facts and get back to the inspector.
- ✓ Do not get unduly pressurised. If uncomfortable, request a break; take legal advice if you feel you are being asked an inappropriate question or may prejudice yourself.
- ✓ Check any written record of the interview prepared by the inspectors – correct inaccuracies and incompleteness before signing.
- ✓ Request copies of any written statements (or audio cassettes from tape-recorded interviews).

After the interview

- ✓ Make a detailed record of the interview.
- ✓ Deal promptly with any information the inspector has requested be provided after the interview.
- ✓ De-brief with your legal advisor and colleagues – consider sending the inspectors any further useful information.
- ✓ If you have concerns about personal liability, seek independent legal advice or ask your organisation if they can help to arrange this.

References

Cullen, Lord. (1990) *The Public Inquiry into the Piper Alpha Disaster*. HMSO, London.
HSE (2003) *Work-related Deaths – A Protocol for Liaison*. www.hse.gov.uk/pubns/misc491.pdf
PACE Codes of Practice (*various*) available at www.police.homeoffice.gov.uk/operational-policing/powers-pace-codes/pace-code-intro/
R v Walthamstow Coroner, ex p Rubenstein [1982] Crim LR 509.
Railtrack plc v Smallwood [2001] ICR 714.
Smith (Fatal Accident Inquiry) [2005] SCLR 355, Sh Ct.

7 Rehabilitation of the injured worker and managing the people aspects of an accident

In this chapter:

General introduction

7.1 A workplace accident can have a major impact on a workforce. The injured parties may feel isolated and abandoned if they are off sick. A plan for rehabilitation, getting the injured person or people back to work where that is possible, forms part of the planning response to the accident. Where colleagues are injured or even killed then there may be a potent mix of grief, guilt and anger in those who are not injured. It is important to manage the developing situation sensitively, considering all affected parties, in order to achieve resolution of the incident and a return or move forward to safer working practices.

The HSE has produced guidance on managing sickness absence and return to work which stresses the benefits to employers of having pro-active policies (HSE, 2004). These include:

- avoiding unnecessary recruitment and training expenditure and maintaining competitiveness;
- reducing Statutory Sick Pay (SSP) and overall sickness absence costs;

- avoiding significant penalties for discriminating against disabled workers;
- improving workplace relations;
- raising the organisation's reputation;
- safeguarding the livelihood of employees, and so benefitting their families and communities.

Rehabilitation in the employment context

7.2 For several years now, major employers in the UK have been taking much more interest in what active interventions they can make for their employees who have been injured or who are off sick in order to get them back to work earlier. The evidence shows that the earlier people get back to work, the more likely they are to stay in employment. The latest figures from the Department of Work and Pensions show that in the UK, one million people go off sick each week of whom 3,000 will still be off sick six months later and 80 per cent of those will not work for at least the next five years. 90 per cent of those who start to claim Incapacity Benefit (IB) expect to work again, but for those on IB for twelve months, their average claim will be eight years and for those on IB for two years, they are more likely to retire or to die than they are to work again. Where the employer can support an early return to work this is likely to benefit the employee.

In the past it has been convenient for employers and especially managers to think of employees as either ill or well, either off sick or fully functioning at work. This has never been the full story, but the current focus on rehabilitation specifically recognises that that employer can play an active role in assisting an ill or injured employee to recover by maximising the opportunities for an early return to work.

The types of interventions that major employers can make include paying for medical investigations and fast track access to medical treatment, arranging assessments to identify what is needed both in the home and in the workplace to get the person back to work and then paying for or contributing to making the necessary adjustments. The workplace adjustments can be physical alterations or at least are likely to be modifications to the type of work or the working hours. Small employers may not be in a position to pay for medical assessments or treatments but a lot can be achieved simply by improving communications between the employer and the relevant health care workers. Any employer is able to consider what workplace adjustments are needed to support an employee to get back to work earlier than they would otherwise be able to do.

The advantages are that employees are retained, their skills are available again to the employer more quickly, they are less likely to lose their jobs and they are likely to feel more positive about their employer. In the case of injury these outcomes may lead to a reduced insurance claim. The approach requires active management, a commitment to invest in ill or injured employees and a flexible approach to managing return to work after illness or injury.

The pressure for this change in approach largely came from trades unions, insurance companies and their legal advisers and some health professionals. It is now reflected in the UK Government's Health, Work and Wellbeing strategy jointly launched by the Department of Work and Pensions, the Department of Health and the Health and Safety Executive in (HM Government, 2006). This strategy sets an ambitious agenda. Broadly speaking it is aiming to break the link between ill health and activity, to advance the prevention of ill health and injury, to encourage good management of occupational health and to transform opportunities for people to recover from illness while at work, maintaining their independence and their self of worth.

There is an increasing focus on the "bio-psycho-social" model of recovery from illness and injury. This model emphasises the importance of many other factors as well as the medical diagnosis in determining the likelihood of a successful recovery. Factors such as the individual's attitude to their illness, what they feel about their work, their supervisor and their colleagues, the processes that the employer uses to support (or not support) the sick employee and the way that the employee is treated by health care professionals will all have a bearing on their recovery (Waddell and Burton, 2006).

Limitations of the physical model of injury

- The physical, psychological and social effects of injury influence each other and should be considered together, not in isolation.
- In approximately 20–30 per cent of cases, the victim suffers disability and distress significantly greater than might be expected from physical factors alone. This is explained by the impact of psychological and/or social factors, which research shows, can have a powerful impact on the outcome of any injury or disease.
- In approximately 5 per cent of cases, the physical and social outcomes are seriously adversely affected to an extent that cannot be explained by initial or remaining injury. This is known as "Apparently Disproportionate Outcome" (ADO) and can have a sizeable effect on the cost of treatment, complexity of case handling and level of compensation. This disproportionate outcome is as likely to be felt in apparently minor injuries as in major ones.

Source: IUA/ABI (2004)

Immediate aftermath of an accident

7.3 Immediately after an accident, any injured person needs to be assessed to decide if they need to go to hospital. If a trained first aider is available, they should be able to make this assessment. If not, whoever attends the accident will make a judgement in consultation with the injured person or people. Sometimes it can be quicker to take an injured person to hospital by car rather than waiting for an ambulance, but for some injuries an ambulance must be called. Table 7.1 gives advice on this decision.

Table 7.1: Requirement for immediate medical attention

People who are injured at work may require immediate medical attention:

- The following are some examples of injuries that will require immediate medical attention: Electric shock, cuts and wounds, trips and falls, fractures and spinal injuries.
- If a person is seriously injured, they should be taken to hospital by an ambulance.
- If there is a risk that the person has had a spinal injury they should always be transported to hospital by ambulance.
- If a person had a significant electric shock at work it is advisable that they go to a hospital Accident and Emergency Department (A&E) to be checked up, as there is a small risk of heart rhythm complications. They can go by car if they feel well and appear satisfactory to the attending personnel/manager/first aiders.
- If there are signs of a serious electric shock, that is the person looks obviously shocked and has suffered burns, then they should go by ambulance.
- If in instances where it is obvious that arrival of an ambulance will be significantly delayed then the casualty can be transported by other means. If there is concern that they may collapse during the journey, they should be accompanied by a member of staff in addition to the driver of the vehicle.
- When a person has a relatively minor injury, which requires medical attention (e.g. a cut that needs stitching), they can be transported to hospital A&E in a car.
- Medical emergencies should be dealt with in the nearest hospital A&E.
- Where a trained First Aider is available they can assist and guide other staff in emergency situations.

Someone will need to contact the next of kin, or preferred contact of the injured person, to let them know that the accident has happened, how seriously injured the person is and where they are. It is useful to keep a list of emergency contact numbers which is easily accessible, but which remains confidential to the relevant managers or administrators.

There is likely to be a lot to do immediately after an accident, but it is helpful to think about people's reaction and how that can be alleviated. The injured people and those who witnessed the accident may be very distressed by it. Once the need for medical attention to the injured person or people has been met, it is useful to consider whether "emotional first aid" will help the injured people, if they did not go to hospital, and for those who witnessed the accident. This is not professional counselling. Indeed in the immediate aftermath of an accident counselling is not helpful, because at first people are shocked by the event and what they need is practical help and comfort.

What does emotional first aid involve? (Spiers, 2001)

7.4 Looking after people involves some or all of the following practical measures:

- Take people to a quiet room, away from the scene of the accident, where there is somewhere comfortable to sit.
- Provide hot drinks for them.
- Check whether people are warm enough – offer blankets if necessary.
- If anyone is shaking or shivering, feeling faint, feeling sick or sweating, explain that it is normal to experience this after being involved or witnessing an accident.
- Encourage people to talk about what happened if they want to, but don't force anyone to talk or listen.
- If they want to go home make the arrangements:
  - It may be necessary to arrange taxis.
  - Check whether anyone will be at home if not, offer to call a friend or relative for them.
  - Check whether they would like someone to accompany them.

Compile a list of all those employees who witnessed the accident, together with their up to date contact details, so that ongoing support can be offered to them if it is needed.

First few days after the accident

Communications with the injured person(s)

7.5 The employer needs to keep in touch to find out how the injured person is getting on, wherever they are. It is a very vulnerable time for the injured person, and it can feel like adding insult to injury if they are injured at work and then their employer takes only a desultory interest in their progress. Keeping in touch will initially often be done by telephone, but can be followed up with one or more visits and regular telephone calls. If the person is in hospital then the first visit can be arranged with a near relative, the hospital staff or the person themselves if they can be contacted by telephone. Judgement will be needed to assess how much the person wants to hear about the investigation of the causes of the accident and what the employer is doing to prevent it happening again. The first focus of conversation should be the injured person and their progress.

Some employers may be able to offer access to private medical investigations or treatment. If so, this can be mentioned to the employee, their relatives and the hospital staff. After a few days the offer of professional counselling may be appreciated, if it can be arranged.

Communications with colleagues of the injured person(s) and all the workforce

7.6 The workforce, especially those who witnessed the accident or worked with the injured person, will want to know the employer's first thoughts on the cause of the accident, what has been done to make working conditions safer and how the injured person is getting on. Some colleagues may be keeping in daily touch with the injured person or their relatives. In a small workplace such communications can be very informal. In a larger company some thought should be given about how to keep people informed. It is useful to involve thosewho know the injured person well and the employee representative(s) in these decisions.

Some of those who witnessed the accident may still be experiencing the symptoms of acute stress reaction (Spiers, 2001). These can include:

- *Intrusive symptoms*: Nightmares, can't stop thinking about it, can't stop seeing what happened, feels as though they are reliving the accident.
- *Arousal symptoms*: Difficulty in sleeping, shaking, tremor, sweating or nausea, overwhelming feelings about the accident, jumpy and easily startled, feeling watchful and on guard, increased anger and irritability, difficulty concentrating.

- *Avoidance symptoms*: Trying not to think about it, trying to suppress feelings about the event, avoidance of people and places connected to the event, feeling numb or cut off from people, feeling it hasn't happened or isn't real.

It is helpful if a manager asks those who were there at the time of the accident how they are feeling. Some of them may wish to have access to professional counselling at this stage, if it is available. The focus of any counselling at this stage should be on reassurance and normalisation; such symptoms are common and normal in the days and even weeks following a distressing event.

The management of the injured person

The injured person(s) is off sick

7.7 Managing this situation is similar to managing any employee who is off sick. Particular sensitivity may be needed about keeping them informed about the investigation of the accident, being aware that they may be very nervous about returning to work, and letting them know if there is specific help that the company can offer them.

Regular contact is needed. It is useful to know how the injured person is progressing, what treatment they are having, when they think they may be ready to return to work and whether they think there are any adjustments that will be needed to their work. Strictly speaking employees are not required to discuss any details of their medical conditions with their employer. It can be in everyone's interests for the employer to have some information in order to plan and where there is mutual respect and trust this is usually forthcoming.

It there is access to an occupational health advisor then they should be made aware of the situation at an early stage. If they are based on site then they are likely to have been involved at the time of the accident. They will be able to give advice on whether reports are needed from medical specialists and if so, how to obtain them.

Advice may be needed from the GP or hospital specialist if the person cannot give a clear indication of when they are likely to return to work and/or whether they will need adjustments. Such advice can be sought through the occupational health adviser, but if there is none then the relevant manager will need to write direct to the GP.

The employer must have the employee's written consent to request such a report, the employee has the right to see it before it is sent and the doctor can charge a fee to write the report. These are the requirements of the Access *to Medical Reports Act 1988* for any medical report requested by an employer or insurer that may affect someone's employment or insurance. An example of a consent form is set out in Appendix III which explains the employee's rights.

Detailed advice about how to seek medical reports is given in many human resource textbooks. The following gives a brief summary:

- Do not ask for medical details, just ask the questions you need answers to.
- Is the person likely to make a recovery to the point where they can return to work?
- If so, when are they likely to be able to return to work?
- Are they likely to need adjustments to their workplace, nature of their work or organisation of their work?
- If so, what kind of adjustments?

An example of such a letter is given in the example in Appendix IV.

It is useful to keep the organisation's legal advisers and claims managers informed of the injured person(s) progress. It may be possible to provide help, such as access to private physiotherapy or to counselling, to speed up recovery and the claims manager may be keen that this is done as it may reduce the employers' liability claim. The manager may also wish to arrange this in order to do their best to care for the person.

The injured person(s) can return to work

7.8 When the injured person indicates that they will soon be ready to return to work, it is helpful to plan using the checklist given at the end of the chapter.

Remember that there are psychological considerations as well as possible physical adjustments that may be needed. The person is returning to work to a job where they have sustained an injury. They may have some terrifying memories of the accident and be frightened to visit the premises again; they may need reassurance that changes have been implemented to prevent the accident happening again; they may be angry with the employer, perhaps with a specific person, that the accident happened. If they now have a permanent disability they may be both angry and ashamed; they may dread their colleagues seeing them now.

It is helpful if discussions about these issues are started gradually, during the regular contacts with the person, rather than mentioning them for the first time just before the person returns to work, or not addressing them at all. The manager needs to be aware of the things that particularly bother the person coming back to work and be prepared to discuss what can be done to manage these.

An occupational health adviser can be invaluable in this process provided they have influence with the managers who can make changes to the person's working arrangements.

Options to consider include:

- An invitation to the person to visit the workplace, perhaps for several visits, for them to see how they respond to it including seeing the site of the accident when they feel ready to do so.
- A return to part-time work or shorter hours, either as an initial option or a permanent option.
- Modified duties or a different job.
- Extra or modified equipment.
- Help in getting to and from work.
- Discussing in detail the accident investigation, the findings and what has been changed as a result.

If there is any ongoing disability then a workplace risk assessment is mandatory to assess if there are any hazards that the individual would be more vulnerable to than they were before, and whether there are any specific preventive measures and adjustments that are required. There is now a statutory duty under the *Disability Discrimination Act 1995* (*DDA*) on the employer to make "reasonable adjustments" where a person who is disabled will be at a substantial disadvantage. This could mean making adjustments to the person's method of working, and/or to equipment or premises. Care should be taken to check relevant guidance contained in the Code of Practice for employment which accompanies the *DDA* (DRC, 2004).

There are agencies that can help if the person now has a disability. Jobcentre Plus (*www.jobcentreplus.gov.uk*) is the UK government's national body that promotes work, provides appropriate health and support for those without jobs and ensures the security of the benefits system. Their priority is to help as many of the people as possible who use job centres into appropriate and sustainable work as quickly as they can. Jobcentre Plus has Disability Employment Advisers (DEA) who will provide advice and help. There is a specific Access to Work scheme to provide help with equipment, other alterations and the journey to and from work if it is needed.

After the return to work

7.9 Once the person is back at work they are likely to need continued support from their manager and colleagues, especially at first. A regular review of their progress undertaken with them is useful. If there are significant ongoing health issues, a regular review with the occupational health adviser is also helpful. A small number of people can develop problems at a late stage related to the psychological trauma they have experienced. Symptoms that may appear are those of intrusion, arousal or avoidance as listed in the previous section on communications with colleagues of the injured person(s) and all the workforce. Although such symptoms are normal in the first few days and even weeks after an accident, when they occur a long time after an accident then the person requires assessment. Part of the duty of care to the individual is to be aware and to make them aware of the small possibility that this can occur. If such symptoms do occur a referral should be made to the occupational health adviser if there is one, or to the GP if there is not.

The person may wish to become more involved in advising on health and safety matters. The possibility of them playing either an informal role or a more formal role (for which training would be needed) can form part of the discussions with them, once they have successfully returned to work. They may want to tell people about what happened and how important health and safety management is. If this can be incorporated into health and safety briefings, it can be very effective. Not everyone, however, will want this role.

The injured person(s) cannot return to work

7.10 If the injured person does not recover either physically or mentally to the point where they can come back to work, then managing their termination of service will benefit from careful planning.

It is helpful to keep in touch with the injured employee regularly. It may be obvious from the start that the injured person will never get back to work. In some cases, it takes a long time before it is clear. Each case will need an individual approach. Where it is obvious, it still may not be appropriate to say so at an early stage. The person may need to believe that they are going to make a full recovery for a time, before they can cope with the reality of permanent disability. If there is an occupational health adviser they will help the manager make these judgements. If not, the manager will need to be guided by their own judgement, the advice of work colleagues, especially the human resource advisor, if there is one and any advice available from the injured person's doctors and their family.

The legal advisors and claims manager will wish to be kept in the picture. Losing employment as a result of a workplace accident increases the claim against the company by a large margin.

The legal aspects of dismissal

7.11 Employers must be very careful in how they manage ill health or workplace injury related dismissals. If the dismissal is mismanaged, then the employer could be vulnerable to claims from the employee for breach of contract and/or unfair dismissal and/or disability discrimination. These can be very costly both in terms of the time taken to defend the claims and in respect of the levels of compensation that might have to be paid to the employee.

Termination can be further complicated if the employer operates a Permanent Health Insurance Scheme and the employee is receiving benefits under that scheme. Dismissal of an employee if he or she is receiving long-term disability insurance could result in a legal challenge.

Faced with these risks, the employer may simply have to continue to employ the employee and find suitable ways to manage the employee's workload (e.g. by redistributing it to other employees or by hiring a replacement).

Clearly, there are many traps for the unwary employer to fall into.

Breach of contract

7.12 If the employer is considering dismissal, it must ensure that it terminates the contract in accordance with its terms. Every employment contract requires an employer to give the employee the requisite period of notice to terminate the employment relationship. The notice period may be expressly stated (e.g. three months) or it may be stated by reference to the statutory minimum period that is required under the *Employment Rights Act 1996* (*ERA*). If the contract says nothing about notice or the express notice period is less than the statutory minimum, then the notice period will be implied as being the statutory minimum period prescribed by ERA.

If the employer fails to give the employee the correct period of notice to terminate the contract then it will have acted in breach of that contract and will be liable to pay the employee compensation for that breach. Compensation is assessed according to a well established principle that the employee is entitled to be put into the position that he or she would have been in had the contract been properly performed.

Compensation would be for the wages and other benefits that the employee was entitled to receive during the notice period. For example, if the employee was entitled to be given three months' notice and the employer terminated the employee without notice (i.e. summary dismissal) then the employee would be entitled to be paid the equivalent of three months' salary and be given the value of any other benefits that he was entitled to. If the employer gives one month's notice, then compensation would be calculated by reference to two months' salary and benefits.

In every case, the employee is required to mitigate his or her loss (e.g. by making reasonable efforts to find alternative employment) which can reduce the amount of compensation payable. However, in the case of an employee who cannot return to any form of work because of his injury or illness, the question of mitigation will be largely academic and he or she can expect to receive the full amount of compensation.

Unfair dismissal

7.13 Many employers believe that if they dismiss the employee with the correct period of notice then that will be the end of the matter. This is a popular misconception and frequently exposes the employer to the risk of further claims because the employee may have statutory employment rights, which could be triggered by the dismissal. These rights operate quite independent of the termination provisions in the employment contract.

If the employee has at least one year's continuous period of service with the employer (or with a previous employer if the current employer agreed to accept this when he employed the employee) then the employee will have acquired a statutory right not to be unfairly dismissed. Once dismissed, the employee can complain to an employment tribunal that he or she was unfairly dismissed. This must be done within three months of the effective date of termination (although there are special rules for allowing claims to be filed later than three months). Once the employment tribunal accepts the claim, it will notify the employer of the claim. If the employer intends to defend the claim, it must lodge its defence at the employment tribunal within twenty one days of receipt of the claim.

Unlike a breach of contract claim the employee does not have to prove anything. This is because it is presumed that all dismissals are unfair until the employer can prove otherwise. The employer must, therefore, satisfy the employment tribunal that it acted fairly when it dismissed the employee. In doing this, the employer can only rely on a range of permissible defences that are prescribed by the *ERA* (known as the "potentially fair reasons"). In cases of ill health or workplace-related injuries, the only potentially fair reason that the employer can rely on is capability.

The employer would argue that because of the employee's injury or illness, he or she is not longer capable of working for the employer.

In dealing with an employee who cannot return to work, the employer will be expected to approach the matter with sympathy, understanding and compassion (*Lynock v General Packaging Limited*). Each case will depend on its own facts in deciding whether the employer acted fairly in dismissing the employee. The employer would be expected to assess the following:

- the nature of the injury or the illness;
- the likelihood of recovery or recurrence;
- the length of the absence;
- the need for the employer for the work done by the particular employee;
- the impact of the absence on others who work with the particular employee.

The employer must, therefore, strike a balance between the needs of the business and the needs of the employee and must be seen to have resolved the matter in a manner that a reasonable employer might have adopted. In doing this, the employer will have to demonstrate that it carried out an investigation to show that it is sufficiently informed of the employee's medical position. This entails consulting the employee and may require more detailed medical investigations with the employee's doctor or a relevant specialist.

Medical information

7.14 The employer would be required to establish from the employee or his or her doctor what the employee's true medical position is. This could entail requesting the employee to attend a medical examination. This requires the employee's consent because an employer cannot apply to a medical practitioner for a report without the employee's consent (*Access to Medical Reports Act 1988*).

Consultation with employee

7.15 It is critically important for the employer to consult with the employee before any decision regarding dismissal is reached. A failure to consult except in very exceptional circumstances will render the dismissal unfair (*East Lindsey District Council v Daubney*). The onus is on the employer to keep in touch with the employee whilst he or she is off work. The employee is under no equivalent obligation to provide information to the employer. What is required is a dialogue between employer and employee so that the situation can be assessed bearing in mind the employer's need for work to be done and for the need of the employee to recover from his or her injury or illness.

Having consulted with the employee it may also be necessary to consult with a doctor. The employee may refuse to be medically examined and he or she cannot be compelled to attend a doctor (unless there is an express provision in the employee's contract). If the employer has a contractual right to require the employee to be medically examined and the employee refuses to co-operate, this may justify a decision to dismiss the employee (on grounds of misconduct) and render the decision fair.

Alternatively, the employer may simply have to act on what information it has and form a conclusion without the benefit of a medical report. This can sometimes mean that dismissal due to injury or ill health will be fair even though had a medical examination been available it would have been unfair.

In most cases where an employer dismisses in accordance with medical advice, the dismissal will be fair.

Considering alternative work

7.16 In proving fairness, the employer would find it difficult to claim that it had acted reasonably if it took no steps to try and fit the employee into some other suitable available work. This would not require the employer to create a special job for the employee and each case will rely on its own facts and circumstances. It may well be that the employer may have available lighter work that the employee is capable of doing. If that is the case, the employer would be expected to at least encourage the employee to consider the chance of doing that work (*Merseyside and North Wales Electricity Board v Taylor*).

Dismissal is only one of a range of options open to the employer and a variety of factors have to be weighed up to determine whether the decision to dismiss was reasonable. These would include the nature of the injury or the illness, how it was caused, its likely duration, the nature of job, the need for the employer to have the work done and the length of service of the employee. The weight given to those factors differs from case to case. In some jobs, the need for robust health may be very strong. If the employer is a small business, the need to find a replacement may be great in circumstances where other employees cannot easily perform the work. The employer also has to consider the effect that the absence has on the other employees. Do they have to work overtime to make up for the absent employee's duties?

The duration of the absence is also very important. The employer must try to establish how long the employee is likely to be off work. If after a reasonable period of time the employee is still unable to say when he or she is likely to return, that fact will weigh heavily in the employer's favour. The employer would not be expected to wait indefinitely for the employee to return to work when his or her work has to be done and there is a need for a replacement (*McPhee v George H Wright Ltd*).

Sick pay and permanent health insurance schemes

7.17 Many employers offer contractual sick pay and/or permanent health insurance schemes. The injured employee may be off work and receiving sick pay or remuneration under the scheme. If those benefits are exhausted and the employee is still unable to return to work, employers often believe that it would be fair to dismiss the employee. The existence of the scheme or contractual sick pay would be one of the factors considered in evaluating the fairness of the dismissal. The fact that the employer had the scheme could be taken to mean that the employee envisaged the possibility of illness or injury related absences for at least the length of time for which the employee can recover under the scheme. This may be taken to have implicitly indicated that it is able to cater for such absences. However, an employer may also be held to have acted unfairly if it waits until the sick pay scheme has been exhausted before dismissing the employee (*Hardwick v Leeds Area Health*). (In this case the dismissed employee indicated that he would be returning to work one week after the sick pay scheme ended.)

An employee may receive benefits under a permanent health insurance scheme. If the employer dismisses the employee who is receiving those benefits, the dismissal may amount to a breach of contract, even if the appropriate notice was given (*Aspden v Webbs Poultry and Meat Group (Holdings) Ltd*). For example, the scheme may provide that if the employee is wholly incapacitated by sickness or injury from continuing to work the employee will receive a proportion of his or her salary until death or retirement provided that the employee continues to be employed by the employer. Even if the contract allows the employer to terminate based on prolonged absences there will be an implied term to the effect that unless there was gross misconduct justifying summary dismissal, the employer will not terminate the contract whilst the employee is incapacitated for work and receiving the insurance benefit.

Disability discrimination

7.18 The employee may be disabled in terms of the *Disability Discrimination Act 1995* (*DDA*). A person has a disability for the purpose of these rules if he or she has a physical or mental impairment, which has a substantial and long-term adverse impact on his or her ability to carry out normal day-to-day activities (*DDA, Section 1(1)*). The European Court of Justice has ruled that sickness is not the same thing as disability for these purposes. Merely being on sick leave and unavailable for work does not afford the protections given under the DDA (*Navas v Eurest Colectividades SA*). (However as we have seen dismissal for sickness could nevertheless be unfair if it for a fair reason and the correct procedure is not followed.)

Conversely, a dismissal may be fair for the purposes of the law relating to unfair dismissal but nevertheless unlawfully discriminatory for the purposes of the *DDA*. All employers contemplating dismissal must be careful to ensure that they do not render a fair dismissal unlawful in terms of the *DDA*. This is especially problematic as discrimination can occur even if the employer is unaware of the existence of a person's disability, and since 2004 the burden of proof has been on the employer to show its action was lawful once a complainant has established prima facie evidence of discrimination.

It is unlawful for the employer to discriminate against a disabled person in the field of employment including dismissal, recruitment or promotion (*DDA, Section 4*). This includes various categories of discrimination: "direct discrimination" (which is discrimination purely on grounds of a disability, usually where there is no consideration of individual's capability and a blanket ban is applied), "disability-related discrimination" (which is where a person is treated less favourably for a reason related to his or her disability), "failure to comply with the duty to make reasonable adjustment" and harassment. Most forms of discrimination by an employer cannot be justified at all. Disability-related discrimination can however potentially be lawful if it meets the test of a defence of justification. In order to justify the dismissal, the employer would have to prove that the reason was material to the particular circumstances of the case and substantial.

An example of less favourable treatment is where the employer dismisses the employee on the grounds of incapacity. Prima facie, this is unlawful discrimination because the employee has been dismissed because of his or her disability. Under these circumstances, the employer could seek to justify the dismissal by reference the employee's inability to perform the standard that it, the employer, requires. Alternatively, the employer might argue that having performed a risk assessment, if it allowed the employee to return to work, this would expose the employer to liability under *Health and Safety at Work Act 1974* (i.e. the duty of an employer to ensure safe working conditions).

These arguments may be sustainable in certain cases but it should always be remembered that the policy underlying the DDA is that the disabled person should be given every chance to continue working. The employer still has to consider making reasonable adjustments to the work or alternative employment for the disabled person. If the employer fails to do that and dismisses the employee, then this will negate the defence of justification and it will be found liable for unlawful disability discrimination.

If an employer is contemplating dismissing a disabled employee, it should pay close attention to the Code of Practice published by the Disability Rights Commission that accompanies the *DDA* (DRC, 2004). The Code of Practice does not have force of law but will be referred to by the employment tribunal as evidence of correct practice. If an employer can show that it has followed the relevant provisions of the Code of Practice, it will generally not be found liable for unlawful disability discrimination.

If the employee has a worsening progressive condition, it could be justifiable to terminate his or her employment if the increasing degree of adjustment necessary to accommodate the effects of the condition (e.g. shorter hours of work or falling productivity) became unreasonable for the employer to have to make.

Need for communication

7.19 Work colleagues need to be informed if the injured person is about to lose their job. The timing of this communication is important. The affected individual should know before anyone else, but as soon as it has been confirmed it will be helpful to let colleagues know. They may want the opportunity once again to discuss what has been implemented since the accident to prevent any recurrence. If any of them are developing or struggling with mental symptoms related to the accident, this news may worsen their symptoms and it is useful to be aware of this.

Saying goodbye

7.20 If the organisation has a culture of a formal or informal goodbye ceremony for employees who leave, then the employee should be asked what (if anything) they would like to take place to mark their leaving. This is often forgotten for people who leave through illness or injury and it can play an important role for them to mark the end of this stage in their lives, and to acknowledge that they are valued by their employer. Not everyone will want anything at all, but for some people it is very important. It also has an effect on their colleagues, who are likely to be pleased to have the chance to say goodbye and that their employer has treated their injured colleague with respect. It can be particularly difficult for the employer in this type of case, where the person is leaving because of injuries sustained in an accident at work. Nevertheless if the employer is able to provide this it is likely to be a far more positive than a negative event.

Communication with the workforce

7.21 As many of the previous sections have indicated, it is important to keep the workforce informed of what is happening. Their reactions to a serious accident at work may be unpredictable, and the managers and health and safety adviser need to bear this in mind. People's immediate safety will, of course, be the first priority. After that has been addressed people will continue to discuss the accident and judge their employer on how the aftermath is handled. If their views are sought, and subsequent actions make it clear that their views have been taken into account, they are likely to be better satisfied with their employer. There will also be close attention paid to the way any injured colleagues are treated. As the aphorism has it actions speak louder than words, and if people see their injured colleagues being, in their view, unfairly or unkindly treated following an injury at work, this is likely to adversely influence their opinion of their employer.

Compensation claims and the Rehabilitation Code

7.22 Rehabilitation issues have been high on the agenda for the government, insurers and the trades unions in attempting to improve the cost and efficiency of dealing with claims for compensation. Many claims can prolong injured employees' feelings of resentment and in some cases can have profound psychological effects which hinder the recovery process (IUA/ABI, 2004). It has been recognised for some time that waiting until claims had been settled or decided in court lost important opportunities to make timely treatment interventions, and increased the likely amount of time in returning to work (and the cost of claims for lost earnings).

In 1999 the main insurers associations and representatives agreed on a Code of Practice intended to promote rehabilitation, such as physiotherapy or counselling, sooner rather than later. The Code has been revised since and, while it is voluntary, it encourages a large measure of co-operation between claimants and defendants where there might otherwise be a conflict over the timing and cost of treatment (BICMA, 2002).

The Code requires claimants' lawyers and insurers to consider from the earliest opportunity whether the injured person would benefit from rehabilitation or other interventions, and to keep this issue under review. If necessary advice will be obtained from their doctor. If necessary there may be an independent assessment of needs paid fro by the insurer. Any funds made available for treatment by the insurer are then treated as a payment on account of damages.

There is also a Bodily Injury Claims Management Association guide to the Rehabilitation Code (BICMA, 2003) which contains more substance on the kinds of issues that the parties ought to consider. It is mainly focused on injuries at the more serious end of the spectrum which will require a significant period of treatment or which may result in permanent disability. It covers:

- *Case management*: Case managers are generally health care professionals acting as independent co-ordinators who can help with assessing the injured person's needs and organising access to various services (standards for case managers are set for its members by the Case Management Society of the UK (*www.cmsuk.org*).
- *Immediate needs assessment*: Applicable to more serious injuries when there may be a prolonged recovery period, the assessment is to determine the short-term support needed, for example care arrangements for after the injured person leaves hospital.
- *Emotional and psychological care*: The guide suggests that an assessment may be needed, not just to help the injured person come to terms with the what has happened but to identify psychological problems that might be triggered and which might prevent other treatments from being effective or which may require support for some time after the physical injuries have been treated.
- *Physiotherapy, osteopathy and chiropractic*: This will be based on a needs assessment, and the guide recommends that there should be a formal treatment plan covering the number of treatments and when they should be given.
- *Accommodation*: This mainly concerns adaptations to the home. Major alterations for the longer term will generally require specialist help and consideration of the availability of grants.
- *Nursing and care*: The guide envisages a detailed report being undertaken to identify all the various needs for assistance, equipment, etc.
- *Social services*: This concerns non-nursing care such as help with house, transport and carers. The availability some services may depend on the local authorities' funding policies.
- *Social security benefits*: The guide states the aim of maximising all rights to means-tested benefits, as well as other non-means-tested benefits such as IB.
- *Mobility*: This cover mobility in the home and driving, but access to public transport should also be taken into account.
- *Vocational*: The guide states that a detailed assessment should be carried out of the injured person's ability to resume their old job, if necessary by adaptations to the workplace, or the alternative options for new employment if this is not possible.

There are a number of practical problems to be overcome in making rehabilitation readily accessible for injured people with; these concern for example clarity of the tax position so it does not risk being treated as a benefit in kind, the overlap with NHS provision of services and when it is desirable to seek private alternatives, and the adequate provision of suitable case managers. However the concept of rehabilitation has now been firmly established in the litigation process by the inclusions of requirements in the personal injury pre-action protocol (see page 263). Any claimant or the defendant must consider as early as possible whether the claimant has reasonable needs that could be met by rehabilitation treatment or other measures and should specifically consider the Rehabilitation Code.

Conclusion

7.23 It is important to plan to manage the aftermath of an accident at the same time as doing everything possible to prevent accidents. Such planning should take into account people's psychological reaction to a serious accident which means the inclusion of plans for managing the psychological as well as physical problems that the injured people may develop and for ensuring ongoing communication with all the workforce. This communication will provide information about the injured as well as the findings and subsequent actions of the accident investigation.

Arrangements to assist any injured people to make a return to work as soon as they are able to do so, providing support and workplace adjustments to achieve this are the hallmark of good management.

Checklist for planning to return to work

7.24

- ✓ Obtain assessment of any adjustments needed at work – use occupational health personnel if there is access to them.
- ✓ Plan the rehabilitation back to work.
- ✓ Involve agencies who can provide help if they are needed.
- ✓ Consider involving family members especially partner if there is one.
- ✓ Consider what and how to communicate to the workforce.
- ✓ Ensure that any recommendations from the investigation following the accident have been implemented.
- ✓ Discuss with health and safety adviser and, if appropriate, the employee representative.

References

Aspden v Webbs Poultry and Meat Group (Holdings) Ltd [1996] IRLR 521.

BICMA (2002) *The Rehabilitation Code, Early Intervention and Medical Treatment in Personal Injury Claims.* The Bodily Injury Claims Management Association. www.bicma.org.uk/rehabilitation

BICMA (2003) *Rehabilitation – A Practitioner's Guide*, Second Edition. www.bicma.org.uk/Guide.doc

DRC (2004) *Code of Practice – Employment and Occupation.* TSO. www.drc-gb.org/PDF/employment_occupation.pdf

East Lindsey District Council v Daubney [1977] IRLR 181.

Hardwick v Leeds Area Health Authority [1975] IRLR 319.

Health, Work and Wellbeing – Caring for Our Future. A Strategy for the Health and Well-Being of Working Age People. HM Government, 2006. ISBN 0-84388-680-4.

HSE (2004) *Managing Sickness Absence and Return to Work: An Employers' and Managers' Guide (HS (G)249).* HSE Books.

IUA/ABI (2004) *Psychology, Personal Injury and Rehabilitation. The IUA/ABI Rehabilitation Working Party.* www.abi.org.uk/Display/File/364/Psychology_Personal_Injury_and_Rehabilitation_July_2004.pdf

Lynock v General Packaging Limited [1988] IRLR 510.

McPhee v George H Wright Ltd [1975] IRLR 132.

Merseyside and North Wales Electricity Board v Taylor [1975] IRLR 60.

Spiers, T. (2001) *Trauma – A Practitioners' Guide to Counselling.* Brunner-Routledge (Taylor and Francis Group). ISBN 0-415-18695-1.

Waddell and Burton (2006) *Is Work Good for Your Health and Wellbeing.* TSO (The Stationery Office). ISBN 0-11-703694-3.

8 Legal processes

In this chapter:

Introduction

8.1 The *Health and Safety at Work etc Act 1974* (*HSWA*) introduced a new approach to the regulation and control of health and safety at work, which is more flexible and prevention-orientated than had been seen before. It has given enforcing authorities wide-ranging powers of enforcement.

More recently, the civil justice system has been shaken up, with the aim of promoting settlements, reducing delays in getting compensation to accident victim, and making the system more cost effective.

This chapter provides:

- an overview of the enforcement processes under health and safety legislation including criminal procedure in England and Wales, and also Scotland;
- a summary of the penalties available to the courts for breaches of health and safety legislation;
- an outline of the ways in which civil claims for compensation may also be brought by injured parties;
- an overview of civil litigation procedure together with the main causes of action upon which claims are based.

Enforcement

8.2 The Health and Safety Executive (HSE) is the principal enforcement body for enforcing legislation in this area. Responsibility for enforcement procedures and prosecutions is shared between the HSE and local authority environmental health officers (EHOs).

Primary responsibility lies with the HSE, which in broad terms has enforcement powers in relation to most industrial premises. The local authorities have responsibility for lower risk situations, principally offices, shops and other commercial premises. (The main exception to the general rules set out above is that the HSE cannot enforce provisions in respect of its own premises and, similarly, the local authorities' premises are inspected by the HSE.) *The Health and Safety (Enforcing Authorities) Regulations 1998* set out the activities for which the HSE and EHOs are responsible (see Table 8.1 below). Enforcement can be transferred (though not

Table 8.1: Situations in which the HSE and local authority are the enforcing authority

HSE is the enforcing authority	*Local authority is the enforcing authority*
• Airports	• Office activities
• Hospitals	• Catering services
• The Channel Tunnel system	• Consumer services in a shop (except dry cleaning, radio/television repairs)
• Offshore installations	• Provision of permanent or temporary residential accommodation, including hotels and caravan sites or camps
• Construction/building site	• Site/storage of goods for retail/wholesale distribution (with some exceptions)
• Schools, colleges, universities, etc.	• Display/demonstration of goods at an exhibition
• Factory premises	• Laundrettes
• Fairground activity	• Leisure facilities
• Any activity in a mine or quarry	• Practice/presentation of arts, sports, games, entertainment or other cultural/recreational activity, except where the main activity is the exhibition of a case to the public

(*Continued*)

Table 8.1: (Continued)

HSE is the enforcing authority	*Local authority is the enforcing authority*
• Broadcasting, recording, filming or video recording and activities in premises occupied by a radio, television or film undertaking	• Care, treatment, accommodation or exhibition of animals, birds or other creatures except in veterinary surgeries, breeding/training at stables or agricultural activity
• Certain works carried out by independent contractors, pertaining to gas, electricity and ionising radiation	• Undertaking, except coffin making
• Use of ionising radiation for medical exposure	• Church worship /religious meetings
• Any activity in radiography premises where work with ionising radiation is carried out	• Car parking facilities within an airport
• Agricultural activities	• Childcare, playgroup, or nursing facilities
• Activity on board a sea-going ship	
• Ski-slope, ski-lift, ski-tow or cable car activities	
• Fish, maggot and game breeding (but not in a zoo)	
• Railway operations	
• Any activity in relation to a pipeline	
• Premises of county councils, local authorities, parish or community councils	
• International HQs and defence organisations visiting forces	
• United Kingdom Atomic Energy Authority	
• Common parts of railway stations/termini/goods yards	
• Manufacturers, suppliers, importers and designers of articles for use at work or any article of fairground equipment, as set out in HSWA 1974, s 6	

in the case of crown premises) by prior agreement from the HSE to the local authority and vice versa.

Selecting incidents for investigation

8.3 The HSE has published criteria for the priorities it will give to most incidents which come to its attention through reports made under the *Reporting of Injuries, Diseases and Dangerous Occurrences Regulations 1995* (*RIDDOR*) (HSE, 2005). Inevitably this involves being selective and some incidents will not be followed up which has led to some criticism, but enforcing authorities have to operate within the limits of their resources.

Briefly, the incidents certain to trigger investigation by inspectors are:

- accidental deaths;
- "major injuries" (see page 144);
- any occupational disease;
- any serious breach of health and safety law which would under the Enforcement Management Model (EMM) (see below) warrant an enforcement notice;
- incidents which concern matters that are included in the Health and Safety Commission's (HSC's) strategic priority action areas, for example falls from height;
- incidents likely to give rise to serious public concern, for example accidents involving children or "near misses" with potential for death or serious injury;
- incidents which could be precursors to major incidents (e.g. chemical leaks).

These are merely guidelines and are not intended to rule out investigations of other situations where there have been significant control measure failures.

Enforcement decisions

8.4 Enforcing authorities are not immune to legal challenges if their approach to bringing legal action is unfair or lacks inconsistency. To address these issues various policies and procedures have been formalised and published. Details of the HSE enforcement approach can be found at *www.hse.gov.uk/enforce*.

The enforcement policy statement

8.5 The HSC, which oversees the enforcement work of the HSE, has issued an Enforcement Policy Statement setting out the general principles that it expects enforcing authority inspectors to comply with (HSE, 2004).

The Enforcement Policy Statement is intended to reflect the HSC's primary aims to be open and transparent, to ensure that a consistent approach is taken in similar circumstances to achieve similar results and to balance enforcement action so that it is proportionate to the risks involved. The policy provides clear guidelines for enforcing authorities to help to achieve and promote increased compliance with health and safety legislation. It also provides guidance on the following:

- Purpose and method of enforcement.
- Principles of enforcement:
  - proportionality,
  - targeting,
  - consistency,
  - transparency,
  - accountability.
- Investigation.
- Prosecution.
- Death at work.

The rationale underlying enforcement action is described in the policy statement as follows:

> "The ultimate purpose of the enforcing authorities is to ensure that duty holders manage and control risks effectively, thus preventing harm. The term enforcement has a wide meaning and applies to all dealings between enforcing authorities and those on whom the law places duties (employers, the self-employed, employees and others).
>
> The purpose of enforcement is to:
>
> - ensure that duty holders take action to deal immediately with serious risks;
> - promote and achieve sustained compliance with the law;
> - ensure that duty holders who breach health and safety requirements, and directors or managers who fail in their responsibilities, may be held to account, which may include bringing alleged offenders' before the courts in England and Wales, or recommending prosecution in Scotland, in the circumstances set out later in this policy.
>
> Enforcement is distinct from civil claims for compensation and is not undertaken in all circumstances where civil claims may be pursued, nor to assist such claims."

The circumstances when a decision to prosecute or recommend prosecution would normally be viewed as being in the public interest and legal action would follow are described in the HSC's policy as:

- death was a result of a breach of legislation;
- the gravity of an alleged offence, taken together with the seriousness of any actual or potential harm, or general record and approach of the offender warrants it;
- there has been reckless disregard of health and safety;
- there have been repeated breaches giving rise to significant risk, or persistence and significant poor compliance;
- serious non-compliance with an appropriate licence or safety case;
- the standard of managing health and safety is found to be far below what is required by health and safety law and to be giving rise to a significant risk;
- failure to comply with an improvement notice or prohibition notice; or repetition of a breach that was subject to a formal caution;
- false information has been supplied wilfully, or there has been an intent to deceive, in relation to a matter which gives rise to significant risk;
- inspectors have been intentionally obstructed in the lawful course of their duties, or they have been assaulted.

There are also some other wider criteria which the Enforcement Policy Statement includes as justifications for legal action:

- enforcement action is appropriate as a way to draw general attention to the need for compliance with standards required by law, and conviction may deter others from similar failures to comply with the law;
- there has been a breach which gives rise to significant risk has continued despite relevant warnings from employees, or their representatives, or from others affected by a work activity.

The Enforcement Policy Statement also requires taking into account the criteria stipulated in the Code for Crown Prosecutors (CPS, 2004) when deciding whether or not to prosecute. In essence, this takes into consideration two questions: (1) is the evidence strong enough for there to be a likelihood of securing a conviction and (2) is a prosecution in the public interest? (This is why the wording of the Enforcement Policy Statement is framed with reference to the "public interest.") One important consideration is that proceedings should not normally be brought if the court would be likely to impose only a minimal fine. Most of the other quoted public interest factors are not relevant to health and safety offences except as regards the issue of delays in bringing proceedings (see 8.14). Sometimes though considerations concerning the personal circumstances of an individual as opposed to corporate defendant do arise. These can include the fact a defendant is elderly or is, or was at the time of the offence, suffering from significant mental or physical ill health, unless the offence is serious or there is real possibility that it may be repeated.

Personal prosecutions

8.6 The HSC policy states that enforcing authorities should identify and prosecute or recommend prosecution of individuals if they consider that a prosecution is warranted. In particular, it suggests they should consider the management chain and the role played by individual directors and managers and should take action against them where the inspection or investigation reveals that the offence was committed with their consent or connivance or to have been attributable to neglect on their part.

The HSE Enforcement Management Model

8.7 The Enforcement Policy Statement serves as a statement of fundamental principles, but it does not provide guidance on enforcement (and at what level) in any particular case. Complaints are often heard that enforcement in practice is inconsistent, haphazard and that too few legal cases are pursued. To meet these (not always fair) criticisms much work has been done to increase openness about the enforcement process. In fact the HSE's decision-making criteria are now published a great deal more than those of other agencies.

The EMM (HSE, 2002) seeks to provide inspectors with more description of more detailed criteria upon which inspectors make enforcement decisions. It involves a structured consideration of facts of individual cases including the effectiveness of compliance arrangements, the "risk gap" (i.e. the degree to which short of what was required in controlling the level of risk), the seriousness of potential injury and strategic enforcement factors (as indicated in the HSE Enforcement Policy).

The EMM needs to be treated with a degree of caution. It is not a mechanical exercise that will automatically produce answers to whether or not action should be taken. Rather, it is a framework which is meant to guide inspectors in reaching consistent and fair decisions, and it allows for a considerable amount of flexibility in applying the principles it sets out. "*The process of making enforcement decisions is complex. Each duty holder is unique, and inspectors must have a thorough understanding of the hazards and control measure associated with each duty holder's activities. It is vital that inspectors have wide discretion to exercise their professional judgement so that action appropriate to each situation can be taken*" (HSE, 2002).

An organisation which is under investigation normally has no direct input or involvement into the way the considerations contained in the EMM are taken into account. It may sometimes however be helpful for organisations under investigation to draw inspectors' attention at the later stages of enquiries – in particular at a *Police and Criminal Evidence Act* (*PACE*) interview (see page 180) – to any positive aspects contained in a set of criteria which are referred to in the EMM as "duty

Table 8.2: Duty holder factors

Descriptor	*Definition*
Does the duty holder have a history of relevant, written enforcement being taken against them?	
Yes	Enforcement action has been taken against the duty holder on the same or similar issues, by notices, prosecutions or letter requiring action.
No	No written enforcement action against the duty holder on the same or similar issues.
Does the duty holder have a history of relevant verbal enforcement being given to them?	
Yes	Enforcement action has been taken against the duty holder on the same or similar issues, by verbally telling them what they have to do in order to comply with thelaw.
No	The duty holder has not been told previously what they have to do in order to comply with the law on the same or similar issues.
Is there a relevant history?	
Yes	The duty holder has a history of related incidents, or that there is evidence of related incidents, for example accidents, cases of ill health, dangerous occurrences.
No	No previous history or evidence of related accidents, ill health or dangerous occurrences.
What is the intention of the duty holder in non-compliance?	
Deliberate economic advantage sought	The duty holder is deliberately avoiding minimum legal requirements for commercial gain. (For example failing to price for or provide scaffolding for high roof work.) advantage sought
No economic	Failure to comply is not commercially motivated.
What is the level of actual harm?	
Serious	A "serious personal injury" or "serious health effect" has occurred as a result of the matter under consideration.

(*Continued*)

Table 8.2: (Continued)

Descriptor	*Definition*
Not serious	There has been no actual harm, or the harm has been no greater than "significant personal injury" or a "significant health effect".
What is the standard of general conditions?	
Poor	There is a general failure of compliance across a range of issues, including those matters related to the activity being considered through the EMM. For example, failure to address risks arising from hazardous substances, machinery, transport, vibration, noise, etc. or inadequate welfare facilities.
Reasonable	The majority of issues are adequately addresses, with only minor omissions.
Good general compliance	Full compliance across the whole range of indicators with no notable omissions.
What is the inspection history of the duty holder?	
Poor	The duty holder has an inspection history of significant problems, copious advice and poor inspection ratings.
Reasonable	The duty holder has an inspection history of nominal or piecemeal problems, where non-compliance has been related to new or obscure duties and where the rating history is in the average range.
Good	The duty holder has an inspection history of good compliance, effective response to advice, consistently high standards and a low rating.
What is the attitude of the duty holder?	
Hostile/indifferent	The duty holder is actively antagonistic or completely interested in health and safety issues. Impossible to establish an effective relationship.
Reasonable	The duty holder is open to discussion and reasoned persuasion and effective communications can be established.
Positive	The duty holder is enthusiastic and proactive towards health and safety issues, actively seeking advice and pursuing solutions.

holder factors". These are factors (positive and negative) to be taken into account which reflect the organisation's blameworthiness or its disregard of health and safety principles – they are set out in Table 8.2. They will not in themselves determine the findings of the analysis under the EMM and the enforcement outcome, but in some circumstances they may be enough to indicate that the level of expected enforcement can move up a level (e.g. just Improvement Notice, as opposed to an Improvement Notice and a prosecution, or vice versa).

Offences and penalties

Principal offences

8.8 Failure to comply with the general duties imposed by the *HSWA, Sections 2–7* is a criminal offence under *HSWA, Section 33*. Breaches of health and safety regulations are deemed to be offences in the same way. In nearly all instances the company or employer or employing organisation will be the main defendant in a prosecution.

Where a person has been injured in an accident it is nearly always the case that the defendant will be accused of one of the main offences under the *HSWA*, usually under *Section 2* for failure of the duty to ensure the safety of employees or the *Section 3* duty not to expose others to risks. In some cases there may also be offences alleged in relation to compliance with more specific obligations such as those under the *Management of Health and Safety at Work Regulation* (*MHSWR*) or other regulations which applied to the activities being carried out at the time of the accident. In exceptional circumstances an employee can be charged alone as a principal offender. The offence would normally be that of a breach of *HSWA, Section* 7 which requires all employees to take reasonable care for the health and safety of themselves and others affected by their activities. (In very rare cases a different offence may arise under *HSWA, Section 8* for deliberate or reckless misuse of safety equipment.)

A prosecution of employees is not common and is only likely to arise where the employer has done everything reasonably practicable to control the risks in question, and where an employee has wilfully or recklessly disregarded the safe systems of work he or she was properly trained to follow.

Secondary offenders

8.9 Prosecutions may also be brought against individuals or others in addition to or instead of a prosecution against the main offender.

Offences by directors and officers (HSWA, Section 37)

8.10 The *HSWA* also makes special provisions for senior managers where offences are committed by companies and other corporations.

HSWA, Section 37 states:

> "Where an offence under any of the relevant statutory provisions committed by a body corporate is proved to have been committed with the consent or connivance of, or to have been attributable to any neglect on the part of, any director, manager, secretary or other similar officer of the body corporate or a person who was purporting to act in any such capacity, he as well as the body corporate shall be guilty of that offence and shall be liable to be proceeded against and punished accordingly."

Most cases brought under this section allege that the defendant was guilty of "neglect", for example in not ensuring that adequate assessments of risks were carried out and that safety systems of work were put in place. It has been held that this refers to proof of a person's "negligence" in the ordinary sense of carelessness amounting to breach of a duty of care. The nature of the duty will be delineated by various factors, including the individual's executive responsibilities, what is contained in his/her employment contract and the duties allocated under the organisation's health and safety policy statement (*Armour v Skeen*). Also relevant will be the HSC guidance on directors' responsibilities (see page 60) and any more detailed codes and guidance the HSE might produce for directors and other senior executives.

Whether a person is of sufficient seniority for the purposes of this section will depend largely on the management of the organisation and how it is run. In particular, the degree of control exercised in any given case will be taken into account. In the case of *R v Boal* brought under the fire legislation which contained analogous provisions to *HSWA 1974, Section 37*, it was held that the offence was intended to deal with "*only those who were in a real position of authority, the decision makers within the company, who had both the power and responsibility to decide corporate policy and strategy*". It is therefore unlikely that managers would fall within the ambit of *HSWA, Section 37*, unless they were also in a position to make corporate decisions. In this case, it was held that a manager of a shop in charge while the general manager was on holiday was not sufficiently senior for these purposes.

Powers also exist under the *Directors Disqualification Act 1986* to bar a person convicted of an offence from holding directorships.

Where companies are managed by their members instead of directors (or where it is shown that shareholders exercise control over the business) are also be subject to another part of *HSWA, Section 37*. Liability under this provision could extend to a parent company if it exercises the requisite degree of control and there has been some neglect, consent or connivance in relation to the offence by its subsidiary.

*Act or default of another (*HSWA, Section 36*)*

8.11 If an offence is committed by one person due to an act or default by some other person that person shall be guilty of the offence regardless of whether any action is taken against the other person. Similar provisions apply when an offence is committed by the Crown due to an act or default of another person. Although Crown immunity prevents legal proceedings being instigated against the crown, the provisions described above permit prosecution of a Crown employee where he has acted in such a way which results in the commission of an offence by the Crown. Crown employees can also be prosecuted for failing to discharge their own employee duties under *HSWA*, *Section* 7.

Enforcement in practice

8.12 The statistics published by the HSE for 2005/2006 (HSE, 2006) indicated that there has been steady decrease in the prosecutions brought by the HSE: since 1999/2000 the number of offences prosecuted dropped from around 2000 per year to about half that number in 2005/2006. The rate of successful prosecutions dropped to 73 per cent. In contrast the numbers of local authority prosecutions, having peaked at just over 500 in 1997/1998, dropped by 2003/2004 to just 332, but with an 85 per cent conviction rate. See Figure 8.1 for the number of offences prosecuted and the number of convictions secured since 1990.

The headline levels of average fines per "offence prosecuted", on the other hand, have increased very significantly to nearly £30,000 as can be seen in Figure 8.2 which tracks changes in the cases taken by the HSE. (These cases tend to be more serious than those brought by local authorities, whose average fine per offence in 2004/2005 was £5,899.) The HSE figures are affected by the inclusion of 13 cases where the fines exceeded £100,000, and without these the 2005/2006 average would have been only £6,219.

In 2003/2004, the HSE prosecuted thirty six individuals, of whom thirty were convicted. Of the cases, ten were brought against directors, with nine convictions.

Figure 8.1: Offences and convictions since 1990 – HSE

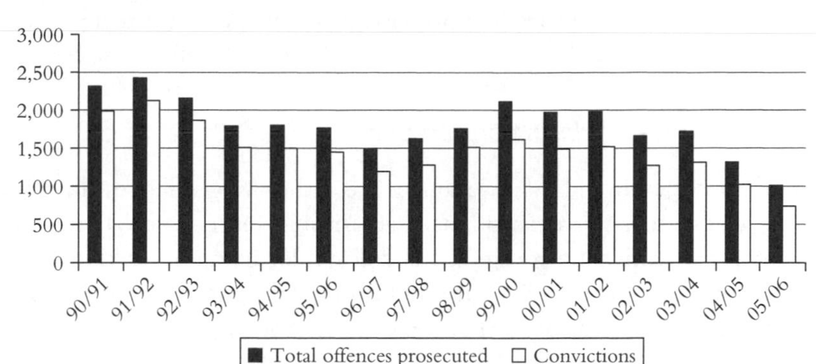

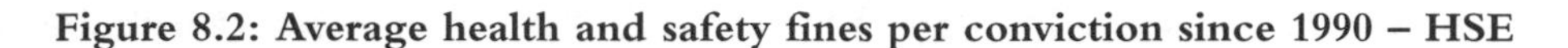

Figure 8.2: Average health and safety fines per conviction since 1990 – HSE

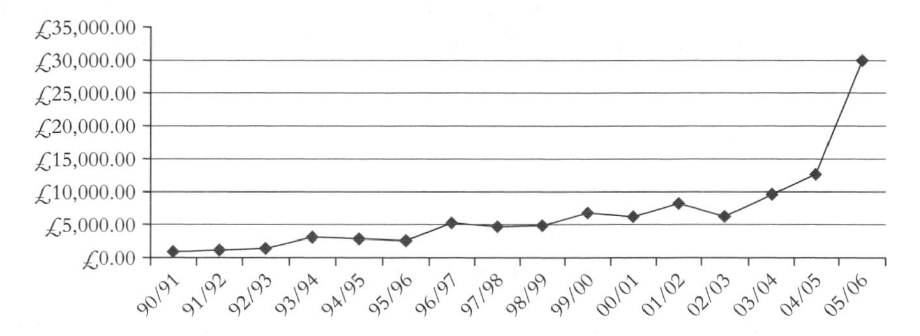

As with the more common prosecutions of companies and other organisations these numbers represent fewer proceedings being brought than in previous years.

Average individual fines were £4,222 for directors and £899 for other employees. In the 2002 case of Austin Brickworks (a fatality relating to the use of a fork-lift truck by an unqualified driver), the director received an exceptional fine of £20,000.

The prosecution process

8.13 Health and safety prosecutions are dealt with in the mainstream criminal justice system. This contrasts with many other jurisdictions in Europe and elsewhere in which the enforcing authorities can impose administrative penalties.

The procedural aspects of prosecutions are described in more detail below first with reference to England and Wales, and then to Scotland which has separate courts and processes, but first it is useful to consider some common questions which anyone connected with proceedings (senior managers, witnesses and accident victims) often ask.

Is there a time limit for bringing prosecutions?

8.14 For most health and safety offences are either way offences the answer is no. Nearly all health and safety offences are triable either way. Time limits of 6 months or a year, which are commonly laid down for many other kinds of regulatory prosecutions, do not apply here. It is not unusual for enforcement authorities to instigate a prosecution a year or so after the offence is actually commissioned.

The issue of delay can be a public interest factor against bringing a prosecution but the circumstances described for this in the Code for Crown prosecutors are quite circumscribed. The Code indicates that a prosecution would not normally be in the public interest where there has been a long delay between the offence taking place and the date of the trial, unless:

- the offence is serious;
- the delay has been caused in part by the defendant;
- the offence has only recently come to light;
- the complexity of the offence has meant that there has been a long investigation.

The only legal arguments open to the accused, when it is alleged that such a delay results in potential unfairness, are under the common law of abuse of process or under human rights legislation. There are no firm guidelines in this respect and each case will be determined on its own facts.

How will I know if there is going to be a prosecution?

8.15 This will not be clear until the latter stages of an investigation and (usually) not until interviews under caution have been completed and a senior inspector has reviewed the file. The investigating inspector will usually write to state that a decision to prosecute has been taken. After this the formal steps to commence proceedings will begin.

In England and Wales, health and safety prosecutions are initiated by a prosecutor laying an "information" before a magistrate (or magistrate's clerk). The prosecutor is usually an HSE inspector or an inspector of a local health authority who, for these purposes, has the same right of audience (i.e. entitlement to appear in court as an advocate) as solicitors and barristers do in the magistrates' courts (*HSWA, Section 39*). Increasingly inspectors do not conduct cases themselves however, and they will usually engage solicitors (who, in turn, may appoint a barrister) to conduct the prosecution. Analogous provisions for instigating proceedings apply in Scotland whereby the procurator fiscal issues a "complaint" under either summary procedure or solemn procedure (see below).

The information is where the details of the alleged offence are set out formally. If the information is correct in detail, the court will authorise the issue of a summons requiring the accused to attend before the court to hear the charges made against him and to enter a plea. The summons is then usually served by post by the prosecution, at the organisations registered office or the home address of an individual defendant.

Is the defendant presumed innocent until proven guilty?

8.16 Generally in most criminal cases the prosecutor is required to prove beyond all reasonable doubt that the defendant is guilty of the offence with which he is charged. However, *HSWA, Section 40* modifies this rule for many health and safety offences:

> "In any proceedings for an offence under any of the relevant statutory provisions consisting of a failure to comply with a duty or requirement to do something so far as is practicable or so far as is reasonably practicable, or to use the best means to do something, it shall be for the accused to prove (as the case may be) that it was not practicable or not reasonably practicable to do more than was in fact done to satisfy the duty or requirement, or that there was no better practicable means than was in fact used to satisfy the duty or requirement."

This is referred to as the "reverse burden of proof". There have been several cases in the UK courts where the question of whether this is consistent with the defendant's right to a fair trial under has been considered. The Court of Appeal decision in the case of *Davies v Health and Safety Executive* has settled for the time being conflicting decisions at crown court level as to whether the legal burden of proof requirements of *HSWA, Section 40* is compatible with *Article 6(2)* of the *European Convention on Human Rights*. This case concerned the death of a subcontractor employed by Mr Davies at his plant hire firm. The defence argued that Mr Davies had done all that was reasonably practicable. Mr Davies appealed against his conviction on the point of burden of proof. The Court of Appeal held that the imposition of a legal burden of proof in *HSWA, Section 40* was justified, necessary and proportionate for the following reasons:

- The *HSWA* is regulatory and its purpose is to protect the health and safety of those affected by the activities referred to in *HSWA, Sections 2–6*.

- The prosecution must first prove that the defendant owes a duty and that the safety standard has been breached. Once these matters have been proved the defence of reasonable practicality has to be raised and established by the defendant. Furthermore, the defence itself is flexible, as it does not restrict the way in which the defendant can show that he has done what is reasonably practicable.
- The reverse burden of proof takes into account the fact that duty holders under the Act have chosen to engage and be in charge of a commercial activity or work which is subject to regulatory controls. It is therefore justifiable to ask the duty holder to show that it was not reasonably practicable for him to have done more than he had in fact done to avoid or prevent a risk.
- The facts relied upon by the defence should not be difficult to prove as they are within the knowledge of the defendant. In complex cases it may well only be the defendant that has the relevant expertise to assume this burden and therefore enforcement in these types of cases might become impossible if the defendant had only an evidential burden.
- The defendant does not face imprisonment and therefore the consequences of a conviction might be newsworthy but are not the same as that involved "in truly criminal offences".

Thus, the current state of the law on this issue is that the interpretation of the reverse burden of proof provisions in *HSWA, Section 40* is compatible with *Article 6(2)*. Consequently all the prosecution needs to prove in most cases is the existence of a risk. The issue prosecution's proof of risk was considered by the Court of Appeal in *R v Board of Trustees of the Science Museum*, a case concerning *HSWA, Section 3* and the duties of employers to ensure that persons who not in their employment are not exposed to risks. It was held here that a risk is to be taken as conveying the idea of a "*possibility of danger*" not an actual danger. (In this case the risk in question was Legionnaires' disease that might be contracted from air conditioning cooling towers, and the risk was held to exist irrespective of there being no evidence of inhalation by persons nearby or even the presence of legionella bacteria in the equipment when it was operated.)

Can we see the prosecution evidence?

8.17 On commencing proceedings for health and safety offences, the prosecution is generally required to advise the defendant of the right to be supplied with certain information. This is known as "advance information" which the accused is allowed to see to make an informed decision as to whether or not to elect trial on indictment. The rules pertaining to advance information are found in the Criminal Procedure Rules 2005. It is important to appreciate the difference between advance information and fuller prosecution disclosure (see 8.23) which may occur at a later stage. Advance information comprises the evidence the prosecution intend to rely upon to prove the offence for which the accused is charged. The rules on disclosure relate to information the prosecution has gathered in the course of investigation and does not intend to rely upon, either because it is inconsequential, or because it might undermine the prosecution case or assist the defence. The prosecutor (usually the lead inspector) can choose whether to supply the defendant (in response to a request for advance information) either with copies of witness statements and other documents relied upon in evidence, or the prosecutor's own written summary of the statements and other evidence. Usually, it will be the former.

What court will deal with the case?

8.18 Normally the court hearing the case will be the closest geographically to where the offence occurred. The real issue though is what *level* of court will deal with the case. This question features prominently at the early stages of any case. It may have a significant effect in terms of the level of the sentencing powers available.

Offences are classified as either indictable offences or summary offences. Summary offences are tried in a magistrates' court before either a bench, normally comprising three lay magistrates, or a single district judge (who is a qualified lawyer). Indictable offences (an example is the offence of manslaughter) are only tried in a crown court before a judge and jury. However, some offences, and in fact nearly all health and safety offences, are classified as being "either way" offences, meaning they may be tried in either the magistrates' court or the crown court. There are no indictable only offences under the *HSWA*.

An either way offence can be dealt in a crown court if either (a) the defendant elects trial by jury, (b) the magistrates choose to send the case to be tried or sentenced in the crown court. Magistrates will transfer a case to the crown court if they believe the case concerns a matter of particular importance or complexity or if the magistrates feel their sentencing powers are not sufficient to provide an adequate penalty for the severity of the offence committed.

When will the case be over?

8.19 This can range from around a month to a year or more. It depends on a number of factors, the most significant being whether the defendant to plead guilty or not, and the gravity of the offence. The reason for this is that the procedural paths can vary (see 8.22) and do not run to any strict pre-determined timetable.

When hearing dates are set these may be purely for case management hearings and the full trial may be some way off still. The prosecution or defence may request adjournments, for example to obtain evidence, and the court will try to accommodate the availability of the parties' lawyers when fixing hearing dates.

What will the penalty be?

8.20 The penalties for offences under these sections will depend upon which court has jurisdiction over the case. Very few offences are subject to imprisonment (the main exceptions being contravention of improvement or prohibition notices) and imprisonment does not apply where the defendant is a company or other organisation.

There are sentencing guidelines laid down by the Court of Appeal for cases in England and Wales (see 8.27). These are however broad statements of principle and do not set a "sentencing tariff" of set amounts for specific types of offences. The judicial policy has been that each case must be looked at in the light of the specific circumstances. This results in widely difficult fines being imposed by different courts across the country. So far the Court of Appeal has declined to impose any principle that there should be consistency, and in fact court have been slightly discouraged from looking at previous cases and treating the fines set there as precedents or guidelines (*R v Yorkshire Sheeting & Insulation Ltd.*).

What involvement in the case do accident victims have?

8.21 In the formal proceedings the role of the victim is generally limited to being a witness as to the facts of his or her accident. The victim is not a party to the criminal case and is not legally represented.

The criminal justice system has been changing in recent years however to give the victim's interests more prominence. Enforcing authorities will normally try to keep the victim (or their family if there has been a death) aware of the process and up and coming hearing dates so they can attend and observe the trial or sentencing of the defendant. There is also a practice of offering victims or relatives the opportunity to make a written "victim statement" to be given to the sentencing court if they wish.

Another important consideration is the powers the courts have been given to make compensation orders against convicted defendants. These involve a payment being awarded on top of the fine, which goes the victims in recognition of their injuries and losses.

Procedure

8.22 When a summons is issued it will state the date of the first hearing, normally four to eight weeks ahead. It is not uncommon for this date to be adjourned while the defence considers the advance information. In straightforward cases where the defendant pleads guilty though the case is sometimes dealt with at this first hearing with the fine being imposed there and then. As can be seen from Figure 8.3 there is a variety of other possible paths which a case can take, including a transfer to the Crown Court for more serious offences or where the defendant chooses to be tried by a jury.

Initial procedures are mainly for deciding whether an offence is dealt with in the Magistrates' Court or the Crown Court, is known as the "plea before venue". If the accused pleads guilty the court will move on to sentencing and, as stated above, either the magistrates will impose a sentence or the case will be transferred to the Crown Court for sentencing.

If the accused is convicted in the magistrates' court or if the accused pleads guilty in the Magistrates' Court and, the magistrates consider their sentencing powers are not sufficient, then the magistrates have powers to commit the case for sentence to the crown court. The Crown Court has much greater sentencing powers than the Magistrate Court (see 8.26).

If the accused pleads not guilty in the Magistrates' Court the court will proceed to consider the "mode of trial" as follows:

- The prosecution followed by the defence make representations to the magistrates as to whether the matter should be tried summarily or whether trial on indictment is more suitable.
- The magistrates consider the representations made by both parties, the nature and seriousness of the offence and their sentencing powers and decide which court would be the most appropriate for the case to be tried.
- If the magistrates decide that the matter could be dealt with summarily then they will inform the accused that they can try him, if he consents. The accused is told that if the accused is convicted in the Magistrates' Court, then the magistrates may still send the matter to the Crown Court for sentencing if the magistrates feel their sentencing powers are insufficient. The magistrates ask the accused which court he elects to be tried. If the accused accepts the magistrates' decision the case remains to be tried in the Magistrates' Court. If, however, the accused elects for trial by jury the case will automatically be transferred to the Crown Court via committal proceedings and neither the magistrates nor the prosecution can object.

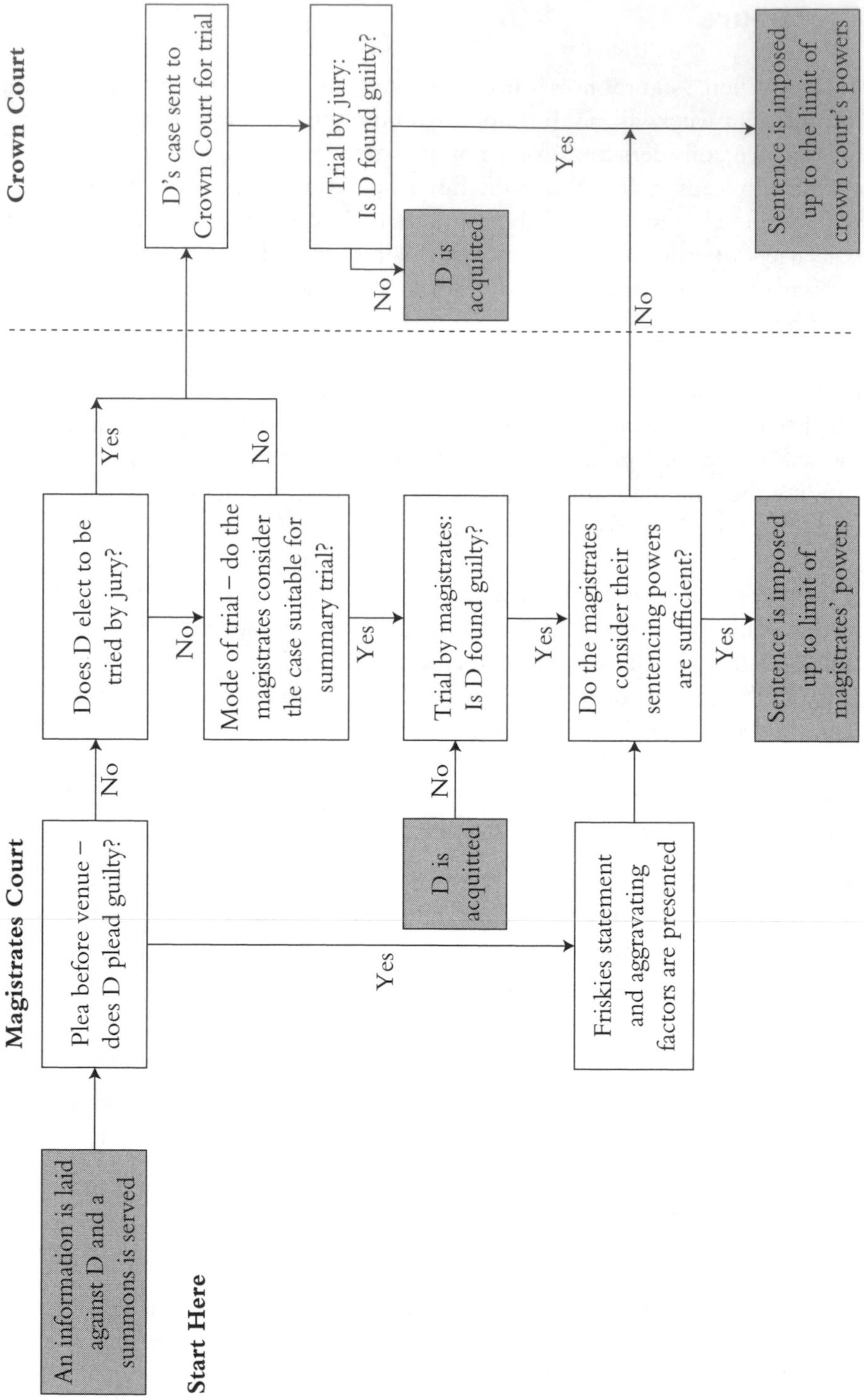

Figure 8.3: Procedures for health and safety prosecution

- If the magistrates choose that the case is one which should be tried in the Crown Court then the accused is informed and the case is sent to the Crown Court. In this case neither the accused nor the prosecution can object.

Disclosure of prosecution evidence

8.23 The duty of disclosure is a key element of fairness in criminal trials in England and Wales – it is believed that the defendant is entitled to know all the evidence, not just the prosecution rely on, but anything else which the prosecution have that is helpful to the defence but it only arises in cases where a not guilty plea is entered by the accused. (In contrast, in guilty plea cases the defendant is only entitled to receive evidence the prosecution regards as helpful and which it does rely on.)

These disclosure rules are now set out in statutes (mainly in *Criminal Procedure and Investigations Act 1996*) but also some important amendments to this Act and associated regulations and Code of Practice. This Act also sets out various other requirements to retain certain types of material (e.g. interview records) as well as material that might cast doubt on the prosecution evidence (e.g. draft witness statements different to final versions). The prosecution will be required to disclose to the defendant anything not already provided which either might be (a) capable of undermining the prosecution case, or (b) assisting the defence case. (If there is believed to be nothing meeting these criteria the prosecution has to positively state this.)

This step is known as primary prosecution disclosure. In cases being tried in the Magistrates' Court the defendant then has the option to serve a "defence statement" on the prosecution setting out its case. In Crown Court cases the defence statement is mandatory. This process is slightly controversial as it is an inroad into the principle of the defendant only having to meet the prosecution case and not to establish its innocence, but in health and safety cases it must be remembered that defendant usually has the reverse burden of proof to demonstrate it took reasonable practicable precautions anyway; consequently the defence statement enables the HSE or other prosecuting authority to consider what evidence it may need to rebut the defendant's arguments.

Finally, if the defence has provided the defence statement the prosecution must then provide further disclosure comprising any additional undisclosed material which might reasonably be expected to aid the defence case.

Expert evidence

8.24 Parties to summary trial and trial on indictment are required to give the court and the other parties advance notice of the expert evidence they intend to adduce. Parties are also entitled to request copies of any tests or other analysis relied on by the expert. This is to allow time to consider the evidence (and obtain expert evidence in response if required) but above all to prevent "surprises" at the trial which might otherwise lead to delays which the other party has to deal with the expert opinion at short notice or otherwise seek an adjournment.

Summary trial

8.25 Where an accused decides to enter a not guilty plea and the Magistrates' Court agrees to accept jurisdiction over the case, the case will proceed to trial. In a summary trial the magistrates are under a duty to hear the evidence in a case and to reach one of only two verdicts – i.e. whether the accused is guilty or not guilty. Where an accused fails to appear at the time set for trial the magistrates can try the accused in his absence. In cases where the accused fails to attend for trial, the accused is deemed to have pleaded not guilty to the offence(s) with which he is charged and the prosecution are permitted to plead their case on the evidence. The process of trial is outlined in Figure 8.4.

Figure 8.4: Summary trial process

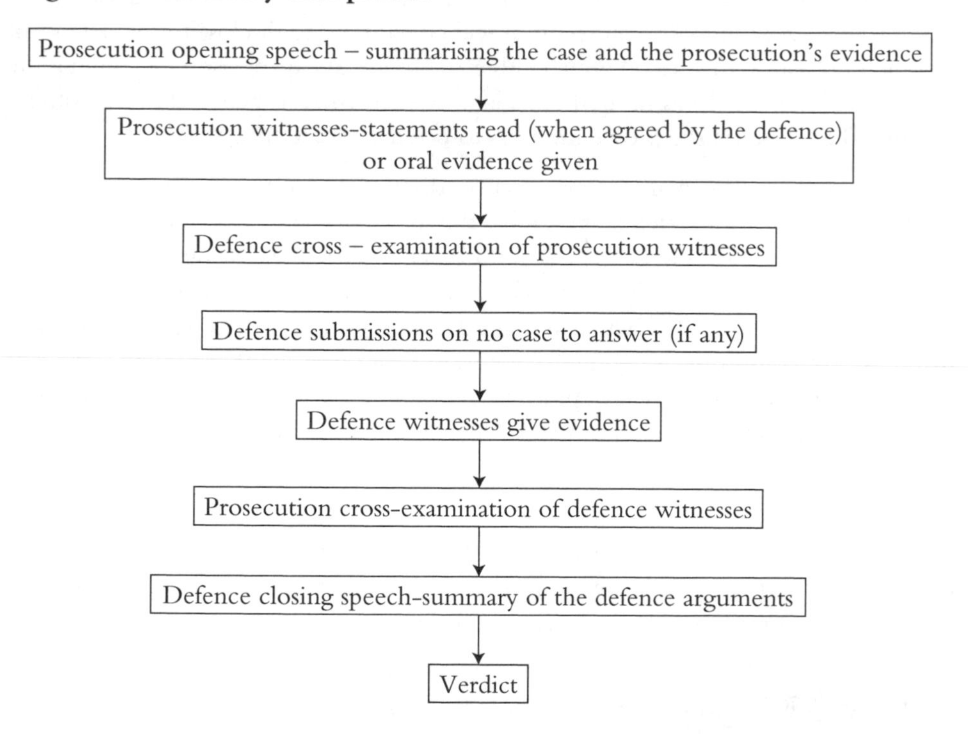

Trial on indictment

8.26 A trial on indictment involves findings of fact and of guilt or innocence being determined solely by a jury. The role of the judge is to decide on the admissibility of evidence and any other purely legal arguments and to direct the jury on the evidence and the issues which they should properly determine. If the judge has concluded that there is no case for the defendant to answer he or she may however instruct the jury that they must give a not-guilty verdict.

Generally, a "plea and directions" hearing will take place before the trial, so that issues can be identified in advance – most importantly the questions of how the defendant will plead and which witnesses the defence intends to call. (On a guilty plea in a health and safety case the court will normally proceed straight to sentencing at this stage.) The prosecution and defence will identify for the court at the plea and directions hearing what points of law and other issues will need to be dealt with at the trial and the likely length of trial. Other preparatory hearings may take place later, for example to rule on points of law.

At the start of a trial on indictment the counts on the indictment are put to the accused. This is known as the arraignment. The accused then makes a plea to each count. Where the accused pleads not guilty to a count the prosecution must prove beyond all reasonable doubt that the accused is in fact guilty on that count. The prosecution must prove each element of the offence in this manner and where the prosecution fails to adduce evidence that meets this standard, the judge will direct the jury to acquit. Furthermore, the prosecution must anticipate and counter any defence put forward. This is particularly significant in health and safety offences where the prosecution will seek to undermine the defence case of having taken reasonably practicable steps to comply with the duty (see 8.16).

The trial process and hearing of evidence is broadly the same as that for summary trial shown in Figure 8.4, except that the judge gives a "summing up" to the jury which recounts the evidence and highlights the legal issues they need to consider in reaching a verdict on that evidence.

It should be noted that at any stage of the trial the accused has the option to change his plea from not guilty to guilty. If the accused wishes to change a guilty plea to one of not guilty, the trial judge's consent is required.

The trial then proceeds in the same sequence as that set out above for summary trial, except that the judge will, at the close of the evidence, give the jury a structured summing up of the evidence and the correct approach they should take in applying the facts to the legal elements of the relevant offence in the legislation.

Sentencing guidelines and penalties

8.27 Penalties for breaching health and safety legislation fall into three main categories: breaches of general duties under the *HSWA, Sections 2–6* and most other offences, including breach of health and safety Regulations. The penalty is a fine or imprisonment (see Figure 8.5 below). Furthermore, the maximum penalty will depend upon which court the sentence is passed. Crown Courts can, upon conviction, impose unlimited fines for any of the offences. The Magistrates' Courts, on the other hand, have limited powers in relation to the penalties they can impose up to £20,000 for breach of general duties under *HSWA*, or up to £5,000 for breach of health and safety regulations. It is important to note that while the Magistrates' Courts have upper limits for the fines which they can impose, the limits relate to each individual offence. Therefore, if a prosecution consists of several separate offences the resulting fine could be considerably higher than the £20,000 limit for a single offence.

Figure 8.5: Summary of maximum penalties for breaches of health and safety legislation

Type of offence	*Court*	*Maximum penalty*
Failure to discharge a general duties under *HSWA, Sections 2–6*	Magistrates' Court Crown Court	£20,000 fine Unlimited fine
Breach of *HSWA. Sections* 7–9 and breach of any health and safety Regulations˜	Magistrates' Court Crown Court	£5,000 fine Unlimited fine
Contravention of the requirements of an improvement or prohibition notice	Magistrates' Court Crown Court	£20,000 fine Unlimited fine and/or up to 2 years' imprisonment

The level of fine imposed will ultimately depend on the court's impression of the facts and circumstances of the particular case, including the gravity of the offence, whether the breach resulted in death or serious injury and any mitigating evidence the accused puts forward. There has, in the past, been a general concern as to the low levels of fines imposed for health and safety offences. In the case of *R v Howe and Son (Engineering) Limited*, the Court of Appeal sought to address these concerns by laying down guidelines to assist magistrates and judges in sentencing health and safety offences. The case concerned an appeal by the company against fines totalling £48,000 and an order for costs of £7,500 imposed in respect of four health and safety offences arising from a fatal accident to one of the company's employees who was electrocuted while using an electric vacuum machine to clean a floor at the company's premises. In reducing the level of fine imposed to reflect the company's limited financial resources the Court of Appeal laid down the following general sentencing principles:

- The level of fine should reflect:
  - gravity of the offence;
  - degree of risk;
  - extent of the breach – an isolated incident may attract a lower fine than a continuing unsafe state of affairs;
  - defendant's resources and effect of the level of any fine on its business.
- Aggravating factors include:
  - failure to heed warnings;
  - if the defendant deliberately flouts health and safety legislation for financial reasons;
  - if the offence results in a fatality or other serious injury.
- Mitigating factors include:
  - prompt admission of liability and early guilty plea;
  - steps taken to remedy deficiencies;
  - previous good health and safety record.

As a general principle the court stated that "*A fine needs to be large enough to bring that message home where the defendant is a company not only to those who manage it, but also to its shareholders*". It will usually be necessary to produce copies of recent accounts to the court if there is any question of a fine being unaffordable.

Sentencing guidelines have continued to evolve since the *Howe* case. One issue that has emerged how to deal with cases is where large numbers of the public have been put at risk. This has arisen particularly in the context of cases railways maintenance activities. The Court of Appeal has indicated that sentencing courts are entitled to take a more severe view of breaches of health and safety at work provisions where there is a "significant public element". This is so particularly in cases where public safety is entrusted to companies and the public has to trust in their competence and efficiency (*R v Jarvis Facilities Ltd*).

This principle raises interesting problems when the defendant concerned is itself a public sector body. On one view these are often large organisations with heavy responsibilities, and they should not be given special treatment. An example of this approach was the fine of £400,000 imposed on Doncaster Metropolitan Council in 2001. The difficulty in taking this line though is that there are no shareholders to punish or to call the senior management to account, and the impact the fine has to be borne by increased taxes or more widely through reduced expenditure of funds of public services. In 2006 the Court of Appeal held that a fine of £100,000 against an NHS trust for a case involving the death of a patient was excessive and reduced it to £40,000. This was intended to strike a balance between not treating public bodies as being immune to penalties and the fact that the public would suffer from the reduced resources the trust would have left (*R v Southampton University Hospital NHS Trust*).

If a plea is entered on a factually agreed basis between the prosecution and accused, the case of *R v Friskies Petcare (UK) Ltd* sets out a recommendation for the agreed plea basis to be put in writing, so that there is no doubt whatever on what basis the court should pass sentence. Where such a statement is prepared the accused must also provide the prosecution with the factors he will be submitting in mitigation, known as the "mitigation statement". If facts are not agreed, the court may, in some cases, require any disputed issues (which are material to the penalty) to be resolved by having witnesses called to give relevant evidence.

The prosecution will normally also be awarded its legal costs of the case, including inspectors' time spent dealing with evidence and case preparation. The courts are encouraged to order the defendant to pay all these costs in full, although this should not result in an award disproportionate to the amount of the fine.

In some instances, courts will consider making a compensation order in favour of a specific accident victim under the *Powers of Criminal Courts Act 1973, s 35.* (Magistrates' powers are restricted to £5,000.) In practice, orders are not often made in health and safety cases, because the evidence has not dealt in detail with issues of the causation or extent of injuries, and because compensation claims are often pending and already being dealt with by the defendant's insurers.

Publication of health and safety convictions and notices

8.28 The HSE pursue a policy of "naming and shaming" companies and individuals convicted of health and safety offences. Convictions are published by the HSE in an annual report and on the HSE website at *www.hse.gov.uk/prosecutions/*. In addition, the HSE website now also contains a register of improvement and prohibition notices issued by the HSE, accessed at *www.hse-databases.co.uk/notices/*.

Similarly, a local authority annual report contains the names of duty holders that have been convicted of health and safety offences following investigation by the local authority inspectors. This report is also accessible on the HSE website at *www.hse.gov.uk/enforce/index.htm.*

Criminal procedure in Scotland

Commencement of proceedings

8.29 Most prosecutions are instituted and conducted by the public prosecutor. In the district court and sheriff court the prosecutor is known as the Procurator Fiscal. In the High Court of Justiciary (Scotland's superior criminal court) the prosecutor will be Crown Counsel acting in the name of the Lord Advocate. Bodies such as the HSE do not operate as a prosecuting authority (as is the case in England and Wales). The HSE will investigate accidents and incidents and then report its findings to the Procurator Fiscal who decides whether to prosecute the case.

There are two types of criminal procedure: summary and solemn. Summary cases are heard by a judge sitting alone in the district or the sheriff court. Solemn cases are heard by a judge and a jury in the sheriff court or the High Court of Justiciary.

Under summary procedure the Procurator Fiscal starts proceedings by serving a complaint on the accused. The complaint sets out the charges against the accused and gives details of the court at which the accused must appear, the date and timing of the first calling when the accused must first appear or be represented or state their plea in writing.

Under solemn procedure the accused is either served with a petition or with an indictment. A petition is a formal notice of the charges that the prosecution intend to bring against the accused. A petition is an interim document and will be superseded by an indictment. Once the case is fully prepared, the Crown serves an indictment on the accused. The indictment contains all the charges which the Crown is making against the accused.

Advance information

8.30 On commencement of health and safety proceedings, the accused has certain rights to receive information about the case against it. These rights are triggered at specific stages of the proceedings. If the case proceeds according to solemn procedure and before receiving the indictment, the accused does not have a right to insist on the prosecutor disclosing documents to it or providing a list of witnesses. In practice, the prosecution will allow such examination as is reasonable as long as this does not interfere with its own inquiries. Once the indictment is served, information about witnesses and documents becomes immediately available to the accused and will be annexed to the indictment in a schedule. The accused may recover the documents and interview witnesses for the Crown. Any document that the Crown intends to rely on must be lodged in court before the trial.

If the case is prosecuted under summary procedure, the accused may petition the Crown to recover documents which are necessary for the defence. There is no requirement on the Crown to serve a list of witnesses on the accused nor is there a requirement to lodge documents in court before the trial. However, because the proceedings are intended to be summary, in practice the parties actively discuss documents to be relied on and witnesses to be cited at an early stage of the proceedings.

Time limits for bringing a prosecution

8.31 If an offence can be prosecuted either under solemn or summary procedure, there is no time limit unless the statute provides otherwise. Most health and safety offences can be tried either way, few are summary only and none are indictable only.

Under solemn and summary procedure there is no time limit for raising proceedings in relation to a common law offence (e.g. culpable homicide – the Scottish term for manslaughter). However, if there is undue delay in starting the case, the accused may challenge the case under the common law relating to delay or as a breach of human rights legislation.

Once proceedings under solemn procedure have commenced, there are time limits within which the accused must be brought to trial.

Types of offences and jurisdiction

8.32 The Procurator Fiscal has the right to decide whether the proceedings are to be summary or solemn. He also has an absolute discretion as to what court the proceedings are to be brought before, subject to the charge being competent in that court and taking account of any statutory restriction requiring summary procedure. The accused has no rights in this matter at all. He cannot demand that his case be taken in any court other than that chosen by the Procurator Fiscal or to insist upon being heard by a jury rather than judge or vice versa. Most prosecutions are taken before a judge alone. In the case of a statutory offence, the Act creating the offence may provide that it must be prosecuted summarily; where no such provision is made, the choice of procedure is determined as for common law crimes.

Procedure

Summary procedure

8.33 Once the complaint has been served, the accused is summoned (cited) to appear in court at a specific time and date. At the first hearing (or diet) the accused or his solicitor (either in person or by letter) will state a plea of guilty or not guilty. The accused can ask for the hearing to be adjourned (continued) without a plea. If the accused wants to make a preliminary plea (e.g. about the relevancy of the charge) he must do this at before entering a plea of guilty or not guilty. The preliminary plea will be dealt with at a separate hearing.

If the accused pleads guilty, the Procurator Fiscal will read a narrative of the offence and, if applicable, details of any previous convictions. The accused or his solicitor then has an opportunity to state any mitigating circumstances. The court will then convict the accused and impose an appropriate sentence. If there are several accused and all of them have not pleaded guilty, it is unlikely that the one who has pled guilty at an early stage prior to the charges against the co-accused having been determined.

If the accused pleads not guilty, the court fixes an intermediate hearing and a trial date. The intermediate hearing is normally held about 2 weeks before the trial. At the intermediate hearing:

- The prosecution and the defence will state to the court whether they are ready to go to trial.
- The defence will state to the court whether it intends to continue with its not guilty plea.
- The court will assess whether there is any evidence that can be agreed by the parties.
- The defence will have an opportunity to change its plea to guilty. The prosecution will also have the opportunity to drop any or all of the charges.
- The court may adjourn the trial date, if necessary.

The prosecution may discontinue the case against the accused at any time before the conclusion of the case. The prosecution may do this either by:

- dropping the case with the right to bring a fresh prosecution on the same charges in the future; or
- dropping the case without the right to bring a fresh prosecution on the same charges in the future.

If the accused does not change his plea to guilty or if the prosecution does not drop the case, the matter will go to trial.

Solemn procedure

8.34 If the prosecution starts with a petition, it will do this at an early stage when the accused is in custody of the police following his arrest. The petition will be superseded by the indictment once the case has been fully prepared.

It is rare for health and safety prosecutions to be started by petition. Most health and safety prosecutions commence with an indictment.

If the case starts in the sheriff court, there will be a first hearing (diet). This is very similar to the intermediate hearing under summary procedure. At the first hearing, the accused has the opportunity to state any special defences. There is no first hearing if the case proceeds in the High Court of Justiciary. If the accused wants to rely on a special defence he must state it at least ten days before the trial.

The case can come to an early conclusion if the prosecution decides not to continue with it or if the accused decides to plead guilty by way of a *Section 76 letter.* This is a formal offer to plead guilty. If the prosecution accepts the offer, the indictment is prepared detailing only those charges to which the accused has pled guilty. A special hearing is arranged for the accused or his solicitor to appear at to state any mitigating circumstances. The court will then convict the accused and impose an appropriate sentence. The case will proceed to trial:

- if the prosecution rejects the offer; or
- the accused continues to plead not guilty.

Trial

8.35 The prosecution starts by calling its witnesses. They give their evidence under oath or affirmation. There are no opening speeches. The sequence for examining witnesses is as follows:

- The prosecution examines the witness in chief through question and answer. Unlike in England and Wales, there are no witness statements for the prosecution to read out to the court.
- Next, the witness is cross-examined by the defence.
- Finally the witness can be re-examined by the prosecution. The prosecution can only re-examine on issues that have been brought up by the defence or for clarification.

Once the prosecution has presented its evidence, the defence may call its own witnesses. Each witness is examined in chief by the defence and its then cross-examined by the prosecution. The defence may re-examine the witness if that is necessary.

At the end of the trial, the prosecution and the defence make closing speeches. If there is a jury the judge or the sheriff will give the jury directions on the law.

In order to secure a conviction the prosecution must prove the charge(s) beyond all reasonable doubt. All of the evidence supporting the key elements of the charge(s) must be corroborated. In many health and safety cases the burden of proof shifts to the accused once the prosecution has established a prima facie case. Thereafter the accused will have the task of shifting the burden of proof back to the prosecution (e.g. by proving that all reasonably practicable precautions were taken to comply with the statutory duties).

There are four potential verdicts:

1. guilty;
2. guilty in part;
3. not guilty;
4. not proven.

In effect the verdicts of not guilty and not proven are the same – the accused will be acquitted.

Sentencing

8.36 Sentencing may be immediate or it may be deferred (e.g. pending receipt of a social enquiry report). In health and safety prosecutions it is rare for sentence to be deferred.

The level of fine will be based on the following:

- The offence and the maximum sentence applicable under statute.
- The plea in mitigation presented by the accused.
- Previous convictions.
- The financial means of the accused to pay a fine.

In limited circumstances in a health and safety case, the court may also impose a custodial sentence. Unlike in England and Wales, there are no formal sentencing guidelines.

Civil liability and compensation claims

8.37 When an incident in the workplace results in personal injury or illness to an employee, the employee may make a civil claim for compensation, usually against his employer. Liability for damages will generally arise because there has been a breach of the common law duty of care owed by the employer (negligence) or following a breach of statutory duty by the employer (under the *Occupiers' Liability Acts 1957 and 1984*).

An outline of the civil procedure for a personal injury claim is given in Figure 8.6.

Figure 8.6: An outline of the civil procedure for a personal injury claim

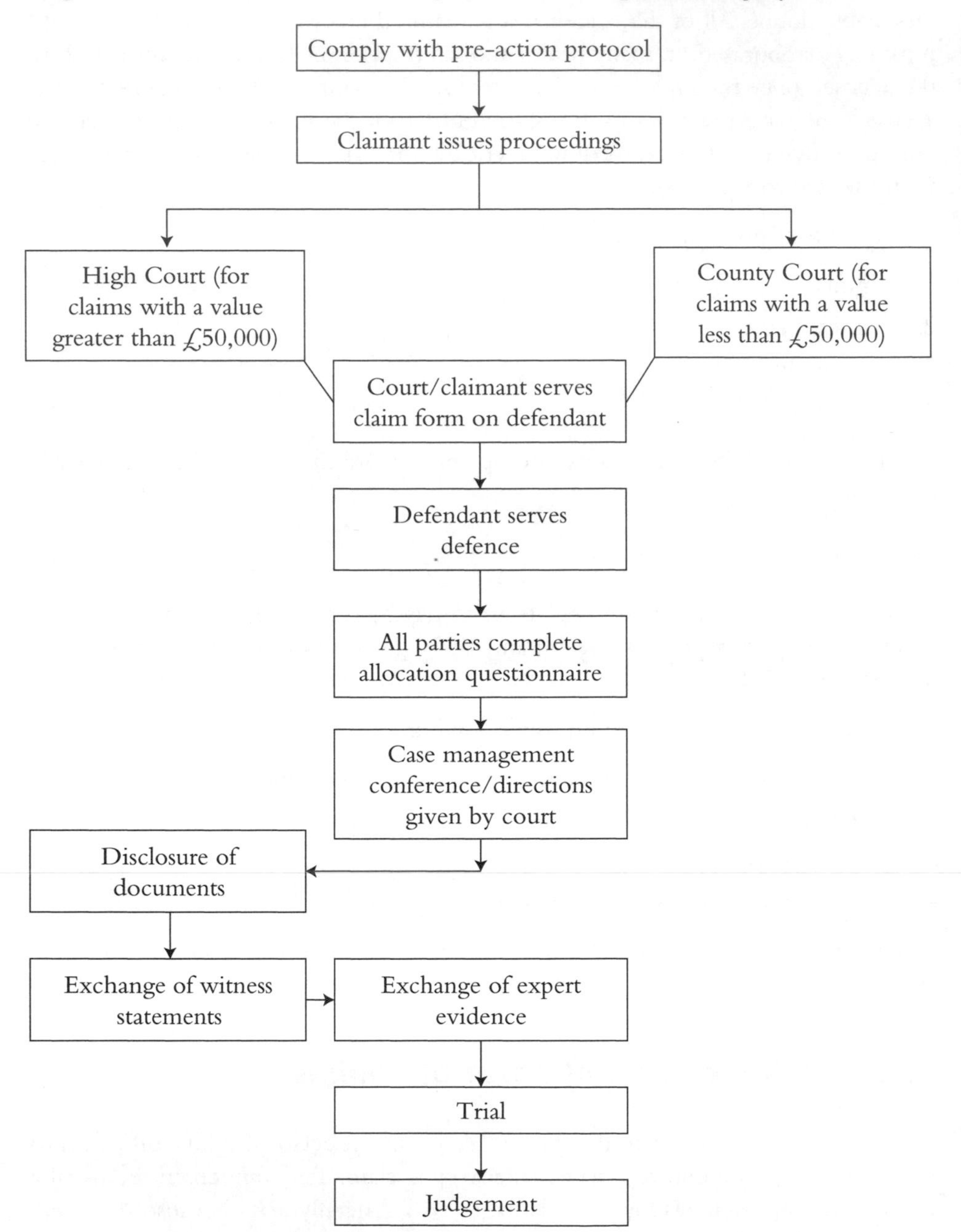

Pre-action protocols

8.38 Following Lord Woolf's final Access to Justice report in 1996 the civil litigation procedure in England and Wales saw the introduction of the *Civil Procedure Rules 1998* (as amended) together with pre-action protocols (PAP). In the context of civil claims arising following breach of health and safety legislation, the relevant PAP one for personal injury claims and for disease and illness claims (which are very similar). The purpose of the PAP is to encourage more pre-action contact between the parties, including better and earlier exchange of relevant information and better pre-action investigation by both sides. The aim of the PAP is to put the parties in a position where they may be able to settle the case early on in a fair manner and without the need to initiate proceedings. The PAP also serves to ensure that, in cases where litigation cannot be avoided, proceedings run efficiently and to the court timetable.

The effect of the pre-action protocol is to achieve a more consistent approach to pre-action conduct. Where proceedings are issued they enable the courts to apply the standards set out in the PAPs to establish whether non-compliance merits any adverse consequences.

The pre-action processes for injury claims are summarised in Table 8.3. It is clear that the courts expect parties to personal injury claims to adhere to and comply with this protocol. Failure to do so will require a full explanation later if there are proceedings and could result in an adverse costs award on the offending party if the court considers that unnecessary expenses have arisen.

Table 8.3: Pre-action process

1. The Claimant may wish to give early notification to defendant that a claim is likely to be made. This is particularly encouraged when rehabilitation may be the immediate need and the defendant might be asked to pay.
2. Claimant must formally notify a claim is intended by sending a letter of claim to the proposed defendant. The letter of claim must include:
   - a clear summary of the facts;
   - the nature of any injuries;
   - an indication of any financial loss;
   - a request for details of the defendant's insure and that where appropriate the proposed defendant should send a copy of the letter to his insurer.

A standard letter format is included in the PAP. This includes references to standard lists of documents the defendant is asked to disclose.

3. The defendant should reply to the letter of claim within twenty one days of the date of posting of the letter identifying the insurer. Failure to reply by the defendant or insurer within this time will entitle the claimant to issue proceedings.
4. The defendant or its insurer is allowed three months from the date of the acknowledgment of the claim to carry out investigations. By the end of the 3-month period the defendant or insurer must reply stating whether liability is accepted or denied and if denied on what basis. (Note that if a defendant admits liability at this stage there is a presumption that the defendant will be bound by this admission for all claims with a total value of up to £15,000.)
5. If the defendant denies liability, it should enclose documents with his letter of reply which are material to the issues between the parties, including any allegations of contributory negligence, which would be likely to be ordered to be disclosed by the court.
6. If the defendant raises allegations of contributory negligence it should specify the reasons, and should specify the reasons, and the claimant should respond to those allegations prior to the issue of proceedings.
7. As soon as possible, especially where liability is admitted, the claimant must send to the defendant a schedule of special damages (this is a schedule of expenses or losses already incurred, such as loss of earnings, pension rights, etc.).
8. If expert evidence is required the parties should try to agree on a nominated expert. The parties may send written questions to the agreed expert on the report that are relevant to the issues.
9. The claimant's solicitor should organise access to the claimant's medical records where a medical expert is to be instructed.
10. Rehabilitation – both parties are to consider this as early as possible and have regard to whether the Rehabilitation Code may help to identify the claimant's needs (see page 226). Any rehabilitation report cannot be used in subsequent litigation without consent as is not subject to the requirement concerning appointment of joint expert.

(The full protocol is provided is available at *www.dca.gov.uk/civil/procrules_fin/contents/protocols/prot_pic.htm*)

An example of a PAP letter from a claimant's solicitor is provided in Appendix V in this chapter. It is worth noting how extensive the requests for disclosure of documents are in cases such as this example which involved unguarded machinery. The early exposure via the PAP process of any deficiencies in these documents is intended to act as a driver for early settlement of claims.

Personal injury claim-procedure

Commencement

8.39 Proceedings in a personal injury claim are commenced by the claimant issuing a claim form. It is important to note that there are time limits within which any claim for personal injury must be made, as set out in the *Limitation Act 1980*. The claimant must normally commence his action by the of issue proceedings within three years from either the date of the accident occurring or the date of the claimant's relevant knowledge. Knowledge for these purposes is defined as the date on which the claimant first had knowledge of all of the following:

- that the injury was significant;
- that the injury was attributable (either in whole or in part) by negligence, nuisance or breach of statutory duty;
- the identity of the defendant.

Once the claimant has all the information with which to commence proceedings he must then decide which court to issue proceedings. In personal injury claims, only claims worth £50,000 or more can be commenced in the High Court, all lower value claims must be issued in the county court. It should be noted however, that a claim issued in the county court may be transferred to the High Court in certain circumstances, for example where the claim is against the police or is for a fatality.

The claim form must be accompanied with the claimant's statement of case, medical report statement of special damages and "response pack" for the defendant to reply. The statement of case sets out the claimant's case and why the defendant is liable. The statement of special damages provides a summary of the losses and expenses already incurred by the claimant together with an estimate of any future expenses and losses.

Defence

8.40 Once proceedings have been served on the defendant he must either file an acknowledgement of service or his defence within fourteen days of the date of service of proceedings. If the defendant files an acknowledgement of service he has a further fourteen days within which to file and serve his defence. However, the parties can agree between themselves, without the need for court approval, to an extension of time of up to a further twenty eight days, within which the defence must be served. It is the defendant's responsibility to write to the court notifying it of any agreed extension of time. Any further extension will require approval via an application by the defendant to the court.

If the defendant fails to file a defence or acknowledgement of service within the required time periods as set out above, the claimant may make a request for a judgement in default. It should be noted that if the defendant has made it clear that he intends to defend the claim and he has simply missed the deadline, any request for judgement in default may, on application by the defendant, be set aside by the court where the defendant can provide a reasonable excuse and he has an arguable defence.

Allocation questionnaires

8.41 The introduction of the *Civil Procedure Rules* (*CPR*) brought with it many changes to the previously long-standing rules of court. One of the greatest changes is seen in the court's case management powers – the courts now retain responsibility for allocating cases to particular tracks (small, fast and multi-track) according to value which in turn controls the timetable of the case:

- *Small claims track*: Work up to £1,000.
- *Fast track*: Worth £1,001–£15,000.
- *Multi-track*: Work £15,000 or more.

Once a defence is received or filed at court, the court will send out an allocation questionnaire to each party in the case. The scope of the allocation questionnaire is fairly broad and requires each party to provide an indication of the track that it considers most suitable for the type and complexity of claim involved. In addition, the questionnaire requires consideration of the following matters:

- an estimate of costs to date and costs to trial;
- confirmation of whether the personal injury pre-action protocol has been complied with;
- details of any expert evidence required for which permission is to be sought from the court;
- an estimation of the length of the trial;
- details of counsel and dates where counsel or witnesses are unavailable;
- suggested directions.

Which track?

8.42 Where a claim is allocated to the small or fast track, the court will usually make directions (taking into account the suggested directions in the allocation questionnaires). This is usually done without a hearing and the parties simply receive a list of directions from the court on paper.

In multi-track claims it is usual for the court to fix a date for a case management conference so that each party can provide a summary of its case and propose a timetable of directions. In many cases the case management conference can take place over the phone and is often specified by the court to be by telephone to save time and costs.

Where the claim is complex and involves many parties, there are likely to be further case management conferences scheduled by the court, or upon request by a party, so that progress can be monitored and the timetable varied as necessary.

Disclosure of documents

8.43 The process of disclosure allows the parties in a case to see the relevant documents held by each party which are purported to support its claim. Documents for case purposes include information stored in electronic form such as e-mails.

In personal injury claims, the forms of disclosure are as follows:

- *Pre-action disclosure.*
- *Standard disclosure*: This is common in most multi-track and fast track claims. Standard disclosure requires the parties to disclose those documents upon which it intends to rely as well as documents that adversely affect that party's or another party's case. Standard disclosure is achieved by each party preparing and serving a list of documents (see below for contents of list of documents).
- *Specific disclosure*: In certain circumstances the court may make an order that specific documents or class of documents must be searched for and disclosed.

It should be noted that documents disclosed may only be used for the purposes of the claim and cannot be disclosed to another third party, other than for the purposes of the claim.

Disclosure in the course of litigation (as opposed to pre-action) is normally given by way of a list of documents. There are three sections to the list. The first part must identify those documents the party has in its possession and that are to be disclosed. The second section identifies those documents which the party is withholding from disclosure on the grounds of privilege (see below) and the last section is for those documents that the party no longer has in its possession.

Typically, in accident cases in the workplace, disclosure by the employer of other defendants would include the following:

- *RIDDOR* Report to HSE;
- accident book entry;

- first aid report;
- minutes of health and safety committee meetings in which accident was considered/discussed;
- pre- and post-accident risk assessments;
- documents identifying previous accidents of a similar nature;
- reports by supervisor/HR/safety representative;
- earnings information;
- copies of an accident investigation report.

There are two key points to remember about disclosure; firstly the list of documents contains a disclosure statement which must be signed to verify that the signatory understands the duty of disclosure. This document is usually signed by the defendant itself (or a director if it is a company) and it is therefore imperative that the solicitor provides the client with comprehensive advice relating to disclosure; secondly, the duty of disclosure is on-going and therefore if more documents come to light after the list of documents have been exchanged these must be immediately disclosed to the other party(s) in the action. Where there are a number of additional documents then a supplemental list of documents should be prepared.

Once lists of documents have been exchanged each party is at liberty to inspect (at a mutually convenient time and place) those documents. It is normally more convenient and practicable for copies of those documents to be provided (at the requesting party's expense).

Some documents may be withheld due to the privileged nature of the document. These documents are often described in a generic way in Part 2 of the list of documents. Privileged documents include solicitor or client correspondence, counsel papers and correspondence, without prejudice letters and documents prepared in contemplation of litigation.

Witness/expert evidence

8.44 The next step after disclosure is the mutual exchange of witness statements and expert evidence.

Witness statements relate to evidence of facts which the parties intend to rely upon and which are to be decided upon at trial. The statements form the oral evidence in chief at trial. It is necessary therefore to ensure that all evidence that is to be relied upon and that is available at the time the statement is made is included in the statement. Where facts contained in a witness statement are uncontroversial then attempts should be made, if possible, to agree the witness evidence with the other party/parties in advance, as this will save both the time and expense of calling that witness at trial.

In most accident cases the expert evidence will be medical evidence and in the majority of cases this will have been complete at the start of proceedings. If further medical evidence is required then, if not already given, court permission may be required to rely on the evidence. Where each party has its own expert evidence, rather than a joint expert, then the parties should prepare a note on the areas of agreement and disagreement between the experts. This will save both time and expense at trial. Experts in other areas may be called, such as engineering or toxicology. Any discipline of expert can be engaged, provided the court gives permission.

Trial

8.45 A brief summary of the order of events at trial is given in Figure 8.7.

Figure 8.7: Civil litigation: trial procedure

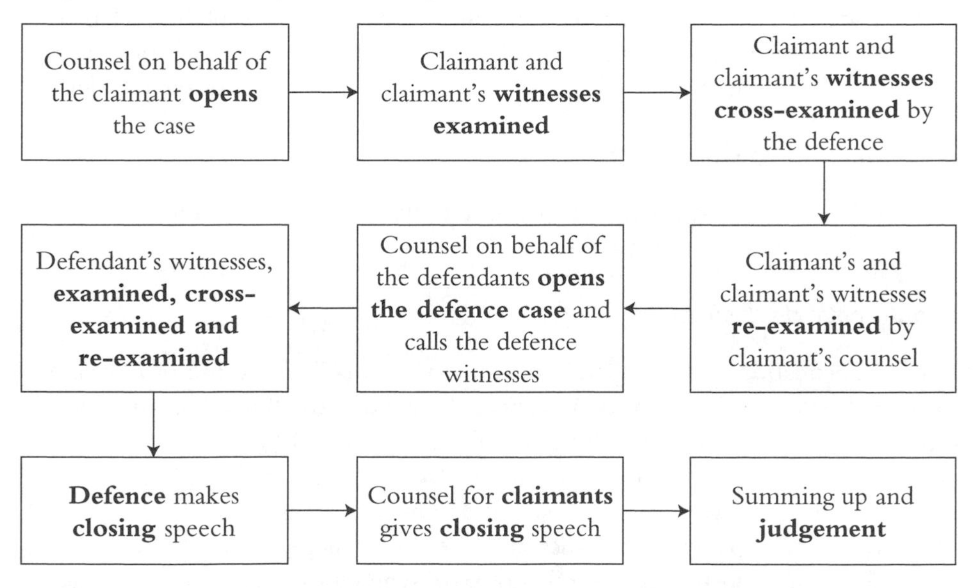

Scottish civil procedure

8.46 Scottish civil procedure differs markedly from English civil procedure and also uses quite distinct terminology. A claimant is referred to as a "pursuer" and a defendant as a "defender". The Woolf reforms do not apply in Scotland, and there has been no similar systematic overhaul of Scots civil procedure. Key differences between Scots English civil procedures are highlighted below.

Time bar

8.47 The *Limitation Act 1980* is not part of Scots law. Time limits in Scotland are governed by the *Prescription and Limitation (Scotland) Act 1973* (as amended). For personal injury actions, the relevant period is, as in England, three years but this is subject to one important qualification. In Scotland, the period does *not* end when the court issues the document initiating legal proceedings, but only when that document is properly served upon the defender in the action. The prudent pursuer will therefore get his writ warranted in good time before the end of the period, to guard against the possibility that there may be difficulties in serving the claim upon the defender.

Detailed pleadings

8.48 In Scotland, the practice has traditionally been for the parties' written cases to be very detailed. The purpose of the practice is (or is supposed to be) to give full advance notice of one's case. However, there has in recent years been a growing perception that – at least in the field of personal injury actions – the effort required to produce such pleadings is out of all proportion to their benefits produced. In the Court of Session, and for personal injury actions only, an abbreviated form of pleadings has been introduced, requiring much less in the way of detail than previously on, for example, the cause of the accident and the legal duties owed by the defender to the pursuer.

No written witness statements

8.49 Written witness statements are not used in Scots procedure. Witnesses instead give evidence orally at the trial.

No automatic disclosure

8.50 In Scotland, unlike England, there is no automatic disclosure of all relevant material to the other side. Instead, the litigant must either use the information which he already has or obtain documents by Specification. This is the process where a party makes a detailed and (as the name suggests) specific call upon the other side to produce named documents or named classes of documents. There are limits upon what can be sought in a Specification principally rules of privilege, which are broadly the same in Scotland as in England – and also upon when a specification can be taken. The rules are enforced rigidly: expansive calls for, for example, "*all relevant documents in [the other party's] possession*" will be dismissed as a fishing exercise.

Procedure at proof: opening speeches

8.51 No opening speeches are made at proof. The pursuer will simply commence by calling his first witness, which will ordinarily be the pursuer himself. Closing submissions are permitted (Figure 8.8).

Figure 8.8: The hierarchy of the Scottish civil court system

First Instance

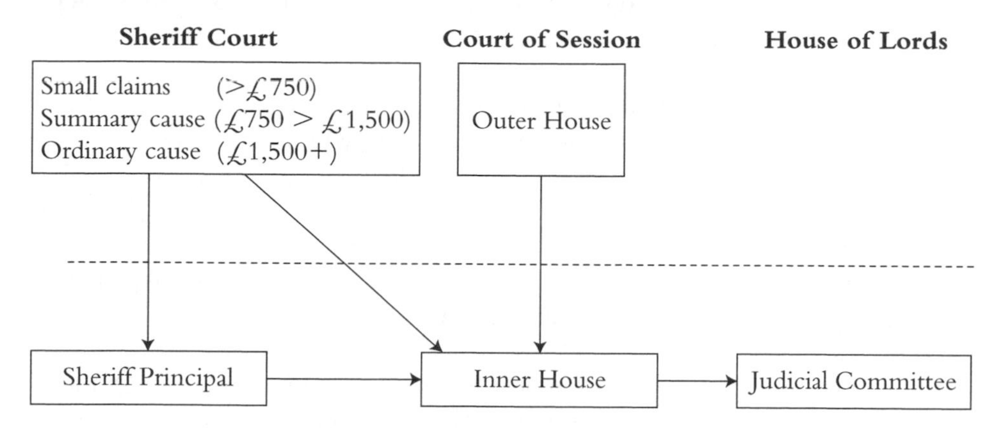

The small claims court and summary cause court

8.52 Disputes worth less than £750 are heard in the small claims court. Disputes having a value of between £750 and £1,500 are held in the summary cause court. There are separate rules of court for each of these processes, but as very few personal injury actions are pursued through these courts they will not be further considered here.

Ordinary cause procedure in the sheriff court

8.53 This is the "mainstream" procedure within the sheriff court, for actions with a value of £1,500 or above. Personal injury actions are frequently pursued through this procedure.

An ordinary cause commences with the warranting by the court of an initial writ and service on the defender. The initial writ contains a detailed account of the pursuer's case and is divided into three parts: a crave (in which the pursuer names the sum of damages he wishes to receive and any other remedy he seeks), condescendences (several paragraphs of detailed narrative, setting out the essential facts the pursuer is offering to prove, and listing the duties he contends the defender has breached) and pleas in law (the legal propositions by which the pursuer contends he is entitled to the sum of money craved).

After receipt of the initial writ, the defender has a period (twenty one days if he resides within the UK or elsewhere in the EU; forty two days otherwise) in which to lodge a notice of intention to defend. This must be followed within seven days by a set of defences, which contain the defender's detailed response to the pursuer's case.

If a notice of intention to defend is not lodged, the case proceeds as undefended and the pursuer is entitled to a decree (or judgement). If a notice of intention to defend is lodged but not followed by defences (or if defences disclose no legal defence), the pursuer may seek decree by default.

Assuming that valid defences are lodged, there will then follow a period of adjustment during which either party can alter their cases without leave of the court. Two weeks after the end of this period, there follows an options hearing. Before the options hearing is reached, however, the parties have to consider if they wish to raise legal points or to contend that in law parts of the other party's written case are irrelevant. If they wish to do so, they must lodge a note in support of preliminary pleas. If no such note is lodged, such points cannot be taken.

At the options hearing , the parties appear before the sheriff, who is charged with the task of "securing the expeditious progress of the cause". This involves hearing representations from the parties and assigning a date for either:

- a proof (a trial where evidence is led and submissions are made following that evidence);
- a debate, where issues of pure law are argued; or
- a proof before answer (a hybrid, containing features of both a proof and a debate).

If a party has a legal point which is either a "knockout blow" or a point on relevance which will seriously restrict the amount of evidence which will require to be heard at proof, then a debate will ordinarily be fixed. If the legal points are less fundamental, but still worthy of consideration, then a proof before answer will be appropriate.

If the case proceeds to a proof or a proof before answer, these will follow a similar pattern to an English trial (other than, as noted above, there will be no opening speeches and evidence in chief will be given orally).

An appeal may be taken on a point of law to either the sheriff principal of the sheriffdom (and thence on to the inner house of the Court of Session, if required) or direct to the Inner House of the Court of Session. A final appeal lies from the inner house to the House of Lords.

Chapter 43 proceedings in the outer house of the Court of Session

8.54 The Outer House of the Court of Session enjoys concurrent jurisdiction with sheriff court ordinary procedure over all personal injury actions valued at more than £1,500. Although generally viewed as being superior to the sheriff court, and as such befitting more valuable or complex cases, there is no particular quality or higher value threshold for raising proceedings in the Court of Session; albeit, the pursuer who does so for no good reason may ultimately be punished in expenses.

The action is commenced by the signetting (warranting by the court) and service upon the defender of the court summons with statement of claim attached. As noted above, the claim will now to be in abbreviated form. In addition, the pursuer will automatically be granted a warrant on an initial specification of documents, calling upon the defender to produce a range of documents generally sought in personal injury actions. The intention is to allow the pursuer to obtain these documents at an early stage of proceedings. However, this is not the introduction to Scotland of a broad disclosure principle – the rules of court provide that only fixed and relatively narrow categories of documents such as medical reports and accident investigation reports may be obtained in the initial specification.

After signetting and service on the defender, the writ must be returned to court for calling (calling is a purely formal matter not involving an appearance in court, it simply denotes that the case has been introduced into the court process). The case must call within three months and one day of having been signetted – if it does not, the warrant lapses and the case falls. After the case is called, the defender has a period of three days in which to note an intention to appear and must lodge with the court a written note of defence within seven days of the cases calling. Thereafter, an automatically generated court timetable will be issued with its end date fixed as a diet of proof (approximately one year after the date of the case's calling). The adjustment period (a fixed period of eight weeks) then follows, at the end of which the court requires to assess whether the case should be assigned to a proof before answer, proof, preliminary proof, a jury trial or for any other specified order it considers appropriate. In tandem with the adjustment period, the pursuer must prepare and lodge a statement of valuation of his claim. This is another important aspect of the new rules and is intended to focus the parties' minds on settlement.

The defender is then required to lodge his valuation of the claim, and, if the case has not yet settled, four weeks before the allotted proof date a pre-trial meeting must be held between the parties to consider whether the case can be settled, or at least, what scope there is for the agreement of evidence or quantum.

If the case proceeds to debate, proof or proof before answer, this will follow the same pattern as a sheriff court ordinary cause.

An appeal may be taken on a point of law to the Inner House of the Court of Session and from there to the House of Lords.

Causes of action

8.55 A brief summary of the main causes of action for personal injury following an accident in the workplace is given below.

Breach of statutory duty

8.56 If an employee alleges a breach of statutory duty by his employer, he is required to prove the following:

- the legislation imposes a duty on the employer;
- that the duty was breached;
- that the injury resulted from the breach and that such an injury was of a type regulated by the statute;
- the employee is within the class of persons to which the legislation applies.

It should be noted that *HSWA* does not confer a right of action in civil proceedings for failures to comply with a general duty imposed under the Act (*HSWA, Section 47*). However, *Section 47* does state that breaches of regulations made under the Act are, except where the regulation provides otherwise, actionable in civil proceedings insofar as the breach caused damage. Most health and safety regulations do in fact give rise to an actionable claim.

Statutory duties may be absolute, for example, if the duty requires that the employer "shall" do something, the employer is required to comply with duty regardless of any particular circumstances (e.g. cost, effort, etc.). However, the majority of health and safety legislation qualifies the duty by the words "*practicable*" or "*so far as reasonably practicable*". Thus, the employer is entitled to balance the costs of averting a risk with the likelihood of injury occurring.

Employers should be of HSE publications such as Approved Codes of Practice and Guidance Notes since compliance with Approved Codes of Practice can support a case to show that all reasonable steps were taken. Conversely, failure to act on hazards identified in published guidance, or to follow the advice, will usually be strong evidence of breach of the relevant statutory duty.

Occupiers' Liability Acts 1957 and 1984

8.57 An employer may also find himself liable for persons other than employees who are injured while working on or visiting premises occupied by the employer.

The Occupiers' Liability Acts (OLA) 1957 and *1984* impose duties on occupiers of premises to all lawful visitors to the premises. The *OLA 1957* state:

> "... the common duty of care is a duty to take such care as in all the circumstances of the case is reasonable to see that the visitor will be reasonably safe in using the premises for the purposes for which he is invited or permitted by the occupier to be there."

Thus, injuries resulting from unsafe premises, for example uneven or slippery floor surfaces or unsafe electrical wiring or defective lifts fall, within the ambit of the *OLA 1957*.

"Control" is an important element in determining if someone is an "occupier" for these purposes. Importantly, an occupier need not have entire control over the premises nor need they be in sole occupation, for instance, an employer may be sharing premises with others and yet it will be sufficient that the employer has some degree of control for a claim to be brought under the *OLA 1957* and *OLA 1984*.

The *OLA 1984* extended the duty of an occupier in respect of trespassers so that a duty of care is owed to a trespasser in respect of any injury suffered on the premises if the occupier:

- is aware of the danger or had reasonable grounds to believe that it exists;
- knows or has reasonable grounds to believe that a trespasser is in the vicinity of the danger concerned, or that he may come into the vicinity of the danger;
- if the risk is one against which, in all the circumstances of the case, he may reasonably be expected to offer some protection.

Negligence

8.58 All persons owe a common law duty of care to those persons who they can reasonably foresee are likely to be injured by their acts or omissions. An employer therefore owes this duty of care not only to his employees but also to others he can reasonably foresee may be affected by the activities of his business, for example visitors, contractors, etc. In relation to his employees, an employer is under a duty to take reasonable care of his employees' health and safety and in particular is required to provide:

- safe premises from which to work;
- a safe system of work;
- safe plant, equipment and tools;
- competent co-workers.

In order for a claim to be brought by an employee against his employer for an injury resulting from an employer's breach of common law duty, the employee is required to prove his case on "a balance of probabilities". The employee is required to show that:

- a duty of care was owed by the employer;
- there was a breach of that duty;
- the breach led to the employee's injury.

The employer's duty is to take reasonable care and therefore the circumstances surrounding the incident will be taken into account in establishing whether there has been a breach of that duty. If there has been a breach of duty, the claimant must then show a causal link between the breach and the injury sustained. Importantly, a criminal conviction for a breach of a health and safety offence is admissible as evidence in civil proceedings and has the effect of reversing the burden of proof such that the defendant in a civil claim has the onus of demonstrating that he was not negligent.

Contract

8.59 The relationship between an employer and employee is a contractual one in which both parties are bound by the terms contained within the contract. An employment contract contains express terms which set out, for example job description, pay, holiday entitlement, hours of work, etc. and these are agreed between the parties. However, there are other terms which are implied into the contract in relation to health and safety. In particular, the common law implies a duty on the employee to perform his work with reasonable care and skill and to abide by the employer's safety rules and the employer is required to ensure his employee's safety. Breach of these duties can invoke rights by both parties. If an employee fails to abide by these duties, the employer can impose sanctions against the employee, the ultimate of which is dismissal.

Failure by the employer to ensure the safety of his employees at work can result in a claim being brought by the employee for compensation for any injury resulting from the breach.

Notices, exclusions and *Unfair Contract Terms Act 1977*

8.60 *The Unfair Contracts Terms Act 1977* (*UCTA*) places restrictions on the use of exclusion clauses for death and personal injury. In particular, an employer cannot by reference to any term in a contract or notice given to persons generally or to particular persons, exclude or restrict liability for death or personal injury resulting from his negligence. Negligence in this context extends to carelessness which as a result amounts to breach of contract. Any attempt by an employer to exclude such liability will be void. However, such a notice may be used by an occupier in defence of a claim of negligent injury to a trespasser under the *Occupiers' Liability Act 1984.*

It is not possible, as a matter of general law, to make provision in a contract to exclude liability for breach of statutory duty with the person to whom that duty is owed. However, the *UCTA* does not expressly address this issue.

In some circumstances, an indemnity clause may be effective as an alternative to excluding liability. For example, if equipment is being hired from a third party with an operator, the employer may take an indemnity from the hirer. The effect of an indemnity clause is to transfer the risk as to who is responsible for damages arising as a result of a particular act or omission. Thus an indemnity clause provides a remedy for specified claims by allocating the risk and consequential damages to another person (compared to an exclusion clause which prevents the recovery of loss at all).

Defences

8.61 Generally, the defence case consists of disputing the allegations of the claimant and seeking to persuade the court that, for example the defendant was not negligent: in essence, what the defence is doing here is saying that the claimant has not proved his or her case. Defendants may also, however, deploy other forms of defence. Outlined below are the primary defences available to an employer following a workplace injury:

- The claim is statute barred – under the Limitation Act 1980, all civil claims for personal injury have a statutory limitation period three years from the date of the incident. Where an injury arises over a period of time before the employee has knowledge of the cause of the injury, the claim must be brought within three years of the date the employee becomes aware that the injury was due to the employer's negligence.
- The employee voluntarily and knowingly exposed himself to a risk. This defence is not available in an action for breach of statutory duty.
- Where an employee has contributed to an injury, for example, if he did not use equipment in the proper manner or where he acted carelessly and against his employer's instructions, the employer's liability can be reduced by the

courts – effectively the employee's own negligence is taken into account. This is only a partial defence. The courts will generally not hold claimants 100 per cent contributory negligent as such a result is not normally consistent with a finding that a breach of duty by the employer caused or materially contributed to the injury.

- Another party (e.g. a negligent contractor) may be liable to the claimant, who should be made to contribute towards the compensation or even to indemnify the employer. To achieve this result, the other party would have to be joined as a party in the same proceedings, or a separate action would proceed in tandem with the claimant's action. Again, this is only a partial defence.
- Indemnity – not a defence as such, but a claim based on an obligation on another party to cover some or all of a defendant's liability. This could be for example an indemnity contained in a sale and purchase agreement for the business the employee worked in favour of the buyer given by the seller.

References

Armour v Skeen [1997] S.L.T. 71.

CPS (2004) *The Code for Crown Prosecutors.* www.cps.gov.uk/publications/docs/code2004english.pdf

Davies v Health and Safety Executive [2002] EWCA Crim 2949.

HSE (2002) *Enforcement Management Model.* www.hse.gov.uk/enforce/emm.pdf

HSE (2004) www.hse.gov.uk/pubns/hscl5.pdf

HSE (2005) *Revised Incident Selection Criteria 2005.* www.hse.gov.uk/enforce/incidselcrits.pdf

HSE (2006) Health and Safety Statistics 2005/06. www.hse.gov.uk/statistics/overall/hssh0506.pdf

R v Boal (199213 All ER 177 (CA).

R v Board of Trustees of the Science Museum [1993] 3 All ER 853.

R v Friskies Petcare (UK) Ltd [2000] 2 Cr App R (S) 401.

R v Howe and Son (Engineering) Limited [1999] 2411 ER 249.

R v Jarvis Facilities Ltd [2005] EWCA Crim 1409.

R v Southampton University Hospital NHS Trust [2006] All ER (D) 181.

R v Yorkshire Sheeting & Insulation Ltd [2003] EWCA Crim 458.

9 Investigating workplace accidents

In this chapter:

The basic principles of accident investigation

9.1 The main reasons why accident investigations are undertaken in the workplace are as follows:

- discovery of accident causes;
- prevention of recurrences;
- minimisation of legal liability;
- collection of safety data;
- identification of trends over time.

The specific aims of an accident investigation will vary depending on the terms of reference and the nature of the accident investigation. However, it should be borne in mind that there are sometimes very difficult balances to strike in terms of objective investigations (see pages 163 and 331).

In the transport sector, accidents investigated by the air, marine and rail accident investigation branches are aimed solely at determining the causes of the accident, in order to make recommendations to prevent recurrence. This underlying principle was succinctly defined by Lord Cullen, when he recommended the establishment of an independent rail accident investigation branch, following the Ladbroke Grove railway accident:

> "The sole objective of the investigation of accidents or incidents should be the prevention of accidents and incidents. It should not be the purpose of such investigations to apportion blame or liability" (Cullen, 2001, Recommendation 62).

Witness statements given to Air Accident Investigation Branch (AAIB), Marine Accident Investigation Branch (MAIB) and Rail Accident Investigation Branch (RAIB) investigators are protected, in that the identity of specific individuals cannot be released without a court order, and individuals know that the evidence they give will not be attributable to them (RAE, 2005). This allows witnesses to have confidence in the investigation and to have faith that the information they provide will not be used against them. Many participants in such investigations give information to investigators precisely because they believe that in doing so, they are directly reducing the chances that such an event will happen again. Individuals involved in transport accidents, either directly or indirectly, often have a very keen desire that lessons from the accident are learnt, and that recurrences are prevented, particularly where there have been injuries or fatalities. Hence, accident investigation can often be perceived as a moral and/or ethical obligation, as well as a safety requirement.

Large-scale transport investigations are intended to be blame- and liability-free, and the transport accident investigating branches cannot take legal or disciplinary action in response to the findings of an investigation. However, it must be noted that some individuals may indeed face such actions once the findings of the investigation have been published, either from enforcing authorities or from their employer, and evidence obtained in the course of the earlier investigation will be highly relevant. This is likely to be a serious concern for any employee who participates in any workplace accident investigation in any sector, whether it is an internal or external investigation. However, the principle of conducting accident investigations without blame or liability in order to determine the causes, and to take appropriate action in order to reduce the probability of recurrence, is nevertheless paramount.

The primary aims of any investigation team are to determine the causal and contributory factors leading to the accident, and to take action to address both the immediate and underlying causes. Recommendations and actions should consider both primary and secondary safety. Primary safety involves remedying any identified deficiencies in the safety management system, in order that the probability of recurrence is reduced. Secondary safety recommendations aim to reduce the probability of injury, fatality and damage to property and assets in the unfortunate instance that a similar event should occur. Determination of liability or disciplinary action for any accident is clearly outside the remit of this type of investigation process. If such matters are likely to be salient in any accident, they should be considered by a separate exercise with legal/disciplinary terms of reference, preferably once the accident investigation is complete, although this may not always be possible.

As far as possible, all investigations of workplace accidents should both follow this principle, and be seen to do so. As such, any internal accident investigation function should ideally remain distinct from normal line management and disciplinary processes. Special arrangements may be necessary to achieve this separation both in reality and in employees' perceptions. For example, it may be possible to have the investigation commissioned or sponsored by a senior manager or director outside the usual management chain. Another option may be to make special arrangements for recruiting an independent chairperson as principal investigator.

"Best practice" in accident investigation

9.2 The Royal Academy of Engineering conducted a review of accident and incident investigation processes in a number of UK industries. The sectors examined were aviation, chemical and allied industries, nuclear, offshore oil and gas, and rail and marine industries. The RAE concluded that the investigation of causes of an accident should have as many of the following attributes as possible (all quoted directly from RAE, 2005, pp. 11–12):

- The establishment of the full causal chain of events, including broader management actions and policies as well as specific practical actions or omissions. The "full causal chain" is intended to include every relevant issue and answer every link in the chain of "and what led to that" with a "why did it happen that way", and reaches as far back in time and as far around the accident as it is reasonably possible to go. It is intended to embrace the concepts of omissions as well as deliberate acts, and those by people indirectly connected to the specific accident as well as those directly involved.
- The identification of changes to policy, procedure, design, operation or use that are used to prevent that or a similar chain of events from recurring in relevant situations.
- The capture, dissemination and absorption of more generally applicable lessons that add to the wider body of knowledge and inform the education of people concerned with future systems and make them safer.
- The communication to interested parties, which may include the general public, of such facts as may be appropriate to their future confidence in the failed system and in others related to it.

- Safety remedies, that whilst being entirely sufficient and appropriate against the assessed profile of risk, are neither excessive nor deliberately penalising in their impact on future operations.
- The recognition of bereaved relatives as interested parties to the accident by way of regular, open and consistent information about the progress of investigations.

Accident investigation in UK industries

9.3 Current accident investigation practices vary widely within the UK. Recent research conducted for the HSE involved a telephone survey and follow-up interviews with companies in agriculture, construction, mining/utilities/transport, manufacturing and services (Henderson et al., 2001). The results revealed that many companies used a very unstructured and ad hoc approach to investigating accidents, while a few used processes that were well supported by relevant procedures, including analysis tools and techniques. However, the majority of companies were found to use a more "traditional" approach to accident investigation. These organisations relied predominantly on identifying active failures – those human and technical failures which were the immediate cause(s) of the accident. Very few organisations investigated underlying and contributory causes, although there was a tendency for larger companies to use more sophisticated methods.

In most instances, investigations were completed by health and safety practitioners (87 per cent), although line management were involved in approximately half of investigations (57 per cent). Senior management were involved in approximately one-quarter (26 per cent) of accident investigations, as were technical specialists. In terms of time spent investigating, 42 per cent of incidents took less than five hours to investigate, while 35 per cent took up to twenty hours and 18 per cent took over twenty hours. The time spent depended on the severity of the incident being investigated, with investigations of more serious accidents taking a greater amount of time, as might be expected.

The HSE report also found that many organisations over-estimated the quality of their investigations (Henderson et al., 2001). With the amount of management time and resources spent on investigating workplace incidents and accidents, and the costs associated with both accidents and their investigation, this is a surprising finding. With traditional, unstructured approaches to accident investigation, an organisation not only wastes time and resources, but it is also vulnerable to the following:

- Failure to identify the underlying causes, meaning that they cannot be addressed in the recommendations. This may allow similar incidents and accidents to occur in the future, since only the immediate cause(s) will have been addressed, and the investigation may not have been completed to an appropriate level.

- Total dependence on the abilities of the person conducting the investigation, leading to inconsistencies in the depth and nature of the investigation between individuals.
- Lack of structure may lead to an over-reliance on opinion and judgement, as decisions made and actions taken cannot be audited.
- Investigations which are highly sensitive to time pressures and other organisational priorities, meaning that minimum requirements standards may not be addressed.
- The organisation will lose the technical skills and experience of the person who conducts the investigation when they leave the company; this is especially problematic since the lack of a structured recording database will mean limited "organisational memory" for incidents and accidents.

Advantages of investigating workplace accidents

9.4 Reporting and investigation of certain incidents, accidents and occurrences may be a requirement in certain sectors under specific conditions (e.g. RIDDOR, medical products "vigilance" systems, Pollution Prevention and Control Regulations, transport accident investigation branches, etc.). However, there is currently no mandated requirement for an organisation to conduct its own internal investigations of workplace accidents, although most enforcing and regulatory authorities would argue that to do so is best practice and is implicit in other obligations. As well as the safety improvement focus of accident investigation, there are a number of other significant benefits to organisations which routinely investigate workplace incidents and accidents.

Organisations which do conduct thorough internal investigations of their workplace accidents may be able to use this to provide evidence of their commitment to health and safety issues. Competent investigation of workplace accidents shows a level of integrity in managing workplace health and safety, and demonstrates a willingness to learn from incidents and accidents. A thorough and objective internal investigation may reduce criminal liabilities, both immediately following the accident and in the future. However, accident investigation is only likely to be effective in reducing criminal legal risk if it forms part of a structured and focused safety management system, and if a documentary audit trail of recommendations and actions is maintained. Hence, tracking recommendations and providing a full rationale where recommendations are not accepted are also important considerations in reducing legal risks.

In addition to prosecutions from enforcing authorities, there are also civil legal risks associated with incidents and accidents. There is a clear and understandable possibility of legal action from those parties directly involved, such as those present, and the family, friends and colleagues of anyone injured or fatally injured. Where an accident has led to loss of income, or has contributed to financial difficulties, the incidence of such claims is likely to rise. This is particularly likely where management contact with the accident involved parties is perceived as being unduly focused on "return-to-work" issues, rather than concern for the employee's well-being and welfare (Cormack et al., 2006).

The probability of a civil claim is also increased where those involved do not perceive that the accident is being investigated properly. In these situations, there may be concerns that the true causes of the accident will not be revealed, that the findings are a foregone conclusion or that the accident will soon be forgotten without appropriate safety actions being taken. Whether involved directly or indirectly, some people can experience high levels of cynicism following accidents, and these are exactly the situations which may be regarded as high risk in terms of individuals pursuing civil claims.

These issues are related to the wider perception of a company which has been involved in a serious accident. Accidents influence the perceptions and actions of customers, suppliers, industry associates and competitors. Ultimately, the way that an organisation manages its recovery from a serious accident can have a profound impact on business continuity and reputation.

Finally, many organisations will want to prevent recurrence as a simple matter of ethical management. Most health and safety managers would agree that it is important to investigate workplace accidents in order to achieve some degree of catharsis and organisational learning. However, many health and safety managers acknowledge that they see the "same" accident time after time – which would suggest that accident investigation reports and recommendations are not always integrated into the safety management system as well as they might be. Thorough accident investigation, supported by a structured analysis framework and associated database, allows an organisation to review trends in incident and accident data, and introduce and evaluate appropriate policy changes and safety initiatives. Accident investigation as part of a pro-active safety management system can be a powerful driver of organisational safety behaviour and can ultimately enhance company performance.

Developing an accident investigation procedure

9.5 A procedure for investigating incidents and accidents should be in place before the event occurs, providing details of the company policy and procedures for investigating incidents and accidents internally. The level of detail in the accident

Figure 9.1: Flowchart of the accident investigation process

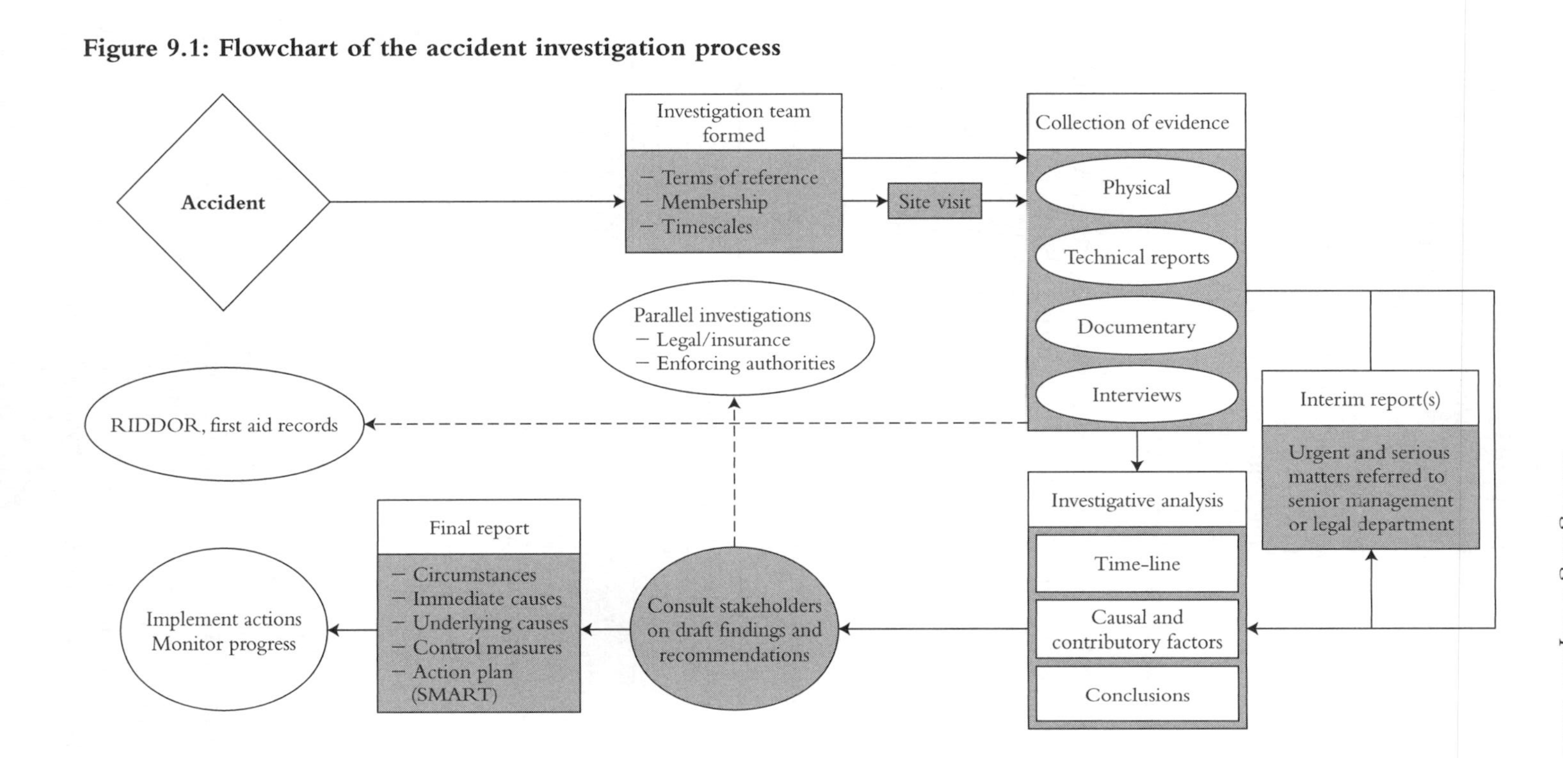

investigation procedure will depend to a certain extent on the nature of the industry and sector. Safety critical industries will inevitably have more rigidly prescribed procedures and requirements than less regulated industries. However, it is important to note that procedures cannot possibly cover every eventuality, and the investigative team must be given discretion to use their professional judgement in matters not covered therein. Processes need to be tailored to individual organisations' needs and management structures. Figure 9.1 outlines a basic model for investigation of workplace incidents and accidents.

What will be investigated?

9.6 The accident investigation procedure should provide guidance on the types of incidents and accidents to be investigated. Further, the level of investigation and the resources to be made available are likely to vary according to the severity of event outcome. Table 9.1 shows the level of investigation that the HSE suggest is suitable in a range of circumstances. This is based on the level of risk of the adverse event, since it incorporates the potential worst consequence of the adverse event, and also the likelihood of recurrence.

Table 9.1: Appropriate level of investigation based on risk. From HSE (2004, p. 10)

Likelihood of recurrence	*Potential worst consequence of adverse event*			
	Minor	*Serious*	*Major*	*Fatal*
Certain	Low	Medium	High	High
Likely	Low	Medium	High	High
Possible	Low	Medium	High	High
Unlikely	Minimal	Low	Medium	High
Rare	Minimal	Low	Medium	High

© Crown copyright. Reproduced with permission of the Controller of HMSO and Queen's Printer for Scotland

The HSE (2004) defines these levels of investigation primarily in terms of the composition of the investigation team. For example, minimal investigations involve the supervisor in seeking to establish what has happened and prevent recurrence. A low-level investigation is a short investigation by a supervisor or line manager which is focused on investigating immediate, underlying and root causes; a medium-level investigation has the same aims, but will typically involve a health and safety specialist and/or relevant employees; while a high-level investigation is more detailed still and will typically be commissioned by a senior manager or director.

The actual severity of the outcome of an incident or accident should not be the only factor used to determine the appropriate level of response. Since luck almost inevitably plays some role in determining the outcome of an accident, seemingly minor events can have potentially extreme consequences. It may therefore be that valuable safety lessons can be learnt from the thorough investigation of less serious incidents and near misses. This is especially likely to be the case where a significant or interesting chain of causal factors is suspected or where the safety issues arising are likely to be far reaching.

As an example of this principle, a critical failure in the mechanism used to secure a helicopter rotary blade may occur a few inches above ground, leading to minor injuries to the occupants and minor damage to the helicopter. The identical failure, with identical causes, hundreds or thousands of feet above the sea, would have far more serious outcomes. In addition to injury or loss of life and the loss of the helicopter, and depending on the established causes, there could be losses to the business and legal risks. Hence, in practice, the *possible* outcome of the event, as well as its actual outcome, will determine the appropriate level of response. However, this general rule should be applied with a good degree of common sense, since one way in which people cope with the occurrence of incidents and accidents is to console themselves with the thought that it could have been far worse. The possible outcome must have had a reasonable likelihood of occurring and not have been an outside chance or remote possibility.

Finally, in very serious accidents, the descent of dozens of media representatives into the accident location can cause real problems in managing the aftermath of the accident and in conducting the accident investigation. Tasks such as finding accommodation, keeping witness and employee details confidential and managing the release of information to the press become much more difficult in these situations. It is important to acknowledge that journalists may try several approaches to obtain information from individuals associated with the accident, including investigators, employees and witnesses.

For example, following a recent train derailment which resulted in five passenger fatalities, an injured passenger was approached on his doorstep by a journalist asking for information. The passenger had left his car in the station car park, and the journalist was able to trace the passenger from the car registration details, since his was the only car remaining in the car park overnight. However, media involvement is not necessarily negative, since the press may be able to provide investigators with useful information. Witnesses with photographs and video footage taken at the time of the event may make this information available to journalists, and a good relationship with the media can assist in gaining access to such materials, if required. Close liaison with the company press office should be maintained, and all members of the investigation team should be briefed accordingly. Media awareness training may also be a necessary consideration, particularly for the principal investigator. Table 9.2 provides detail of the type of factors to be considered in compiling an accident investigation policy.

Table 9.2: Points to consider in compiling an accident investigation procedure

- Consider how the incident and accident investigation process will fit into the organisational structure, and consider its consistency with the existing safety management system, reporting lines and existing procedures on emergency response and disaster management.
- Define levels of incidents and accidents, for example low-level investigation, medium-level investigation, high-level investigations, and the resources generally available for each.
- Provide general guidelines for the conduct of the investigation, such as who can convene or commission an investigation team, under what circumstances, terms of reference, and the timescales for reporting different levels of investigation.
- Outline criteria for the composition of the investigation team for different levels of incident/accident and outline the key responsibilities of different team members.
- Consider what practical support will be made available to investigation teams, such as liaison with enforcing authorities and regulators, on-call legal advice, press office support, arranging site visits, booking accommodation and travel.
- Provide training for all managers and specialists who might be called upon to chair or act as an investigation team member. Training will need to include the key skills of investigative interviewing techniques, methods of analysing investigation data and report writing techniques.
- Provide detail of the procedures for conducting any interviews, such as whether the interviewee can be accompanied and by whom, how the information will be recorded (tape recorder and verbatim transcript or summary statement from contemporaneous notes), who the information will be shared with, whether a copy of transcripts or the summary statement will be provided to the interviewees (and when) and whether such transcripts and statements are to be signed.
- Provide information on gaining access to other sources of information, such as site visits, maintenance and personnel records, etc. In addition,

provide information on how any necessary forensic testing, research and/or technical analysis required to support the investigation can be commissioned if it is required.

- Provide specific guidelines on the retention and use of written documents or files, and electronic information (including digital photographs), generated as a result of the investigation. For example, consider who is responsible for these items, who may have access to them and how long they should be kept for.
- Specify any particular data analysis or classification techniques to be used, and explain how this is important for conducting the investigation, for database storage, search and retrieval of investigation results, the investigation of trends over time, and use in risk assessments.
- Include interim reporting procedures if necessary, such as where urgent and serious matters may need to be referred to senior management or the legal department for decision and/or immediate action.
- Provide some detail of the required structure and style of the report, and who the audience for this report might be (internal and external). Provide an indication of how recommendations will be approved and tracked, and issue guidelines on how actions are to be allocated and resourced.
- Provide some detail on how the incident and accident investigation process itself is to be reviewed and monitored, and how investigators may provide feedback on the policy, procedures and the process itself.

Who controls and co-ordinates investigations?

9.7 In a small organisation where lines of communication are short it will be obvious who takes charge of the investigation. In larger organisations there can sometimes be confusion or disagreement about who should launch and conduct an investigation. This may even result in more than one investigation commencing. The accident investigation procedure should make clear where the responsibilities lie for setting the terms of reference, the reporting lines during the process and other lines of communication. These should differ depending on the appropriate level of investigation (see Table 9.2). The procedures should encourage co-ordination between various departments with differing responsibilities including insurance, health and safety management, human resources and legal.

Consideration should also be given to how the internal investigation team operates in relation to parallel investigations being conducted by enforcing authorities. Some organisations might find it valuable to have a liaison officer specifically to deal with any external investigation. This individual can be made responsible for dealing with all requests from enforcing and regulating authorities, such as the HSE, the police, the Environment Agency and the Department for Trade and Industry, and can also co-ordinate the response. Ideally, this person should be independent of the internal investigation team, allowing a degree of separation between internal and external investigations. This will allow the internal investigation team to focus on completing their own investigation, without being distracted by requests for information from external sources.

The investigation team

9.8 An investigation may often be conducted by a single individual, but more significant events are likely to be investigated by a team or panel. An investigation team will normally be led by a principal investigator or chairperson. This person may be an operational manager, a health and safety manager, a manager from another department or directorate, or in very serious cases, a director or independent expert. Whatever the background of the principal investigator, his or her role will be to take overall responsibility for the conduct of the investigation, including duties such as:

- Allocating roles and co-ordinating activities between investigation team members.
- Ensuring that the day-to-day management of the investigation runs smoothly in terms of resources, timescales and allocation of new tasks.
- Holding regular briefings and meetings to share information, findings, hunches or concerns and to plan further activities.
- Representing the "face" of the company while leading the investigation in a professional and ethical manner throughout, and particularly when dealing with accidents involving employees, other witnesses and any external parties.
- Taking a key role in analysis of the events and causal and contributory factors, while also ensuring that the voice of all team members is heard.
- If necessary, agreeing the allocation of recommendations and actions to the relevant departments when the draft report is released.
- Ensuring that a cohesive and comprehensive report is produced on time and with appropriate recommendations.

It can sometimes be helpful to have the principal investigator, or another investigation team member, come from a different technical background or functional area to that in which the accident occurred. Firstly, it can increase the independence and perceived independence of the investigation. Secondly, it can help in conducting interviews, as someone from a different field or technical area is less likely to make assumptions, or assume something is "obvious", based on their experience in the field. An independent investigator is more likely to be willing to ask "why do you do it that way?" This may help to ensure that things which might never be questioned, or which might seem common knowledge, are not overlooked.

Investigation team members are normally recruited to an investigation team at least partly for their specific technical expertise or professional experience, although a team member may be allocated to an investigation from a different division in order to provide a greater degree of independence. The advantage of this independent approach is that a non-specialist will be able to ask legitimately simple questions which require interviewees and witness to explain the obvious, without losing credibility. If the accident is believed to have involved a significant element of human factors, such as where there are questions over why someone behaved the way they did, or why a particular decision-making process was followed, it may be useful to employ a human factors specialist on the team – some organisations employ human factors expertise on a retainer basis.

Briefing meetings

9.9 In order to share an understanding of the background to a particular incident or accident and help to build the investigation team, an initial briefing meeting is essential. The aims are to discuss the terms of reference for the investigation, and to confirm that all statutory reporting has been conducted. In addition, the roles and responsibilities of team members may also be discussed, although it may only be possible to allocate specific tasks once the likely scope of the investigation becomes apparent. Information to be shared will typically include:

- Who has commissioned the investigation and whether the investigation is conducted under legal privilege in anticipation of litigation.
- Introduction for all team members, including information on technical experience and expertise.
- A description of the accident, including time, location, weather and so on.
- The well-being of those involved, including injuries, hospitalisations and fatalities.
- Brief descriptions of background of those involved (whether employees, members of the public, bystanders, passengers, etc.).
- Legal reporting requirements and responsibilities.
- Involvement of enforcing authorities and contact details for liaison officer.
- Terms of reference for internal investigation, including timescales.
- Roles and responsibilities of team members.

- Practical arrangements such as travel, accommodation.
- Any media interest generated, press conferences/press office details and company policy on press comments or interviews.

It will help to run through the company accident investigation procedure, particularly for the benefit of team members who have not previously been involved in investigations.

Regular team meetings are a valuable method of sharing information and updating colleagues on data collected and questions generated. However, they can be time consuming and costly if there are many participants, and attendees are widely geographically dispersed. For smaller investigations, they are an ideal means of updating all team members and allowing the principal investigator to keep track of investigation activities and findings. For larger investigations or investigations over longer timescales, greater distances, or on multiple sites, an alternative method of regular communications may need to be established.

Site visits

9.10 Site visits are always valuable, since they allow all members of the investigation team to clearly visualise the location of the accident, and to better understand the factors which might have been involved. A site visit can help investigators to place all evidence in context and can assist with any engineering investigations that are necessary. The site must be made safe before this visit takes place. If enforcing authorities have control of the site, then the company liaison officer will need to negotiate or arrange site access for the internal investigation team. However, enforcing authorities may have disturbed the site, and/or removed physical sources of evidence such as damaged items, which will of course reduce the amount of information available to the internal investigation team.

In making a visit to the site, it will be helpful to have a representative from local management to explain the nature and context of the site, as well as to talk through the accident. Viewing the site will also allow for easier interpretation of any building plans or site maps, where these are available to the investigation team. However, care must be taken to prevent site guides, be they managers, employees or witnesses, from demonstrating how they think the accident happened or how any injuries were sustained. Such activities can sometimes include moving switches and guards, or trying to demonstrate a slip or a fall, before the investigator even realises what is happening. This type of activity could destroy valuable investigative information or lead to further accidents. It is also important to bear in mind that it is the purpose of the investigation to establish the sequence of events and the causes of the accidents, and therefore comments and suggestions from others should be treated as hypotheses in the early stages of the investigation.

On site, it is useful to take as many photographs as possible from different angles and perspectives. Bear in mind that objects may have been moved, and lighting and weather conditions may have changed since the accident occurred. In addition, sketches can be extremely valuable, particularly where they include detailed measurements of distance, angles and so on. Diagrams of the layout and configuration of particular items of equipment or machinery can also be made, although it may be necessary to call in a technical illustrator if the items are complex. Manuals, procedures and instructions may also include useful diagrams, and these can often be made available for referral during a site visit, with copies made available to the team on an ongoing basis where necessary.

Sources of evidence

9.11 The amount of information available to an investigation team can be almost without limit, but typically will include both physical evidence, including any relevant testing and engineering analyses, written or electronic data or documentary sources, and evidence gained at interview. The key challenge for the investigation team will be to obtain all relevant information without becoming overloaded.

Physical evidence

9.12 Physical evidence includes all items of infrastructure, plant, equipment, materials, etc. that may have been involved in the accident. Before disturbing such evidence or removing it from the site for any kind of testing or analysis, photographs and sketches should be obtained of the items in situ, including details of the surrounding area. Depending on the nature of the equipment involved, it can be very helpful for an engineer to examine a fully functional item, before disturbing and examining the piece that was involved in the accident. For equipment, plant and substances, specialist forensic or engineering testing and/or analysis may be required in order to establish the nature and possible causes of any damage. Full reports should be obtained of any such tests in order to document the analysis methods, results and/or professional opinion provided. In addition, a reconstruction or simulation of the accident circumstances may be conducted under certain circumstances. For example, recreating the configuration of a piece of machinery may provide information on the failure modes; re-staging a lifting operation may provide useful information on how the operation failed. Where an accident is reconstructed rather than simulated, these activities should be carefully planned and controlled, in order to avoid a replication of the accident itself. The methods and results of any reconstruction or simulation should be fully documented in written reports.

In the case of serious incidents, the relevant emergency services will be in attendance and the first priority will be the care and treatment of those involved in the accident. In such instances, the enforcing authorities will determine their own response to the accident. This may include taking control of the site and preventing further access to physical evidence with a "direction to leave undisturbed", or by detaining items. In these cases, it may be difficult to negotiate access to physical evidence, although it may be possible to obtain copies of any photographs, analyses and tests conducted by the enforcing authorities.

Documentary evidence

9.13 Documentary evidence describes written or electronic data sources. Typical sources of documentary evidence are provided below.

Personnel records: May provide detail of educational qualifications, length of time with company, disciplinary issues and records, medical information and occupational health issues (e.g. occupational illness, accident records, use of medication/drugs), next of kin and family details.

Training and performance records: May provide details of occupational training courses attended, including qualifications and clearances, training content and the date and level of certification; safety briefings on the safe system of work; appraisal information including performance targets and whether or not these were met, and details of performance monitoring and mentoring.

Rostering, rotas and shift patterns: May provide information on duties/shifts completed immediately before the accident, which could provide an indication of fatigue issues, last-minute changes of staff and availability of more experienced colleagues/supervisors at the time of the accident. Overtime records may give some indication of whether staffing levels were appropriate, and whether adequate personnel were available to meet operational demands.

Equipment manuals, product data sheets, procedures and instructions: Can be useful for clarifying known risks and precautions and the procedures which are *intended* to be used in a given situation. However, accident investigators should recognise that a number of caveats are associated with procedures, manuals and written instructions, since they could bear little resemblance to the way that a task is accomplished in practice. People may not refer to the written procedure for completing a task, especially if the task is normally completed successfully on a frequent or routine basis. Custom and practice in the workplace may determine how a task is completed rather more than a written document, especially where a procedure specifies a cumbersome or time-consuming way of completing the task.

Work permits, method statements and safe systems of work: These normally provide a clear and auditable permission for a task to be completed or a site to be accessed. Any records of monitoring and audit of the safety procedures should be collated. Departures from permitted arrangements can be important in identifying the immediate and underlying accident causes.

Risk assessments: Written risk assessments may be available for specific items of equipment, tasks and jobs, or sites. They will typically identify hazards, risks and precautions. Informal risk assessments might be undertaken where there is a change of plan, or an event or circumstance arises that was unanticipated in the original assessment. These assessments of risk are often undertaken on a dynamic (and verbal) basis and may not adequately reflect the real risks of the situation. Further, no documentary evidence for a dynamic risk assessment will exist, and this information will need to be obtained from interview, especially where there has been a change of plan or it seems that the work deviated from the method statement or agreed procedure.

Written briefings and checklists: Written briefings can provide some detail on the specific plan and instructions for a particular task or job, particularly where there was something unusual about the situation. However, supervisors and first line managers will often provide a verbal briefing, so a written record may not always be available. Checklists can serve as valuable memory aids, but they are only likely to be used routinely in highly regulated industries where there is a clear culture of following checklists in a prescribed order. If they are not referred to routinely, then they are of very little use.

Maintenance schedules and records: These should provide details of installation and commissioning, and should include details of routine and non-routine maintenance. This may include replacement of parts, engineering analyses and information on faults and problems. They can also be cross-referenced to manufacturer supplied information on maintenance and operation.

Automatic recording devices: Depending on the industry/sector, these could include telephone records, tachometers, CCTV footage, voice recordings, equipment monitors and fault tracking programmes. While these data sources are routine in certain sectors, such as transport, their potential can sometimes be overlooked in other industries. It is always worth considering whether useful data could be obtained from any of these sources if they are available. They can provide valuable information which can corroborate that obtained from non-documented sources, such as witness interviews. They may also provide the data necessary to create a computer simulation of the accident, as might be the case with flight data recorders and air traffic control radar replays.

Production targets and schedules: May allow for checks on the efficiency of scheduling and production, and highlight any bottle-necks, over- and under-runs, timing problems and other production pressures.

Committee paperwork: This covers items such as terms of reference and minutes of meetings, and can be a useful data source for tracking the progress in resolving known problems, or taking particular actions and decisions with regard to a particular concern.

Site maps and building plans: Depending on the context, these can provide useful information on the workplace layout. However, such diagrams are not always reliable, since their accuracy and level of detail are generally related to the reason they were produced in the first place. In addition, older plans are less likely to be accurate simply because of the accumulation of changes over time.

Working environment: With accidents outdoors, it may be necessary to obtain weather records. The indoor working environment may also be relevant, for example temperature records.

Proposals, contracts and costings: Depending on the context, these may provide key summary information on the way work was planned and the resources and equipment available, and who was responsible for such items. The use of such items is often particularly relevant where there is a high level of sub-contacted work, since the paperwork will normally specify the obligations of the contractor and any subcontractors.

Previous accident records: Report archives, accident records and near-miss reports may need to be reviewed, for example for patterns of unsafe practices or weaknesses in previous recommendations and remedial actions.

First aid records: These may be relevant to issues of adequacy of the emergency response and whether first aid arrangements can be improved.

An important caveat is that accident investigations in many sectors have shown that documentary evidence does not always reflect the reality of corporate or operational situations. Instead, they may only serve as an "organisational baseline" against which to evaluate actual customs and practices, or to establish how what happened in a given incident or accident differed on this one occasion. Documentary evidence sources are often relied upon heavily in legal circles, since it is difficult to refute the written word. While the authors of written documents are knowingly creating an indelible record that does not mean that the record created will reflect reality.

Any documents created immediately after an accident should also be collated and placed on file for the investigation team, including *RIDDOR* report forms, internal notifications, statements taken from those at the scene, communications with inspectors from the enforcing authorities and any enforcement notices served. In producing written documents in the course of the investigation, whether e-mails, memos or interim reports, authors should bear in mind that the potential audience may be wide. The precautionary notes provided in the section on written investigations reports also apply here (see page 329).

Interview evidence

9.14 In addition to physical and documented sources of evidence, there will be evidence gained from interviews. In essence, this is simply information which is gleaned from asking people. It is not unusual for a very high proportion of the information in an accident investigation to be derived from interviews, although of course this does vary depending on the nature of the event.

Potentially, an interviewer can gain information on a wide range of salient issues: actions and decisions, including perception and diagnosis of the accident situation; what was done once the event had occurred; what verbal workplace instructions were given and received; workplace norms and values; custom and practice activities; relationships between colleagues, supervisors and subordinates involved; use of normal and emergency procedures; competence and confidence; and safety culture, climate and behavioural safety information. In addition to providing verbal answers, the interviewee will use non-verbal communication, such as body language and tone of voice, to convey meaning to the interviewer. Tellingly, sometimes what is left unsaid reveals as much as what is voiced.

Given that human error is implicated in such a large proportion of accidents, it is vital to discover what people did, how they reached their conclusions and why they took particular actions. This can require investigation of workplace acts and decisions from the immediate cause at the front-line, through to latent supervisory and management errors. The relevant decisions and actions need to be examined in context, to find out *why* an individual made a particular decision or behaved in a certain way. As discussed in Chapter 2, it is the intent of an act that determines the nature and type of error, and knowing this is important in making recommendations that will reduce the probability of it happening again. The only way to understand why someone did what they did in a given situation is to ask them, since information on human error and human factors cannot be obtained from a maintenance schedule or engineering report. However, as nobody likes admitting to having made an error, gaining accurate information can be a very delicate matter.

Unfortunately, interviews can potentially provide some of the most unreliable and misleading information in an accident investigation. The information gained from an interview can often *only* be obtained from interview and cannot be corroborated by other information sources. Investigators must therefore have a clear understanding of the limitations of eyewitness testimony in order to accurately evaluate the information obtained from an interview, to obtain corroborating evidence from other interviews where possible, and to assess the accuracy, validity and worth of information obtained in this way.

Perception, memory and deception

9.15 The quality and quantity of information gained at interview is dependent on (1) what the interviewee saw or heard in the first place, (2) on how much of the event can be remembered and (3) how much of the memory of the event the witness is willing to share.

Raw sensory information enters the body via one of the primary sense organs: eyes (sight), ears (sound), nose (smell), mouth (taste) and skin (touch). However, not all sensory information is processed by the brain in a meaningful way, and not all of it reaches conscious awareness. People are only conscious of a mental interpretation of sensory information and this perception is constructed by the brain. However, in order to experience this perception, a witness must generally first have paid attention to the sensory input. If sensory information is not attended to, then the brain will filter it out, and it is unlikely to be processed further.

A striking example of the influence of attention comes from research into inattentional blindness. This is where a visible event or object is not noticed simply because the viewer is not paying attention, and not because the event or object is obscured or masked from view. One study in a flight simulator required pilots wearing head-up information displays to land their aircraft on a runway. Some of the pilots in the study landed their aircraft by actually flying *through* aircraft which were parked on the runways. They simply did not perceive these unexpected aircraft, because they were paying attention to the information displayed on their head-up helmet units (Haines, 1991; cited in Crundall et al., 2005). The simulated aircraft were clearly visible but simply not perceived by these pilots.

Even where an interviewee has paid attention to the event or object of interest, it is important for the interviewer to be aware that their perception of the event or object may not necessarily be an accurate representation of reality. What people think they have seen is not an infallible record of what actually occurred. Because the brain interprets sensory information to make it meaningful, discrepancies between reality and perception do arise. For example, car drivers often misjudge the speed of trains at level crossings, resulting in a large number of train–car collisions and associated injuries and fatalities. The sheer size of the train means that the driver will normally perceive it to be travelling more slowly than is actually the case. Further, the perception of distance which is generated by railway tracks converging towards the horizon can lead drivers to believe that the train is further away than it actually is, and that they will have more time to cross the tracks than is actually available (NTSB, 1998). Because of this common inability to accurately estimate sizes, distances and angles, interviewers are well advised to use models, plans and photographs when interviewing witnesses and to check any estimates following the interview.

In addition to attention and perception, witness memory will also have a bearing on the quality and quantity of information obtained at interview. Most memories relating to an accident will be held in long-term memory. Long-term memory lasts indefinitely, as some long-term memories last a lifetime, and it has an enormous capacity. However, it can be unreliable, since, as with perception, people generally attempt to "make sense" of information held in long-term memory, and interpret their experiences in the context of knowledge, experience and expectations. For example, most people do not have accurate memories of familiar items, even though they may use them every day. You probably do not remember what the ATM machine you last used actually looks like. Instead, you use a mental schema of an ATM. A mental schema is a representation or story, which includes details of what you usually do when you use an ATM machine. The problem is that schemas are general – they contain information on the usual or typical, and not the specific or particular.

Asking an interviewee what they did on the day of the accident may well evoke a schema, in which case the information provided may reveal what it is they usually do, rather than what was different on the day of the accident. The longer the period of time between the accident and the interview, the more opportunities for the interviewee to remember and talk about the event, and the more likely it is that the event will be replayed around a relevant schema, and detailed information will be lost. The use of schemas, along with elaboration, enhancement and gap-filling is simply a function of the way that human memory encodes, stores and retrieves information (see Table 9.3).

In addition, it is important to remember that no one expects an accident to occur. They are unpredictable and unexpected, occurring without advance warning. Accidents are an unusual and unique combination of events, but in isolation, each event is probably perceived as normal, routine or ordinary – at least until the witness realises the significance of what is occurring. It is most unlikely that a typical interviewee will have been making a deliberate and conscious effort to notice and remember everything.

Table 9.3: Types of long-term memory

Semantic memory is memory related to rules and knowledge. If you were asked what car Shakespeare drove you would quickly be able to answer that he did not have a car. Your semantic knowledge indicated that Shakespeare was dead long before motorcars were invented.
Episodic memory is autobiographical; it is a narrative memory of personal experiences and events. Most information in episodic memory is not stored in great detail, and it often requires interpretation to be meaningful. For example, if asked what I did last Saturday, I might say that "I went to El Machos, and then popped into the King's Head for a swift half". In order to make any sense of my answer, you would need to know that El Machos is a Mexican restaurant, that the King's Head is a pub, and that a swift half is a drink of beer.
Flashbulb memory is a particular type of episodic memory. Flashbulb memories are particularly strong memories of unexpected events, and they often have an emotional component. Examples might be memories relating to the assassination of Kennedy or the death of Princess Diana. Accident involvement may well involve flashbulb memory. Although flashbulb memories can usually be recalled in detail, this does not necessarily mean that they are accurate. The experiences they relate to will be remembered and discussed often; the tale can grow and change in the telling.

Even where an interviewee saw the event and has remembered useful material about it, it is important for interviewers to be aware that there may be some reluctance to share that knowledge. Factors which increase reluctance to disclose include fear of reprisal, concern about disciplinary or legal action, perception of intimidation from the interviewer or other sources, protection of colleagues, embarrassment about decisions and actions and uncertainty or lack of knowledge about the event. In these circumstances, interviewees may provide incomplete information, or be somewhat evasive in responding to questions.

In contrast, deception is a deliberate attempt to provide the investigators with information which the interviewee knows to be untrue. The main difficulty with deception is that it is almost impossible to detect, since the behaviours that most people associate with deception are also indicative of other states (Vrij, 2000). For example, many people believe that avoiding eye contact, touching or obscuring the mouth, and increased hesitancy of speech are behaviours typically exhibited when lying. However, these behaviours might be expected of someone who is nervous during an investigative interview. Research also shows that they are not reliable indicators of deception.

Provided that the interviewer is able to take these limitations into account, interviews can provide the richest, most detailed and most revealing information in any accident investigation. The interviewer's competence will directly influence the quality of the investigation, the reliability of the findings and the appropriateness of recommendations and actions generated as a result. Reliable, high quality and comprehensive information is only likely to be elicited if the accident investigator uses reliable and well-researched interview techniques. Suitable techniques are those which elicit as much accurate information from the witnesses' memory as possible, without producing an increase in the amount of erroneous information obtained.

While an outline of approaches and techniques is provided here, it is important that investigators are properly trained in the psychology of conducting interviews for incident and accident investigation. Ideally, such training should incorporate practical accident investigation exercises and investigative interviews, and should include constructive feedback, tailored to each individual's own interview style.

Investigative interviewing

A controlled conversation

9.16 An interview is a controlled conversation conducted to obtain information for a specific aim or purpose. In an investigative interview, the aim is to obtain as much information as possible from the interviewee in order to determine the causal and contributory factors which led to the accident. It is vital that the interviewer takes all possible steps to encourage disclosure – a task that can be made very difficult with a reluctant interviewee. The principles of investigative interviewing apply to all interviews, but their real strength lies in the way that they can help the interviewer to initiate and maintain a conversation with even reluctant witnesses.

Only enforcing authorities can *require* answers to questions. For the most part, investigators within an organisation will be relying on goodwill and rapport to obtain full and honest answers to the questions which are put. The willingness of an interviewee to discuss their experience of the accident has an influence on the approach taken. For reluctant interviewees, the interviewer should treat the interview as a highly structured conversation and should simply persevere politely and persistently. For willing interviewees, conversational approaches can be supplemented with the use of principles from the cognitive interviewing methodology (see Table 9.6). However, cognitive interviewing techniques are of little or no use with an uncooperative witness.

Advance planning is vital, since the interviewer must have an indication of what information is being sought, and the likely perspective of the interviewee, before arranging the interview.

Interviewees

9.17 Potential interviewees include anyone likely to have information of value to the investigation. For example, depending on the accident, workplace interviewees could include:

- Employees and contractors.
- Team leaders and supervisors.
- Junior, middle and senior managers.
- Health and safety advisors.

If appropriate, members of the public and other witnesses (such as passengers, bystanders, etc.) may also be approached. In many instances, contacting such individuals may be difficult, unless their contact details are obtained at the scene. Whenever interviewing anyone from outside the organisation, it is ethical practice to explain why the investigation is taking place, and to provide information on the uses to which the information obtained will be put. Members of the public and bystanders will normally want reassurances regarding the use and release of personal/confidential information, and it is a good idea to provide these details in writing and to obtain written consent.

The range of potential witnesses is sometimes underestimated in an investigation, with the result that valuable information is overlooked. People who were outdoors at the time the event occurred can act as "eye" or "ear" witnesses: consider breaktimes at schools and colleges, sporting events, road and rail work gangs and milkmen and postmen. Passengers in particular are often overlooked, and yet in transport accidents they provide the most valid and useful information on secondary safety. While passengers can typically say very little about how and why the accident occurred, they can often say a great deal about the emergency plans, procedures and equipment provided for their use, and can provide evaluative information on whether they functioned as intended. This information can clearly help to reduce injury and fatality in future accidents (Thomas & Rhind, 2003; Thomas & Muir, 2005).

For anyone under the age of eighteen, or who may be regarded as vulnerable for reasons other than age, the permission of a parent or guardian should be sought prior to the interview taking place. Similarly, employees may need to be provided with independent legal advice if they are to be interviewed by an enforcing authority, since the information they provide may be used against them individually in certain circumstances, as well as against the organisation.

Table 9.4: Conducting group interviews

Individual face-to-face interviews are the most effective type of interview, and they should be conducted wherever possible. Individual interviews should also be conducted with people who are unlikely to open up in a group interview situation. Examples would include a reluctant interviewee, anyone who may have made a critical error or mistake, anyone who may have exacerbated the accident situation, anyone who appears to be less popular with their colleagues and anyone who is particularly distressed or upset by the accident.
Sometimes it may be necessary to conduct group interviews, but it is very difficult to manage these situations. Interviewees will interact with each other, and bounce ideas and answers around. This can sometimes be useful, but it also makes it very difficult to obtain individual accounts of who saw or did what, and it also makes it very difficult for the interviewer to remember who said what. Because people within a group hear each other's accounts of the event, individual perceptions and beliefs may change during the interview.
Some individuals are less likely to share information in a group situation, particularly if they are socially shy or sensitive about speaking in a group. Also, while dominant and talkative individuals tend to speak most and speak loudest, they do not always have the most useful or accurate information.
If you have to conduct a group interview, arrange seats in a ring around a table, and try to have several interviewers, one of whom is the "lead" and one who will take notes. Ensure that all individuals get the opportunity to give their own account of the event, and try to avoid obtaining a "group" decision on the definitive answer to every question.

Making interview arrangements

9.18 The first stage of any interview is to make appropriate arrangements. It is recommended that interviews are conducted individually, in person. Individual interviews allow investigators to focus exclusively on a single witness, and face-to-face communication is more effective than any other form of communication. It is also important that the interview is conducted as soon as possible after the accident, although due consideration must be given to people who are distressed and who may need some form of counselling or psychological support.

Generally, the reliability of information gained decreases the longer the time delay between the event and the interview. Conducting the interview between twenty four and forty eight hours after the event would probably be ideal, but this is unlikely to occur in practice for many interviewees. It will certainly not occur for every potential witness, so investigators will need to prioritise and interview key personnel as soon as possible.

Practical arrangements will include considerations such as the time and date, the location of the interview, and who will accompany the interviewee, if anyone. If possible, interviews should be conducted in a neutral venue. Conducting interviews in a senior manager's office may add to the prestige of the interviewer, but it will do little to enhance the quality and quantity of information obtained from an accident involved employee. Interviews may be conducted at a venue of the interviewee's choice, but be aware that the interviewer loses some control of the situation if the interview is conducted at the interviewee's home. People "just popping in" and other interruptions can make for an uncomfortable experience for the interviewer, no matter that the interviewee is relaxed and talkative.

Although the interview is a formal occasion, it is preferable that the furniture is not arranged too formally. Armchairs are too cosy, but chairs arranged in a panel layout across a vast desk will be intimidating for the interviewee. Chairs at a round table, or across the corner of a square table, represent a good compromise. The interviewer should also ensure that any items such as site maps and plans, photographs or models are available to hand. These can be excellent "props" for discussing the event with interviewees.

It is important that the interviewer has control of the venue, if at all possible. This will mean that interruptions and distractions can be minimised, if not eliminated. It should be possible to ensure that telephones are switched off and that a "do not disturb" sign is placed on the door. Colleagues can be informed that an interview will be taking place and can be asked to prevent disturbances from other people. The availability of refreshments can assist with introductions and "settling-in" an interviewee, but the interviewer will need to bear in mind that it will be difficult for the witness to talk while eating biscuits. Tissues should always be available; it is better to have them and not need them than to have no tissues to offer an upset interviewee.

Introducing the interview

9.19 The introduction to an interview sets the tone for the whole conversation, and first impressions are vital in creating and maintaining rapport with the interviewee. The interviewer should introduce the interview honestly in an open and

Table 9.5: Conducting telephone interviews

It is much more difficult to establish rapport over the telephone than it is in a face-to-face interview because of the lack of non-verbal communication. Although not ideal, telephone interviews may be convenient when interviewees are geographically dispersed. They are most useful when only a few factual details need to be confirmed; they should not be used to collect detailed information or with reluctant interviewees. They should also be avoided where an interviewee is likely to be upset, as the interviewer will not be on hand to deal with the interviewee's response. They should always be conducted on a one-to-one basis, avoiding any conference call facility:

- Consider the interviewee's privacy and convenience, and arrange the phone call for a mutually convenient time.
- Invest some time in developing rapport. Begin by discussing a neutral background topic, then verify background and basic factual information to open the interview.
- Appeal to the interviewee's helpful streak by asking for assistance with clarifying a points or confirming a few details.
- Avoid negative statements ("I know that this is stressful for you") or accusatory statements ("I have reason to believe"); these may antagonise the interviewee.
- One of the few advantages of a telephone interview is that the interviewer can refer to written information without the interviewee knowing; use the hands-free facility to do this where necessary.
- Stay tuned to the interviewee throughout, and be aware that they may become guarded, tired, stressed during the interview.
- Try not to leap ahead to the next question, or anticipate what your interviewee might say next – stay focused on their actual answer.
- Take notes throughout the interview and review them immediately on completion. Where was information volunteered? Where were questions evaded? What hunches do you have?
- Telephones may assist in gaining less guarded information because of the perception of distance.

impartial manner. The aim of the interview is to elicit information from the interviewee, and it is important to lay the ground rules before the interview opens.

In particular, the interviewer should provide information on the names and roles of the interviewer(s), the purpose of the interview, which is normally to establish cause and prevent recurrence, whether notes will be taken contemporaneously, or voice recordings made, and whether a summary or recording will be provided, in what timescales. Issues such as anonymity and confidentiality should also be dealt with at this stage, as should the role of any accompanying party (such as a colleague, or union or health and safety representative).

Questioning techniques

9.20 Questions are used to either open a topic or to probe for further information on a subject already being discussed. There are many different question types, and the actual nature of a question can normally only be determined when both the question and answer are heard in context. However, most question types are a variation of either an "open" or a "closed" question. Open questions are those which encourage a full and frank response, while a closed question elicits a very narrow response, often merely a yes/no answer. Some of the more common types of open and closed questions include the following.

Broad questions: These are open questions which do not impose a limit or parameter on the interviewee's response. Examples are *Tell me what happened*, *Describe your day at work in as much detail as you can*. Broad questions are good for opening the interview, since they allow a willing interviewee to give their own account of the sequence of events. This can then be followed up with more detailed questioning.

Bounded questions: These are open questions that place boundaries or parameters on the interviewee by asking for information between specific points in time or events. For example: *What did you do between clocking on and checking the machinery guard?* They are useful where more details of specific events and timescales are required.

Simple or probing questions: These are open interrogative questions which are used to probe the interviewee. Examples are *who* …, *what* …, *where* …, *when* …, *why* … and *how*…. Probing for further information in an interview will often use these questions.

Invitations: These are open questions which invite the interviewee to share information, such as *Could you please tell me*…. They are polite questions which are good for opening the interview or for dealing with reluctant witnesses.

Mirror or echo questions: These questions reflect a just-received answer, encouraging the interviewee to provide additional information. These questions show empathy with the interviewee and encourage further disclosure of information. For example:

A: "I kept pulling the emergency stop lever"

Q: "You kept pulling the emergency stop lever?"

A: "I tried five or six times but the lever was jammed".

Summary or trailer questions: These are questions which contain an element of summary as a run-up to the main question. They can be used to ask for a response on pieces of information from different sources or to move the interview to a different topic. An example would be: *Your supervisor asked you to put on your harness. You were seen climbing the rig without it. Why did this occur?*

Prompts: Prompts are used to encourage continued responses. Verbal examples include *and* ..., *go on* ... and *uh uh*.... Non-verbal examples include nodding in silence and pauses while still maintaining eye contact. Silence is a psychiatrist's trick which encourages an interviewee to continue talking.

Narrow questions: These are closed questions which allow a very narrow response. An example would be *What colour was the indicator?* They are useful where a definite and specific answer is required, but are often overused and generally of limited use.

Direct questions: These are very restricted closed questions that allow a yes or a no answer only: *Did you* ..., *Were you* ... or *Had she*.... There are occasions where a direct yes or no answer is required, but generally, such situations will be rare, and it will often be better to use an open question instead.

Multiple or multiple concept questions: These questions involve asking for several pieces of information at once: *Was that in the procedures or didn't you check?* or *What did they do next?* These questions generally confuse the intervieweewho won't know which piece of information to provide: the interviewer will have to ask another question to clarify the response.

Negative or double negative questions: Examples of negative question types include "You don't know why she didn't do that, do you?" and "Didn't you know that he wasn't there?" They confuse the interviewee, and the interviewer doesn't know whether a yes or a no is a positive or a negative response without further clarification.

Leading questions and leading tags: These are questions which lead, mislead or suggest a particular response, since they include an implicit or explicit statement of the interviewer's beliefs or values, or the desired answer. An example of a leading question would be *So how fast was the lathe going when it span wildly out of control?* An example of a question with a leading "tag" would be: *But you really should have checked that display first, shouldn't you?* Leading questions are sometimes asked with the sound intention that they will assist the interviewee in remembering. Unfortunately, research has shown that they consistently generate inaccurate information. Leading questions and leading tags have no place in an investigative interview. It is better to require the interviewee to remember something "from scratch" than attempt to assist in this way.

Whether open or closed, all interview questions should be ethical, meaning that they should be fair, and they should not contain any element of threat, promise or insult. A question which contains an allusion to disciplinary action or other retribution is likely to be counter-productive. Questioning can be robust, searching and persistent, but there must be no danger that it may be interpreted as bullying.

Questions should be brief and concise, and theoretical and hypothetical questions should be avoided. Each question should have a specific purpose in the context of the investigation: "filler" questions are mere padding and should be avoided. Every question should either seek new information, seek clarification or further detail on information obtained from other sources, or should check that the interviewee has understood the process and is willing to continue with the interview.

Interviewers need to pay attention to the way that a question is phrased, since the way that a question is asked will determine the quality of response. It can be very easy to unwittingly reveal opinion, judgement, emotion or possible consequences in asking a question. This is often done merely by tone of voice or choice of word. Such questions are often unintentional or unconscious: they just seem to "come out wrong". Examples may include mentioning the name of a colleague of the interviewee in a negative context or mentioning a negative trait (such as carelessness or incompetence) in association with a possible cause of the accident. Such phraseology may easily make a witness reluctant to continue with the interview and may increase the number of evasive answers. Interviewers have to be on their guard at all times and make every effort to demonstrate that they are neutral and impartial. A conscious effort should be made to phrase questions in simple and non-emotive language, and to tailor them to the interviewee's ability and vocabulary.

It is also important that the interviewer pays attention to the interviewee's motivational and emotional state throughout the interview, since it may be necessary to take a break if an interviewee is becoming unduly upset, confused or anxious.

In terms of structuring the interview, it is preferable firstly to gain a full and open account of the event from the interviewee's perspective. This will be a relatively simple matter provided the interviewee is willing, since one or two open broad questions can start off the account. These can then be supplemented by open bounded and simple probing questions to gain more detail and probe for further information. Mirror and echo questions, along with verbal and non-verbal prompts, will aid the interviewer in obtaining a full account of all relevant details. Finally, summary and trailer questions can be used to tackle inconsistencies and to ask for responses to contradictory information obtained from different sources. Closed narrow and direct questions should be used sparingly and only where well-defined responses are required. At all times, care must be taken not to appear judgemental or to antagonise the interviewee.

With a less willing interviewee, open questions are also most useful, since closed questions permit a higher proportion of evasive answers. However, it is unlikely that asking one or two broad questions will elicit much detail, and therefore the interviewer must be prepared to ask many follow-up questions. Simple probing questions are the workhorses of such interviews and can be used for the bulk of the interview. It is best to inform a reluctant interviewee at the outset that the interview will follow a chronological structure, since this will make it easier for the interviewee to follow the sequence of events.

Where an interviewee is reluctant, the interviewer must take care not to allow frustration, irritation or annoyance to show – it is important to maintain as much rapport as possible and keep the conversation going for as long as possible. This will also require attention to body language and other non-verbal communication, such as gestures and eye contact. A stock of generic follow-up questions, such as *What happened next?* and *Why?* can also be useful with reluctant witnesses. Regular summarising will help to maintain the conversation and may even elicit further information if the interviewee is asked to comment on the accuracy or sequencing of the account. Unless the interview is being conducted by an officer of an enforcing authority, it pays to remember that answers to questions cannot be required. Trust in the investigation process and a just and fair organisational culture are the best mechanisms for achieving co-operation with internal accident investigations, and these can only be developed in the long term.

Active listening

9.21 Effective interviewing involves not only asking the right questions in the appropriate format at a suitable time, but also actively listening to the information given in response. Listening is a skill which is seldom taught, in contrast with other communication skills such as reading and writing. However, it is a vital component of an investigative interview, as with any conversation. Ineffective listening will damage rapport, increase the number of repetitive or inappropriate questions and reduce the chances of obtaining quality information. In contrast, effective listening allows the interviewer to demonstrate empathy, deepen rapport, and enhance the quality and

Table 9.6: Cognitive interviewing techniques

A number of cognitive interviewing techniques have been developed by psychologists to enhance the information that can be obtained from willing and co-operative witnesses (Geiselman & Fisher, 1997). Research has shown that these techniques can assist the interviewee in generating additional relevant information, without increasing inaccuracies, as long as they are used appropriately. These methods are based on the psychology of memory and remembering, and they are intended to be used by skilled interviewers who have been trained in the use of cognitive interviewing methods.

Reconstructing the context: This means asking the interviewee to reconstruct as completely as possible the context of the event in the interview. This can be done by asking questions relating to all sensory, emotional and environmental factors. Examples would include asking the interviewee what they could see, hear, smell and feel, what the weather was like and what kind of mood they were in. It may sometimes be possible to return to the accident site and walk through the accident while talking to the interviewee. These approaches assist in putting the interviewee back into the situation they were in and can provide additional cues for retrieving information from memory.

Reporting everything: Asking the interviewee to report everything eliminates the possible bias which may arise from an interviewee thinking that a piece of information is irrelevant or unimportant. The interviewee may not have a complete picture or complete knowledge of the event, and may be unable to assess the accuracy or relevance of the information they hold. If they only report information which they are confident is true, valuable information may be lost. Hence, asking interviewees to report everything, while also providing a caution against fabrication and reassurance that a "don't know" is acceptable, should increase the amount of information elicited.

Re-ordering the events: While an interview normally follows a forward chronological structure, asking someone to recall a sequence in the opposite order can assist in retrieving incidental detail. Remembering and recounting an event as it unfolds produces memories which rely on what we think we saw or what we believe to have been the case. Recalling the sequence backwards can help to reduce the influence of these expectations and exaggerations, and retrieve more of the actual detail. It may be possible to begin with the event that surprised, impressed or concerned the interviewee most, rather than with the event that occurred first.

Recounting from a different perspective: This involves asking the interviewee to try and recall events from a different perspective, which can be a physical stand-point or position in space, or another person's character. Trying to recall an event from a different location, or trying to recall an event as another person or as a colleague may have seen it, can again increase the amount of information obtained.

quantity of information obtained. Active listening is more than merely hearing, since it involves demonstrating attention and interpretation to the interviewee.

Active listening comprises four key stages: hearing, understanding, evaluating and responding. Hearing is fundamental, and it is vital that distraction and background noises are kept to a minimum in order that responses to questions are audible, and also so that the interviewee correctly hears the questions posed. This can be a particular problem with telephone interviews, since a poor reception or background noise can greatly diminish both the interviewer's and the interviewee's ability to hear.

Understanding requires that the interviewer correctly interprets the information provided. It is important that the interviewer does not race ahead, jump to conclusions or react emotionally to what has been said. Deviation, irrelevant answers and technical jargon may all influence the interviewer's ability to extract meaning from an interviewee's answer. Seeking clarification, summarising or paraphrasing the answer, or repeating the question in a slightly different format (such as using an open mirror or echo question) are all ways that the interviewer can check interpretation.

The information then has to be evaluated, meaning that the investigator has to place the answer in context, bearing in mind the information that has been obtained from other sources. Where are there discrepancies? Where are the gaps in the information and what knowledge is missing? Does what has not been voiced provide any information, over and above what has actually been said? Is the information volunteered or is there some evasion? What non-verbal communication accompanies the answer? Is the answer consistent with information obtained from other sources? Does the answer conclude this topic? What new questions does the answer raise? What else needs to be covered before the interviewer moves onto another topic of conversation? Evaluating information is an ongoing stream of thought, continuing throughout the conversation.

In addition to evaluating information and generating the next question, the interviewer also has to respond appropriately to the answer that has been given. Sometimes, the response will be another question, perhaps repeating the original question, or using a simple question to follow up or probe for further information. At other times, the response may be an open mirror or echo, to reflect the answer just given and seek expansion or development of the answer. Prompts, such as use of phrases like *yes*, *go on* and *uh uh*, may also be used, along with silence, eye contact and pauses to encourage further disclosure.

"Stock" responses are habitual phrases used repetitively in response to several interviewee answers – examples include *good*, *right* and *okay*. These can be irritating to the interviewee, and they can also be inappropriate if the answer contains negative information. They should be avoided, although it must also be noted that interviewers generally only become aware of such habits if they receive feedback on their interview performance.

Sometimes the interviewer may voice a hunch, which can be useful where the interviewer feels there is more to the matter than is being told. Sharing these hunches with the interviewee must be done tactfully, since hunches can sound accusatory if voiced as a direct challenge. It is best to phrase a hunch as an educated guess, perhaps in the format of a summary or trailer question. Normalisation can also assist in voicing hunches. Normalisation is where the interviewer makes it clear that the behaviour or decision being asked about is a perfectly normal one. Examples include statements such as *accidents like this do happen* and *nobody is perfect, we all make mistakes*. Such normalisation statements can be useful when investigating human errors. Interviewers must bear in mind that few witnesses will admit to having made a critical error in an interview which they perceive as hostile.

Throughout the interview, the interviewer must also be on the lookout for "drop-it" cues, which are indications that the interviewee does not wish to speak further on this topic. "Drop-it" cues can involve defensive body language, such as folding arms, backing away from the table slightly, avoiding eye contact, providing evasive answers or changing the subject of the conversation. "Drop-it" cues should be respected, and if the matter is important, the interviewer should return to the topic later in the interview. If one of the issues of interest is a human error, such as a deviation from procedures, a mistake or a slip, then blunt or repetitive questioning may be counter-productive. It is of course necessary to find out why a given action or decision make sense at the time, but this must be approached subtly as it is easy to antagonise an interviewee. Human error can be a very sensitive issue, and it is easy for an interviewer to unwittingly appear judgemental or to seem to be "testing" interviewees on their knowledge of procedures and instructions.

If an interviewee is angry, bitter, resentful or has other strong emotions regarding the event, it is important that the interviewer does not take this personally. Allowing the interviewee to "vent" gives them an opportunity to get the matter off their chest. The interviewer can then try to address the problem by naming the issue, and allowing the interviewee time to respond. For example, *It seems that you are very angry about what happened* or *It seems that you are very reluctant to talk about this*. Following such statements with silence will give the interviewee a chance to respond. If necessary, the interviewer can call for a break. If the interviewer is uncomfortable continuing, it is advisable to arrange for another interviewer to conduct the interview another time.

On a related note, interviewers should always have contact details for the occupational health department or employee assistance programme to hand. Employers have a duty of care to any employee who appears to be suffering physical or psychological ill health as a result of the accident. Where this might be the case, the individual should be referred to the appropriate service. The interview is not a counselling session, and interviewers have to remain impartial and neutral throughout, without becoming emotionally involved in the event. It should be noted that it is not unheard of for interviewers and investigators to seek such assistance themselves, and they should be encouraged to seek such support where they feel it is warranted.

Closing the interview

9.22 In closing an interview, it is important not to leave the interviewee "hanging on", and also to emphasise to interviewees that they are only one part of the puzzle. Interviewers should always thank interviewees for their contribution, even if it turned out to be less informative than was originally hoped. It is good practice to ask the interviewee whether there is anything they would like to add, to obtain additional remarks or opinions. Also make clear whether there may be a request for follow-up information, and whether another interview will be sought if this is the case. Where the interviewee is from outside the organisation, any requests for permission to follow-up should be agreed to in writing on an interview consent form, completed before the interview takes place. Interviewers should also provide their own contact details in case the interviewee wishes to add further information or make contact in the future.

Some interviewers may read out a summary of the notes to the interviewee, to check for accuracy and omissions before closing the interview. Some organisations request that the interviewee sign the notes or a summary statement actually at the interview. Others may make available a summary of the notes some days later or provide a transcript of the interview. If the interviewee is able to provide commentary or request amendments to such items, they should be informed of this fact.

Finally, and especially where an interview is recorded, be prepared for off-the-record statements as the interview closes. As recording equipment is turned off and notebooks are put away, some interviewees will make apparently throwaway or facetious comments – these can sometimes be illuminating and should be jotted down as soon as the individual has left.

It will be impossible to wait until all of the evidence is available before generating some hypotheses, hunches, theories or ideas about what might have possibly caused

Table 9.7: Investigative interview checklist

Planning the interview

- Identify what information is sought from this interviewee and whether the interviewee is willing or reluctant.
- Draft a list of principal topics and documents to ask about.
- Key interviewees should be interviewed as soon as possible after the event.
- Will the interviewee be accompanied, and if so, by whom?
- Arrange for the venue to be neutral and under the interviewer's control if possible.
- Confirm date, time and venue with interviewee.
- Arrange the furniture appropriately, for example no glaring spotlights, relatively informal seating arrangements, removal of telephone.
- Availability of refreshments, tissues and props such as plans, models, etc.

Introducing the interview

- Introductions – interviewer(s), interviewee and accompanying observer.
- Summarise purpose of interview, include note-taking, recording, confidentiality.
- Outline structure of interview – chronological, beginning with interviewee's own account, then followed by questions on points of interest.
- Turn off mobile phones, OK to ask for comfort break or pause if needed.

Body of interview

- Use broad questions to initiate free account if interviewee is willing.
- Use simple probing questions if the interviewee is reluctant.
- Closed questions used sparingly, avoiding counter-productive questions.
- Use verbal and non-verbal prompts.
- Actively listen to the answers provided, and evaluate and respond appropriately.
- Use a range of questions to obtain more detailed information.
- Keep asking "why" to find out about critical acts and decisions (human factors).
- Use cognitive interviewing techniques if appropriate.
- Use hunches and normalisation where appropriate.

- Use summary/trailer questions for inconsistencies with other accounts/ evidence.
- Pay attention to drop-it cues, returning to issue later if important.
- Deal with interviewee's emotional reactions without becoming involved.
- Ensure your verbal and non-verbal communication are consistent.
- Maintain neutral, open and non-judgemental manner throughout.

Closing the interview

- Summarise key points and ask for confirmation of account and summary.
- Explain what will happen next (further interviews, testing and analysis, provision of statement and/or transcript).
- Thank interviewee for contribution, mention possibility of follow-up.
- Provide your contact details in case of query or further information.
- Be prepared for any off-the-record statement as interview closes.

the accident. During information gathering, and certainly during interviewing, investigators will find that these thoughts and ideas will inevitably develop further. However, it is important that these possibilities are recognised as merely possibilities. If they are elevated to the status of a pet theory, or to an account of how the accident happened, then the risk of confirmatory bias enters the investigation. Confirmatory bias is a natural tendency to seek to confirm one's own ideas rather than disprove them. If investigators only seek evidence that supports their notion of what occurred, then that is all they will find. It is important for investigators to keep an open mind throughout the proceedings, but especially the interviews, and to maintain a willingness to be shown wrong.

Investigative analysis

Constructing a time-line: asking "how"

9.23 There are many theories of accident causation as outlined in Chapter 2. Early theories focused on immediate causes and understanding *how* the accident occurred. In effect, they emphasised the need to establish the sequence of events that led to the accident. Later theories emphasised that this was not sufficient: thorough accident investigation involves seeking to establish *why* an accident occurred and not just how it happened. The *how* is simply a chronological reconstruction of the sequence of events, whereas finding out the *why* involves an analysis of the causes of acts, events, omissions and failures.

Establishing the sequence of events is fundamental to determining how an accident happened and is the first step in the analysis process. Use of an analytical framework assists in moving beyond this, to ask why accident occurred. There are many analytical frameworks available, some of which are grounded in theory, and others which have developed from a more pragmatic approach (Thomas & Rhind, 2003). HSE "best practice" guidance on investigating and analysing workplace accidents and incidents notes that organisations need to decide what framework to use for analysis and decide how to apply it in practice (HSE, 2004).

In practice, the sequence of events can be reconstructed by drawing a time-line. Using post-it notes is a good way to sequence events in the order in which they occurred. These can be moved, changed, edited and amended as further information is sifted. The important point is that the sequence of events should be chronological and should be actual. That is, the sequence of events should be based on time and ordered in the sequence in which events actually occurred. Developing a sequence of events based on when people became aware of events, or when people decided to do something about them, is misleading. These factors are important, but they should be represented as separate events in the time-line. In addition, it is important to note that the time-line may stretch back some considerable time, to possibly months or even years before the accident actually occurred. This is because latent errors and failures may lie dormant within an organisation for many years before causing or contributing to an accident. For example, a pipeline may have been laid too close to other utility ducts some twenty years before being involved in a gas explosion. If the location of the pipeline was a significant factor in the accident, and the pipeline should not have been laid where it was, then its placing in that location was an error that was undetected and uncorrected for two decades.

Analysing causes and contributory factors: asking "why"

9.24 There are several levels in investigating why an accident occurred, and there are always many "why" questions that can be asked. Indeed, the role of an investigator is to keep asking why (Table 9.8 and Figure 9.2).

Once a time-line has been constructed, one method of analysis is to examine the causal and contributory factors for each event. The factors associated with any accident may be causal or contributory. The key issue is whether removing the event

Table 9.8: Asking why, again and again and again. Adapted from Hopkins (2006)

A worker descended a set of steps carrying his tools, and fell and was injured. In investigating the accident, we ask the first "why?" He was not using the handrail, although using the handrails provided was a requirement of company policy. This is the immediate cause of the accident. To establish the underlying causes, ask *why?* He was using both hands to carry his work tools. *Why?* He wanted to use both hands to carry his tools, because if he hadn't, he would have had to make two trips up and down the stairs to bring his tools to the worksite. He wanted to avoid having to do this. *Why?* Because that would have taken more time, and the supervisor had put him under pressure to get the job done quickly. Another line of enquiry might involve the fact that the stairs were too steep, steeper than would be allowable in the relevant building code. Asking why this was the case, it might be because the designers had not adequately considered who the stairs would be used by, and what the users would be carrying. Why? Because the regulators may not have required the designers to design out (eliminate) the hazard at source. Clearly, moving back to the design of the staircase would involve considering latent errors, which may have been made some years previously. These factors may be illustrated as shown in Figure 9.2.

Figure 9.2: Analysis of causal and contributory factors (adapted from Hopkins, 2006, p. 4)

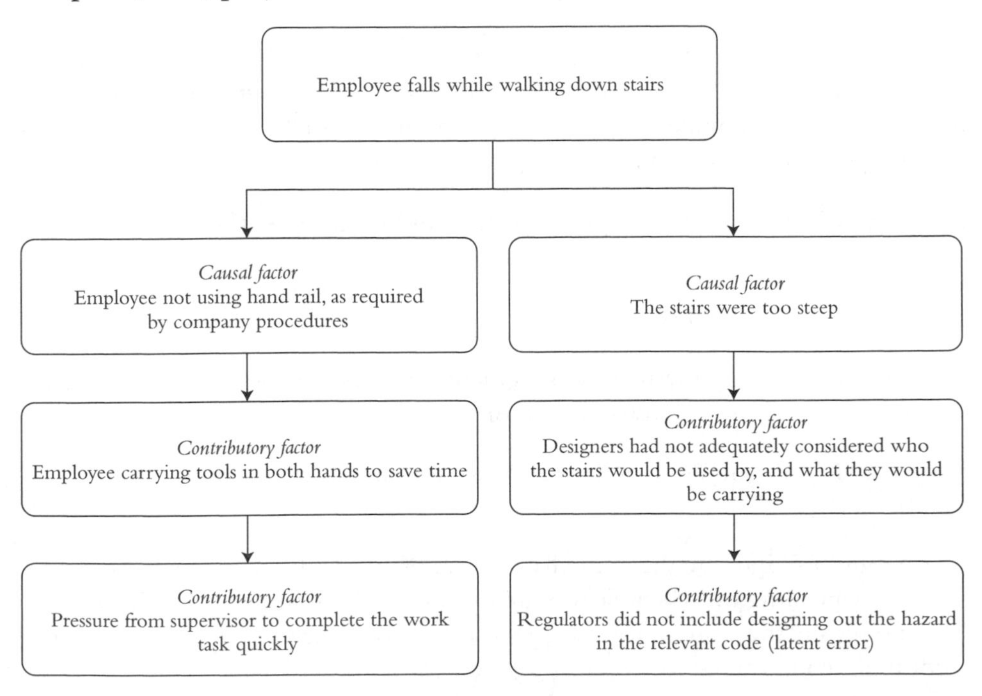

or occurrence, be it a technical or a human failure, would have prevented the accident. If removal of the event or occurrence from the chain of events would certainly have prevented the accident, then it is a causal factor. If removal of the event or occurrence would only have reduced the probability of the accident occurring, then it is a contributory factor. In other words, causal factors are directly involved in the chain of events leading to an accident, whereas contributory factors simply increase the probability of the accident occurring. Both causal and contributory factors can be added to the time-line, allowing for a more detailed analysis of the causes of events.

As a case study, consider the following construction accident adapted from Haslam et al. (2005). An apprentice carpenter, working with his mentor, was installing a rafter. The weather was forecast to become wet and windy, and the job involved a complex crane lift in limited space. The apprentice walked backwards along the working platform of the scaffold, and stepped through a void where two scaffold battens had been removed, seriously injuring his ribs. Although the apprentice was aware that the battens were missing, he continued working, against the requirements of the safe system of work outlined in the method statement. Assuming that additional information is obtained from interview, the basic analysis might begin as shown in Figure 9.3.

A HSE analysis framework

9.25 An analysis method published by the HSE is described in Appendix 5 of HS(G)65 (HSE, 1997), and it provides a framework for investigators to examine both immediate and underlying causes. In examining immediate causes, the guidance suggests that these can be:

Procedures (including access, egress, weather and environment);

Plant precautions and substance control (e.g. equipment and relevant controls, such as guarding and ventilation);

Procedures (including the method statements and systems of work, and standard operating procedures and instructions); or

People (whether they did or failed to do anything which contributed to the event).

These four "Ps" indicate the immediate causes of an accident; one or more can be selected depending on which factors are deemed to be significant (see Table 9.9). Depending on the immediate cause(s) selected, the analysis then moves to the eight underlying causes.

Figure 9.3: Illustration of a time-line and analysis of causal and contributory factors (case study adapted from Haslam et al., 2005)

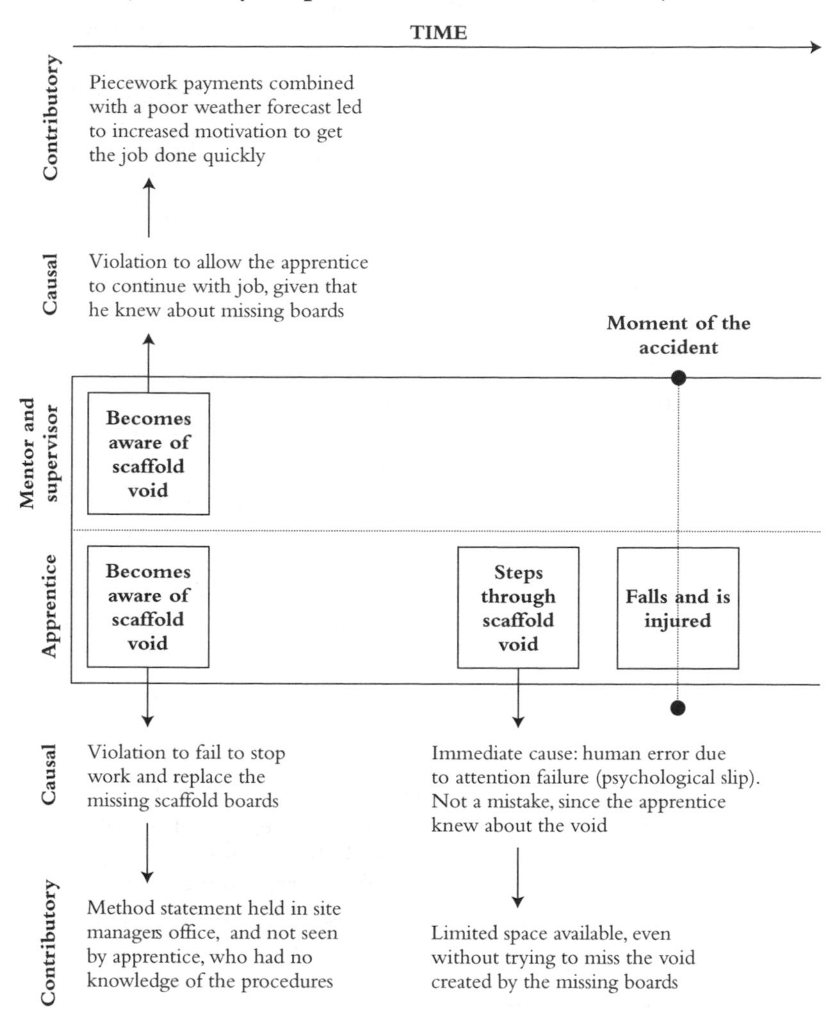

According to the guidance provided in HS(G)65 (HSE, 1997), underlying causes are related to management failures in the risk and management control systems of an organisation (see Table 9.10). The first of these is planning, which involves controlling the risks associated with the buildings, sites, equipment and plant. The second is risk assessment and involves an analysis of whether risk assessments were present and sufficient. Several factors associated with organisational culture are also regarded as underlying causes: these are control, co-operation, communication and competence. Finally, the management functions of monitoring and review can also be underlying causes to the accident.

Table 9.9: Immediate causes. Adapted from HSG 65, Appendix 5, HSE (1997)

A: Premises	*B: Plant precautions*
• Not significant, go to Box B • Adequate premises provided but not used, go to Box C • Adequate premises once provided but not maintained, go to Box E • Adequate premises never provided, go to Box E	• Not significant, go to Box C • Adequate precautions provided but not used, go to Box C • Adequate precautions once provided but now lapsed, go to Box E • Adequate precautions never provided, go to Box E
C: Procedures	*D: People*
• Not significant, go to Box D • Suitable procedures available and in use, go to Box D • Suitable procedures devised but not used: consider accuracy and clearness of instructions (Box I), supervision (Box G) and behaviour (Box D) • Suitable procedures devised but now lapsed, go to Box K • Suitable procedures never devised, go to Box E	• Not significant, go to Box A and reconsider • Person unsuitable for the job due to health or physical ability: consider whether person was never suitable (recruitment and selection, go to Box J) or was once suitable (go to Box F, occupational health risks) • Suitable person but not competent: if never competent, consider training (go to Box J), if once competent but now lapsed consider supervision (Box G) and monitoring (Box K) • Suitable competent person who did the wrong thing: slip, lapse, mistake or violation). Consider training (Box J), control and monitoring (Boxes G and K), communication (Box I), planning (Box E) and co-operation (Box H)

Causes and conclusions

9.26 In determining causes of accidents, and coming to conclusions about the investigation, it is clear than human error will often be implicated. Human error is cited as a cause in up to 85–90 per cent of accidents (Sanders & McCormack, 1987; Wickens & Hollands, 1999; see page 29 for detailed information on human error). However, it is not sufficient to conclude an accident investigation by saying that it was caused by human error.

Table 9.10: Underlying causes. Adapted from HSG 65, Appendix 5, HSE (1997)

E: Planning	*F: Risk assessment*
Planning failures represent failures to anticipate the risk control systems necessary for premises, plant and procedures • Risk control systems absent or inadequate: go to Box F	This category relates to methods for identifying hazards and managing risks • Absent hazard identification and risk assessment, go to Box G
E: Planning	*F: Risk assessment*
• Risk control systems not used: consider risk assessment (Box F), Communication (Box I), Control (Box 7), People (Box D) and Monitoring (Box 11)	• Inadequate hazard identification and risk assessment: consider the competence of those choosing them (Box J) • Adequate methods not used: consider control (Box G) and monitoring (Box K) • Satisfactory methods used but with inadequate results: consider the competency of those using them (Box J), the technical standards used and clarity of results (Box I) and the extent of employee involvement and consultation (Box H)
G: Control	*H: Co-operation*
This category relates to methods of supervision and the ability of management to control work methods, systems and procedures in a manner supportive of health and safety Are roles clearly defined, with responsibilities and authority outlined? If not, consider competence (Box J) and whether senior management have sufficient visible commitment to health and safety	This category relates to employee co-operation in health and safety activities, including: • Involvement in risk assessments • Devising workable procedures • Representation on health and safety committees If there is inadequate co-operation, consider competence (Box J) and whether senior management have sufficient commitment to health and safety
I: Communication	*J: Competence*
Consider whether there is sufficient current knowledge on how to manage risks: • Are there adequate instructions, standards and procedures based on this knowledge?	Consider the adequacy of arrangements for competence assurance, including: • Medical and eligibility criteria, on entry and throughout employment

(*Continued*)

Table 9.10: (Continued)

• Are instructions, standards and procedures clear, concise and available to those who need to follow them? Is there sufficient visible senior management commitment to health and safety	(occupational medicals, eyesight tests, etc.). • Use of reliable and valid assessment methods, such as psychometric tests, assessment centre exercises, competency based interviews, on appointment and promotion
I: Communication	*J: Competence*
	• Provision of training and recurrent training (normal and emergency) to maintain and enhance competence • Assessment of health and safety competence of contractors as part of tender/selection process • Planning to cover staff shortages such as holidays, maternity cover, flexible working practices • An occupational health policy, with adequate monitoring
K: Monitoring	*L: Review*
Routine checks and inspections, conducted frequently and rigorously: • Absent – go to Box G • Inadequate – go to Box F • Adequate but not completed: consider control and review (go to Box G and Box L) • Consider the monitoring function of health and safety departments; if similar incidents and accidents have occurred, then go to control (Box G) and competence (Box J). If recommendations and actions have not been addressed, consider control (Box G) and review (Box L)	Arrangements for evaluating health and safety actions. • Consider control (Box G) and competence (Box J) Also consider whether the senior management commitment to health and safety is sufficient and visible.

Whatever the nature of the error (slip, lapse, mistake or violation), investigators must always look further and find out why the error occurred. Although it may appear that people did something "stupid", "careless" or even "dangerous", investigators must remember that for the most part, *the act or decision seemed acceptable and appropriate at the time*. Individuals do not know in advance that their act or decision will directly result

CASE STUDY 11

9.27

Analysis of an accident using the HS(G)65 framework

A maintenance gang of seven men were working a straight section of railway track at Kelvedon: the track chargeman, the controller of site safety (the COSS), the lookout (to watch the tracks and warn of approaching trains) and four trackside workers (RSSB, 2006). After several hours, the lookout called to warn of an approaching train, and the gang moved from their position to a place of safety. Three of the men sat on a concrete foundation, one facing the track, and two at right angles facing the oncoming train. The fourth man stood nearby with the track chargeman, the COSS and the lookout.

The train passed by, travelling at close to 100 miles an hour. One of the seated trackmen then noticed a threaded steel plug on the ground, 4.5 centimetres long and 3.6 centimetres in diameter. Following this, the lookout noticed that another of the seated trackmen had a bleeding nose and an injured hand; he could not stand without assistance and could not speak coherently. As he could stand with assistance, the injured man was taken to the nearest accident and emergency unit. The worker had a fractured skull and badly injured fingers; he was hospitalised for six days.

Investigation of the accident revealed that the plug had come from a bogie on the locomotive, which had become detached as the locomotive passed the group of workers.

According to HS(G)65, the immediate cause would therefore be associated with plant, and the necessary controls and precautions (Box B). Although these plugs are subject to significant vibration, the correct torque and fitting were not specified in any inspection, maintenance or servicing instructions. The plugs were not recognised as a safety risk, and missing plugs were not regarded as a safety related defect because they could not affect the integrity of the locomotive. Plugs had been going missing since the relevant bogies had been used on the railways and had not been subject to any monitoring, review or reporting requirements.

Adequate precautions had never been provided, meaning that one underlying cause is within Box E. This suggests that there were planning failures involved, since adequate risk control systems were not in place (Box F). As a whole, the

UK railway industry has very sophisticated risk assessment methods, but in this case, the technical standards used did not reflect reality. This meant that knowledge of the potential danger from metal plugs becoming loose was not available, and/or not communicated across the sector (Box I).

It is entirely possible that a different person, analysing the same evidence from the same investigation, might draw different conclusions using the framework. In other words, the reliability of the outcomes obtained from HS(G)65 is not quantified. However, the flowchart does allow investigators to move through a clearly defined process to identify immediate and underlying causes. Each cell requires investigators to make a decision as to an immediate cause and to consider other possibilities as specified. Hence, at least the process by which investigators reach conclusions is consistent and structured. Many health and safety practitioners like the model, since it has clearly defined start and end points, and it is easy to determine when the end of the analysis has been reached.It should be noted that the process does not include analysis of the so-called "root causes", which the HSE defines elsewhere as the failures from which all others stem (HSE, 2004). If root causes exist, they are perhaps best identified within an organisation's health and safety department, where trends across different accidents over time may uncover wider organisational issues, as well as issues associated with specific injuries or items of equipment. If accident data is not routinely stored in a search-and-retrieve format, this information will be lost. While the HS(G)65 framework has clear start and end points for analysis, the same is not true for other analysis models. Alternatives to HS(G)65 may involve a more iterative process to analysing the causal and contributory factors, and require the investigator to move through a cycle of analysis several times before determining conclusions.

in, or contribute to, the accident. If they had this knowledge, then they would avoid taking action on the decisions they made. Because of this, it is important that investigators consider the reason, or combination of reasons, why an individual did not perceive the situation correctly, misinterpreted the risks or carried out the "wrong" action. Examining the human factors in context is the key to understanding their contribution to accidents.

Accident investigators have the benefit of hindsight by virtue of their role, but this can often be a millstone rather than an advantage (Dekker, 2002). In reconstructing a series of events, it can sometimes seem as if the conclusion – the accident – was inevitable. If they are not careful in their analysis, investigators can "cherry-pick" evidence to support their interpretation. In fact, in many accidents the behaviours that immediately precede the accident are normal workplace behaviours – people doing what they would normally do in every day occupational situations – after all that is exactly what defines a behaviour as typical in the first place. Taking this "new view" of human error, it would be possible to argue that human errors are symptoms of a systemic failure, rather than the direct cause of the accident per se. Somewhere is an organisational failure to manage and reduce the errors occurring everyday, in normal workplace behaviour.

In identifying human errors in accidents, it is necessary to consider the implications with regards to negligence and associated disciplinary action. The phrase "human error" is often leapt on as evidence of negligence, or as justification for disciplinary or legal action. However, to discipline an individual for a slip or a lapse which caused or contributed to an accident may not be constructive. If that individual was working on a difficult or complex task, in less than optimal conditions, then the chance of operator error was increased by virtue of the job and the environment. The error is not solely attributable to the individual; organisational factors also shaped that behaviour.

Similarly, if an individual commits a violation which causes or contributes to an accident, disciplinary action is not constructive if that person was only doing what most of his or her colleagues would do in the same situation. Routine and necessary violations often arise from expedience or convenience, and a desire to get the job done. In some cases supervisors and managers even encourage them, either implicitly or explicitly. Disciplining someone for a routine or necessary violation only punishes them for being accident involved, for being caught out. It does not address the wider behavioural issue, which is that these violations had been permitted to continue on a habitual basis without management action.

Another potential problem with determining causes and drawing conclusions is the "confirmatory bias". This is the natural tendency in human decision-making to seek information which would confirm an opinion or belief, and to avoid seeking information that would disconfirm it. The confirmatory bias may be particularly prevalent in interviews, because of the immediacy and credibility of face-to-face communication. However, it is a risk in analysing investigative evidence from any source. To prevent the bias from having undue influence on the final stages of analysis, it is important that investigators make a conscious effort to list all of the information which both supports and refutes each possible cause of the accident. Deliberately challenging evidence, and identifying evidence which does not support the theory on the table, assists in strengthening the attributions of cause and the conclusions derived from the investigation (Table 9.11).

Table 9.11: Common errors in attributing causes

There are a number of common errors that can occur when attributing causes to particular events. An awareness of these may help when analysing evidence and drawing conclusions about why an accident occurred.

Underestimating coincidence: Events do co-occur purely by chance, and people tend to underestimate how often this occurs. When similar events occur simultaneously, or within a short-time frame, we often perceive them as being related when in fact there may be no significance in the co-occurrence.

Spatial proximity: Events that occur close together in space are often associated, whether or not this is actually the case. Sometimes people assume that one event caused another because they occurred physically near to each other, even though they may be unrelated.

Temporal proximity: Events that occur close together in time are often seen as associated. Sometimes people assume that the first event caused second or subsequent events when this is not actually the case.

Relative size: Sometimes people unconsciously assume that major events must have major causes, and small events must have small causes. However, a series of small events can have major, even catastrophic, effects.

Representative causes: There is a natural tendency to seek a cause which is most commonly or most often associated with an event. Unusual or rare causes of an event, even though rarer, are logically also possible.

Singular causes: People tend to prefer to attribute singular causes to events, rather than multiple ones. In seeking the reasons for a human or technical failure, people will often fail to realise that there can be multiple causes, working in combination or interaction.

Always remember that the evidence available may be incomplete, and not all of it will be completely reliable. Engineering analyses are likely to be more accurate than interview accounts from eye-witnesses, but may still contain subjective elements. Interviewees may be extremely confident that their accounts are accurate and complete, but confidence does not necessarily equate to accuracy. Investigators need to be aware that they are using their judgement in considering all of the evidence. In cases where some witnesses disagreed with others, investigators should make it clear that there were different accounts of the event. Where some of the evidence may be open to interpretation, this must be borne in mind and clearly stated. Subjective opinions should be avoided, and all conclusions should be based on evidence obtained in the investigation or given by experts who are qualified to provide a professional judgement.

Writing reports and recommendations

9.28 This is normally the responsibility of the principal investigator or chair of the panel, although large sections may be delegated to team members and technical specialists. This is the one document that will sum up all of the evidence from the multiple sources. The report will provide a factual account of the sequence of events, and an analysis of the causes and consequences of those events, with appropriate conclusions and recommendations. It is important that the investigation team can provide a clear and well-substantiated account of their analysis of evidence in the report. However, the account should also be persuasive. The written report is often the standard by which the quality of the whole investigation and the feasibility of its recommendations are judged.

In writing the report, it is sensible to start with purely factual information. A summary of the accident can be followed by a brief history, and description of the background information. The sequence of events should be laid out, starting as far back in time as is necessary. Conjecture, opinion and speculation should be avoided.

Only once this basic information has been provided should the report turn to investigation and analysis. This section of the report should provide a clear account of the evidence collated, including an account of how it was obtained or accessed and how it was analysed. Where there are limitations to the evidence, a balanced account of the limitations should be provided. The analysis is the account of the investigation team's evaluation of factual information, and it should therefore include a logical argument as to why the accident occurred. Conflicts in the evidence should be highlighted in a balanced manner, and gaps in the evidence should be pointed out. The aim is to explore all possibilities, settling persuasively on the account determined by the team.

The conclusions of the report should be firmly grounded in the analysis – conclusions which have no basis in the analysis will surprise the reader, as they will appear to have come from nowhere. The conclusions can take either a "bottom-up" or a "top-down" approach. A "bottom-up" approach starts with the immediate cause(s) of the accidents and works through to the high-level causes at supervisory and management levels. A "top-down" approach starts with organisational failures and management and supervisory levels, and works downwards towards the "front-line" of the accident. Conclusions should also highlight causal/immediate and contributory/underlying causes. Latent and active failures should also be identified. A suggested structure for accident investigation reports is provided in Table 9.12.

Table 9.12: Suggested structure for an accident investigation report

Title: Including details of the time, location and outcome(s) of the accident.
Summary: Providing background information, which will include factual and historical information. Conclusions and recommendations can be copied and pasted to finish the summary, which will then provide an overview of the investigation.
History of accident: Including detailed factual and historical information, the sequence of events, and any actions taken in immediate response to mitigate or make the site safe, including emergency response and attendance of first aid or medical personnel.
Investigation and analysis: Including evidence from interviews, engineering and forensic testing and/or analyses, other sources of information.
Conclusions: Including the most likely or probable immediate and underlying causes, and an indication of the extent to which the conclusions were unanimous. Where there are discrepancies in the evidence, this should be pointed out.
Recommendations: These should address causes and should have a clear link with the investigation and analysis.

In terms of the style of a report, the accident investigation procedure should contain an outline of the report structure and the organisational style required (see Table 9.2). It is normally the case that the report is written formally in professional style. All technical terms, abbreviations and acronyms should be explained and defined the first time they are used. It is also good practice to include a glossary of terms and abbreviations at the beginning of the report. The report should be written in the past tense, as the investigation will be complete when the report is issued. A logical sequence and a clear structure will assist in keeping the reader focused on the point and will avoid confusion. Try to reduce the need for the reader to cross-reference information, but also avoid unnecessary repetition. Keep the purpose of the report in mind throughout, and avoid showing off technical knowledge if it is not relevant.

With regards to language, dramatic, emotive and unnecessary words and phrases should be avoided, as they can convey meanings or opinions not rooted in the evidence. A useful exercise with draft reports is to question the use of every adjective and adverb. For example, terms such as "approximately", "about" and "roughly" should be avoided, as they are ambiguous and require interpretation. Redundant words and phrases, as in "*totally* destroyed", "*entirely* completed" and "*true* facts" should also be avoided.

Try to edit long sentences, and use active rather than passive phrasing. The aim is to be concise, and yet to provide a full account. Precision and accuracy are the key – so provide full references to sources of evidence and analysis, and place supplementary and supporting information in an appendix. Get someone to proofread the report thoroughly, several times. It can also help to have a legal representative check through the report before it is issued, to help reduce the chance that your intended meaning could be misinterpreted.

Confidentiality and anonymity will need to be considered before the report is published or made publicly available, especially if the report contains personal details or other information that could identify a specific individual.

In writing recommendations, investigators need to focus on the deficiencies identified in the analysis. Recommendations are intended to address these deficiencies and reduce the probability of a similar accident occurring in the future. As with the conclusions, they need to be based on the analysis of evidence, and not be a list of "good ideas" that apparently come from nowhere. In devising recommendations, it may be advisable for the chair of an investigation panel, or the principal or lead investigator, to advise senior management of the proposed recommendations and the potential resource implications in advance. This will allow the team to gauge whether other considerations may render recommendations redundant or impractical. This approach may assist with securing the resources and/or funding necessary for implementation of accepted recommendations.

In writing recommendations, try to phrase them with the SMART acronym in mind. This will help to make recommendations specific, measurable, agreed, realistic and time-bound. *Specific* means particular, avoiding sweeping generalisations. *Measurable* means that the recommendation should be tied to an outcome, so that progress can be assessed. *Agreed* implies that the action is discussed with the responsible manager or department in advance of the issue of the report. *Realistic* means that the recommendations should be feasible. *Time-bound* means that a timescale for achievement should be included. Table 9.13 contains some sample sections from accident investigation reports along with some commentary on why certain phrases and conclusions would have been best avoided.

Table 9.13: Sample sections from accident investigation reports

Summary: A roofing contractor and his mate were fixing the roof on a farm building. To save time in securing his boss' harness, the mate used the works van, throwing the ropes over the apex of the barn roof. To ensure that no one moved the vehicle, the contractor put the van keys in his jacket pocket on the back seat before going onto the roof. The farmer needed to move the van in order to open a cattle gate; the mate took the van keys from his boss' jacket pocket and moved the van several feet. His boss was pulled some distance up the roof, and received several minor injuries as a result. It was fortunate that he was not pulled over the roof, as he could have sustained more serious injuries. The cause of the accident was that the mate was negligent because he knew the harness was attached to the van.

Suggested improvements: The author of the report has attributed blame rather than identify the immediate and underlying causes. The immediate cause was a failure of memory, a human error. This was compounded by the time pressure, since the contractors were in a hurry to complete the job. Use of the term "negligence" could raise difficult issues.

Summary: A saw miller was selling three fence posts to a customer, who asked that the posts be sharpened ready for use. As a favour, the miller agreed to do this as there were only three posts. Resting a post across the top of two logs, the miller secured the first post with his foot before using the chain saw to sharpen the fence post. He slipped when sharpening the second post and cut into his foot. Although wearing steel-reinforced boots, the wound needed fourteen stitches. The injury was caused because the miller hadn't done a proper risk assessment when the customer made a non-standard request.

Suggested improvements: Again, the author of this report has not addressed the cause. The saw miller was trying to do the customer a favour with the best intentions. Had the request been for more posts, then the miller may have refused the request, or set up a safer work system. However, attributing the accident to a lack of risk assessment is not accurate, since the risks of working with a chain saw, and the procedures to follow, were well documented.

Summary: A businessman opened a newly built and designer decorated restaurant. On opening night, three waitresses sustained burns as a result of collisions with other staff when using the swing door between the kitchen and the dining room. The restaurateur was extremely annoyed that the cause of the accidents was the waitresses not being careful, as they had been told that someone else could have been trying to use the swing door from the other side. The chef had previously suggested that putting a window in the door would help stop the waitresses being so careless. However, the restaurateur had dismissed this suggestion as he did not want to spoil the designer ambience of the dining room.

Suggested improvements: The author of the report has implied agreement with the restaurateur, in that the waitresses should have been aware of the potential for someone else to be using the door at the same time. Further, the fact that the risk had been identified by the chef, who had also identified a suitable remedy, means that the restaurateur was perhaps too hasty in dismissing the suggestion. Had the businessman had more experience, this hazard would have been easily identified at an early stage.

The range of activities which follow an accident and its investigation means that the potential audience for written information produced by the investigation team could be very wide indeed. Enforcing authorities could request information using their own statutory powers; requests can also be made of enforcing authorities as public bodies under freedom of information legislation. Even where an investigation is conducted under legal privilege in anticipation of litigation, the restrictions may not apply indefinitely. Care should be taken to ensure that all documents pertaining to the investigation contain only factual information and evidenced reasoning. Emotive, judgemental and dramatic language should be avoided in all investigative documents. Lay opinion can be highly charged and extremely damaging when taken out of context and should be avoided entirely. In producing written documents throughout the investigation, cool and analytical objectivity is the order of the day.

As noted at the beginning of this chapter, accident investigation should focus on the risk management and accident prevention benefits that can accrue from the process, while treating issues relating to liability or blame as subsidiary. This can result in a number of difficulties and potential conflicts, as it is difficult in practice for accidents to be investigated in an entirely blame-free context. Accident investigations can – if there is not careful consideration given to legal and disciplinary implications – prejudice not just the organisation but the position of individual employees and managers who may be subject to actual or implied criticism.

Monitoring, dissemination and feedback

9.29 All recommendations and their associated actions should be tracked and audited in order that progress can be demonstrated. An account of recommendations that are not accepted should be kept, with a reasoned argument showing why they were not accepted. It may be better to have a smaller number of achievable recommendations that stand a realistic chance of being implemented than a "wish-list" of improvements that are more idealistic. This will allow for an "audit trail" of incident and accident investigations and recommendations, and a nominated person can review the outstanding actions on a periodic basis.

Periodic review and evaluation of previous accident investigations can also be very useful. For example, radical recommendations which involve an extensive change may not be accepted if they are based on a single accident. If additional cases or instances are found where the same issue arose, then there may be a clear case for reconsidering the recommendation. Where recommendations are rejected because of significant cost or resource implications, it may be possible to periodically revisit the issue. Recurrent issues will only be identified via periodic review and monitoring. This process will also help to identify issues of concern, such as repeated issues with particular items of equipment, similar injuries occurring again and again or a number of similar accidents in the same location or site.

It is also important that findings from evaluation and monitoring activities are fed back to the workforce. This applies firstly to anyone interviewed in any particular investigation, since interviewees often want to know the outcome. However, feeding back also relates to wider analysis activities undertaken within the health and safety department. If organisations are keen to learn from incidents and accidents, then monitoring and review needs to engage employees. They need to know what the current trends in safety and accidents are, what the causes of injuries and accidents are, what recommendations for change have been made, what changes are being implemented and why. If these issues are not communicated through an organisation, then one of the main drivers of accident investigation – the principle of reducing recurrences – is significantly weakened. Making an accident investigation report available is often not the best way of communicating the lessons learnt, and strategies and initiatives which are more relevant and engaging to employees may need to be designed and implemented. Too often, disseminating the findings of an accident investigation is limited to rewriting a procedure, emphasising that disciplinary action will be taken in cases of non-compliance, or providing a brief paragraph for the employee magazine. On their own, these actions are not likely to be efficient or effective.

Finally, it should be noted that accident investigation processes and procedures can themselves be improved. After investigating an accident, the team can provide details of what was difficult, what practical issues were faced and how they were overcome. This type of feedback can be particularly useful when considering the way that evidence is analysed and causes are classified. Feedback in these areas can assist the organisation in designing and further developing database systems and taxonomies to better measure and manage safety risks.

References

Cormack, H., Cross, S. & Whittington, C. (2006) *Identifying and Evaluating the Social and Psychological Impact of Workplace Accidents and Ill-Health Incidents on Employees*. Research Report 464. Health and Safety Executive, Sudbury, Suffolk.

Crundall, D., Chapman, P., France, E., Underwood, G. & Phelps, N. (2005) What attracts attention during police pursuit driving? *Applied Cognitive Psychology*, 19, 409–420.

Cullen, Rt Hon Lord (2001) *Ladbroke Grove Rail Inquiry, Part II*. HSE Books. HMSO, London.

Dekker, S.W.A. (2002) Reconstructing human contributions to accidents: the new view on error and performance. *Journal of Safety Research*, 33, 371–385.

Geiselman, R.E. & Fischer, R.P. (1997) Ten years of cognitive interviewing. In Payne, D.G. & Conrad, F.G. (eds) *Intersections in Basic and Applied Memory Research*. Lawrence Erlbaum, New Jersey, USA.

Haines, R.F. (1991) A breakdown in simultaneous information processing. In Orrecht, G. & Stark, L.W. (eds) *Presbyopia Research: From Molecular Biology to Visual Adaptation*. Plenum Press, New York. Cited in Crundall, D., Chapman, P., France, E., Underwood, G. & Phelps, N. (2005).

Haslam, R.A., Hide, S.A., Gibb, A.G.F., Gyi, D.E., Pavitt, T., Atkinson, S. & Duff, A.R. (2005) Contributing factors in construction accidents. *Applied Ergonomics*, 36, 401–415.

Henderson, J., Whittington, C. & Wright, K. (2001) *Accident Investigation – the Drivers, Methods and Outcomes*. Human Reliability Associates Report for the Health and Safety Executive. Contract Research Report 344/2001. HMSO, Norwich.

Hopkins, A. (2006) What are we to make of safe behaviour programs? *Safety Science*, 44, 583–597.

HSE (1997) *Successful Health and Safety Management*. HS(G)65. HSE Books, Sudbury, Suffolk.

HSE (2004) *Investigating Accidents and Incidents*. HSG 245. HSE Books, Sudbury, Suffolk.

NTSB (1998) Safety at passive grade crossings. *Volume I: Analysis*. Safety Study NTSB/SS-98/02. NTSB, Washington, DC.

RAE (2005) *Accidents and Agenda: An Examination of the Processes that Follow from Accidents or Incidents of High Potential*. The Royal Academy of Engineering, London.

RSSB (2006) *Kelvedon: Report and Recommendations*. Rail Safety and Standards Board, London. Summary available online from www.rssb.co.uk/pdf/reports/Kelvedon-report-and-recomendations.pdf; accessed 22 August 2006.

Sanders, M.S. & McCormick, E.J. (1987) *Human Factors in Design and Engineering*, Sixth Edition. McGraw-Hill, New York.

Thomas, L.J. & Rhind, D.J.A. (2003) Human factors tools, methodologies and practices in accident investigation: implications and recommendations for a database for the rail industry. Cranfield University Report for the Engineering Link. In Cokayne, S. (ed.) (2004) *Data to be Collected for Investigations of Railway Accidents.* AEA Technology (Rail) Report 09/T122/ENGE/003/TRT. Rail Safety and Standards Board, London.

Thomas, L.J. & Muir, H.C. (2005) *Summary of Passenger Accounts of the Collision and Subsequent Derailment of the 17:35 Paddington to Plymouth Service at Ufton Nervet, Berkshire, 6 November 2004.* Cranfield University Report for the Rail Safety and Standards Board, London.

Vrij, A. (2000) *Detecting Lies and Deceit: The psychology of Lying and the Implications for Professional Practice.* Wiley & Sons, Chichester.

Wickens, C.D. & Hollands, J.G. (1999) *Engineering Psychology and Human Performance,* Third Edition. Prentice Hall, Englewood Cliffs, NJ.

10 Management tools: accident data and performance monitoring

In this chapter:

Recording accident data

10.1 Preceding chapters have demonstrated the need to generate and retain accident (and ill health) data for a variety of purposes. Aside from the regulatory and legal imperatives to record details of individual accidents, diseases and dangerous occurrences, all organisations need to have databases of information which are accessible for maintaining the integrity of the overall safety management system. As we have seen with POPIMAR model (see page 73) performance measurement is a key element of these systems, and evidence of performance would be deficient if it failed to include all events where the system has failed. Accident data is also relevant though to other elements of these systems – for reviewing overall performance, for auditing and for planning improvement strategies.

Looked at from another perspective, there are also sound wider risk management reasons for capturing and evaluating this type of data. Much of the "raw" data will be known first hand or anecdotally to sections of the workforce, although not necessarily its implications and the full potential of the dangers will be understood. From this there emerges a "collective sense" (intangible but nevertheless measurable e.g. by staff surveys) within organisations of how much importance is expected to be attached to safety, and the commitment and attitude of senior management. Organisations which fail to meet legitimate expectations among employees that management maintains a keen awareness of accidents leave themselves open to an eroding safety culture, poor industrial relations, enforcement interventions by regulators and potentially serious reputational or commercial losses.

CASE STUDY 12

10.2

Herald of free enterprise

The Herald of Free Enterprise, a roll-on/roll-off passenger and freight ferry, capsized and partially sank in shallow waters off Zeebrugge on 6 March 1987. Around a third of the 459 passengers perished, along with thirty eight of the ship's crew. The cause was found to be that the vessel had left port with her inner and outer bow doors – used for vehicle embarkation and disembarkation – left open, the person whose job it was to close them having been asleep when the order to make ready for departure was given, and no one else having assumed responsibility for the doors. As the ship accelerated out to sea the vehicle deck flooded, making it unstable and leading to the capsize.

The government ordered a formal inquiry into the accident, conducted by a judge, and which exposed many of the issues affecting the accident including design, operation, life saving appliances and rescue services (Sheen Report, 1987).

The inquiry discovered information that, rather than these being exceptional circumstances that led to the accident, there had been no less that five previous occasions when one of the company's vessels had set out to sea with the bow or stern doors still open. Some but not all of these were known to the management but they had not been drawn to the attention of masters of vessels in the company's fleet. Masters who were aware of the incidents did not change their procedures for how checks were carried out on bow or stern doors. Also, it emerged that on one of the occasions (around four years before the accident) the responsible crew member on another vessel had fallen asleep and had not heard the "harbour stations" order, with the result that he also neglected to close both bow and stern doors.

There was also a paper trail that emerged from documents obtained by the inquiry that showed three masters employed in the fleet had requested several times between 1985 and 1986 that indicator lights be fitted on the bridge to show whether or not these watertight doors had been properly shut. The managers involved at the time did not deem this necessary.

The inquiry concluded that proper checks and monitoring procedures for the doors would have avoided the accident. *All concerned in management, from the Board of Directors down to the junior superintendents, were guilty of fault in that all must be regarded as sharing responsibility for the failure of management. From top to bottom the body corporate was infected with the disease of sloppiness.*

The company and some of its senior managers were prosecuted for manslaughter, but they were ultimately acquitted.

Accident data is not the be-all and end-all of performance monitoring, and there is a growing recognition that other measurements are needed to supplement it. The value of accident records however can be expressed in terms of meeting four basic requirements:

- Historical evidence which facilitates forward planning and setting of targets.
- Detailed management data derived from monitoring what is happening in the workplace.
- Accurate statistics which can be relied upon for reviewing and auditing the safety measure system.
- Information which interests and motivates people at all levels – that is, builds/re-enforces the safety culture.

Fundamentally, this is about having a healthy and functioning "corporate memory"; learning from what has gone wrong, remembering why one has particular controls in place and having current awareness of the levels of safety in an organisation.

Meeting different needs for information

10.3 There is much room for debate about measuring health and safety performance using accident data opposed to other quantifiable data and key performance indicators (KPIs) as we will see later in the chapter. However, even within the narrower confines of accident data there is no straightforward rule of thumb for what to record and measure. The infinite variety of activities, risks and available resources that exist in different organisations measure that there can be no one-size-fits-all solution. Certain simple techniques of measurements of incidence and frequency of accidents and ill health allow a common approach to be taken and comparisons to be made both internally and externally; the level of details and the presentation of information still needs to take account of different need within an organisation.

Internally, data should ideally be available in different forms such as for:

- board scrutiny (see page 58);
- health and safety managers' detailed monitoring, benchmarking and production of reports (see below);
- evidence for performance targets;
- safety representatives' reviews (see page 170);
- statutory record keeping requirements (see page 173);
- ascertaining cost of accidents.

Increasingly too, some organisations are finding data needs to be "cut-and-sliced" for wider consumption. Pressures on the insurance industry for example translate into demands for historical accident and ill health data to better inform the underwriting process; customers and clients in some sectors may demand evidence of competence and performance records for health and safety procurement criteria or their own monitoring; regulators may demand the maintenance of performance data as part of the safety case in high risk industries; corporate social responsibility programmes (driven in part by institutional investors) and targets set for public-sector bodies also increasingly promote transparency and publication of accident and ill health data as part of wider reporting requirements. The need has been growing steadily to have data which is accessible, sometimes in a form that is readily verifiable by independent auditors (see examples in Table 10.1).

Reviews of accident investigations

10.4 A number of major companies, which have seen health and safety performance stubbornly fail to decline after other improvements have been introduced in safety management, have turned to looking at the quality and analysis of accident investigations to drive further improvements. This takes accident investigations further than the rationales already identified for them (see page 279). The process is

Table 10.1: Example accident record Q2 2007 Accident record for Newco Ltd

Record number	*Date*	*Location*	*Injury classification: fatal/major/over-3-day/minor*	*Status of injured party: employee/agency/contractor/visitor/member of the public*	*Injury type*	*Other damage: premises/plant/vehicles/materials*	*Cause*
070014	02/04/07	Warehouse	Minor	Employee	12	–	I
070015	18/04/07	Warehouse	Minor	–	6	Racking broken	D
070016	30/04/07	Driveway	3+ day	Employee	12	–	I
070017	02/05/07	Clean room	Minor	Employee	10	Filter batch lost	C
170018	15/05/07	Reception	Minor	Visitor	12	–	I
170019	25/05/07	Warehouse	Major	Employee	2	Forklift damaged	F
170020	29/05/07	Factory	Minor	Employee	6	–	J
170021	13/06/07	Warehouse	Minor	Agency	12	–	I
170022	17/06/07	Factory	Minor	–	6	–	D
170023	30/06/07	Factory	Minor	Employee	12	–	K

Injury type codes

1. Amputation
2. Fracture (not fingers/toes)
3. Other fracture
4. Dislocation of joint
5. Eye injury/hearing damage
6. Sprain/strain
7. Electric shock
8. Smoke/other inhalation
9. Burns (fire)
10. Burns (chemical) or scalding
11. Toxicity
12. Cuts/bruises
13. PTSD
14. Others
15. No injury

Immediate cause codes

A. Explosion/pressure release
B. Fire
C. Chemical/biological agent
D. Collapse of structure
E. Electrical contact
F. Vehicle movement
G. Falling object
H. Fall from height
I. Slip or trip
J. Manual handling
K. Contact with moving parts
L. Assault
M. Trespass
N. Alcohol/drugs
N. Victim's prior medical condition

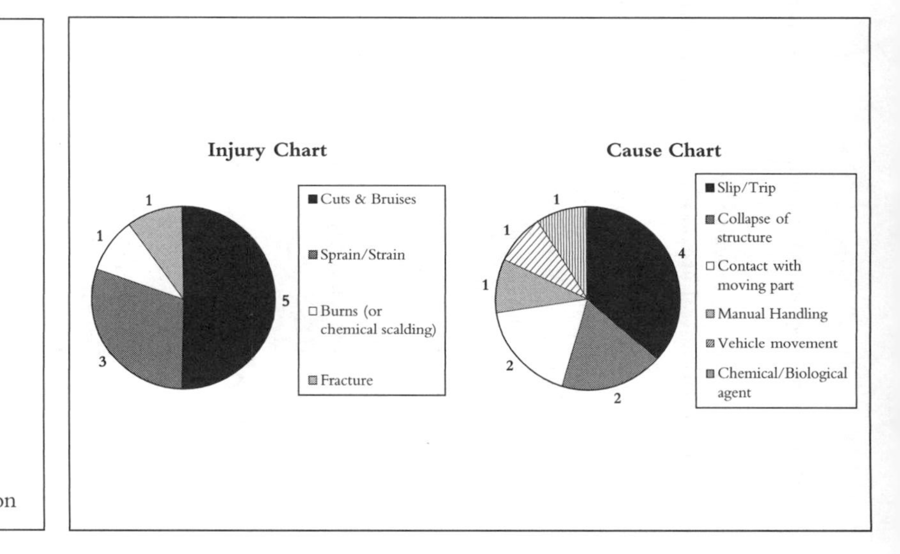

seen as providing clues to underlying causes and under-performance as compared to more successful areas of the organisation or other organisations.

Among the direct and indirect costs of accidents (see page 9) the cost of accident investigation can itself be significant. Members of an investigation team may commit anything from a few hours to a week or more on gathering information and producing reports, or substantially more on complex or serious accidents. Typically this will involve those whose time is viewed as most valuable – senior management, HR heads, health and safety specialists – but there will also have been time incurred by line managers, witnesses (including victims), victims' colleagues, safety representatives, technical and other experts and so on. As well as being an investment of resources in further accident prevention there is value to be derived from this exercise.

Collating accident investigation reports

10.5 Research evidence based on surveys indicates that while around 42 per cent of organisations use paper-based systems to store accident investigation data, 15 per cent use primarily computer-based systems and 41 per cent use combinations of the two. Only 60 per cent of them considered that the data was used to identify trends in underlying causes.

It is straightforward to develop electronic storage of accident investigation reports, either using proprietary database programmes or by recording key findings in spreadsheet formats. Some companies are taking this approach further and by developing electronic accident reporting software which incorporates and centralised the subsequent accident results (Tables 10.2 and 10.3).

Measuring health and safety performance

10.6 Measurement is an important element of any safety management system. *What gets measured gets managed.* This is an area that needs senior management oversight, so that policy objectives can be monitored and information can be generated for management review and identifying targets and priorities. For larger organisations this is a critical element of corporate governance and the board's understanding of business risk management.

Getting to grips with measuring health and safety performance data can prove difficult in many organisations, large and small. The information is not necessarily easy to collect or confirm in ways that can be converted in numbers. Also, if accidents and ill-health are perceived as rare there may be scepticism about committing resources to

Table 10.2: Sample measures of actions taken following incident investigation

Action taken	*(%)**
Risk assessment revised	24
Procedures modified	27
Equipment changed	34
Environmental change	8
Work organisation change	13
Process changed	12
Process eliminated	4
Person disciplined	6
Training	26
Person changed roles	7
Awareness raising	28
Other	19

* These figures do not total 100% because the survey responses were not mutually exclusive
Data derived from: HSE, 2001a

Table 10.3: Sample of common recommendations made following an incident investigation

Recommendation	*(%)**
Be more careful/aware	23
Reinforce safe behaviour	18
Training/refresher training	22
More safety communication	8
Review procedures/instructions	5
Review risk assessment	11
Change in equipment/work organisation	8
More supervision	3
Industry-specific recommendation	2
Range of responses/difficult to answer	8
Not known	9

* These figures do not total 100% because the survey responses were not mutually exclusive
Data derived from: HSE, 2001a

what is viewed as a non-issue. Much of the guidance written on the subject in Health and Safety Executive (HSE) and academic literature treats the issue as one of abstract management which can reinforce negative attitudes. In practice it is important to identify simple and easily accessible database of accidents and incidents (*reactive monitoring*) alongside other measures from which there is a consensus and buy-in from senior management that inferences can be drawn about the effectiveness of the controls intended to prevent accidents (*active monitoring*). Usually this involves finding ways to monitor and count activities that show the extent to which people across the organisation have followed procedures or adopted positive attitudes to health and safety.

For many people the most obvious and appealing information by which to judge safety performance is by the accidents numbers, on the basis that the "proof of the pudding is in the eating" of accident prevention. This data does actually demonstrate the ultimate success of risk assessments and control measures if they are working well. It does so in a way which is easy to convey to senior managers, and it accords with how the workforce perceives its safety in terms of accidents happening frequently or rarely among its number. It is also how regulators and governments worldwide judge their countries' performance and safety records. (This basis of measurement for instance underpins the *Revitalising Health and Safety at Work* strategy of the Health and Safety Commission (HSC) and the government (DETR, 2000).) Reactive monitoring in this way is valid just as long as those who rely on it for assistance and decision making do not fall into the trap of believing that low accident rates are sufficient evidence that risk assessments and control measures are working well. Other measures are needed with which to demonstrate that.

For most organisations accidents are infrequent occurrences and particularly in small workforces it is extremely difficult to derive statistically meaningful patterns. If a company has only say ten employees, one of whom is injured an accident in 2006 but had no other accidents reported in 2000–2005 this does not necessarily indicate a dip in safety management efficiency. It could, as we have seen in Chapter 2, be pure chance that more accidents or injuries have not occurred before.

Even within larger organisations where statistically significant incidence and frequency data may be derived because of the population size of the workforce, this can only provide a narrow field of view to the management. It does not explain the nature of changes that are taking place over time. Certain types of injury or dangerous occurrences may be increasing while others are being reduced. Research has shown for example that success in tackling accidents such as slips and trips is not necessarily a good indicator that other more complex risks are being well controlled. Also, as we have seen with the Swiss-cheese model (see page 38), the final injury event may be a purely fortuitous outcome and it tells us little about the point at which the controls have failed.

Common units of measurement

10.7 A variety of performance measures are found in practice, which count accidents or ill health with varying degrees of sophistication. Each of them has certain advantages and it is a matter of choosing ones which best suits the organisation's needs.

Numbers of accidents or injuries per month/year

10.8 This is easy to monitor from accident book entries and *RIDDOR* reports, and can be a useful measure in small and medium size organisations if there is very low staff turnover and a large degree of homogeneity in the work done by them. It can be a poor indicator of safety performance for example if there are part-time or seasonable workers, or there have been changes in the organisation's operations.

Breaking the numbers down into accidents of different severity and "near misses" or by reference to operations or location will render it more valuable in terms of the inferences that can be drawn from changes over time and comparisons between the contexts. As an example, Royal Mail Group measures KPIs which focus on its biggest impact which is the total number of *RIDDOR* reportable accidents, while at the same time also monitoring total numbers of accidents and some of the other measures described below.

Research has been carried out which has shown that there can be predictable relationships between serious and minor injuries or near misses, such that the study of actual numbers in one's organisation should provide some useful indicators of wider performance. The HSE's studies of accidents in oil, food, construction, health and transport sectors concluded that there was an "accident triangle" (as shown in Figure 10.1) which comprises the ratio between more serious injuries and less serious outcomes. (Individual triangles are available for the data studied in particular sectors HSE (1997, 1994).) This reflected earlier research (Bird, 1974) which also

Figure 10.1: Accident triangle (1)

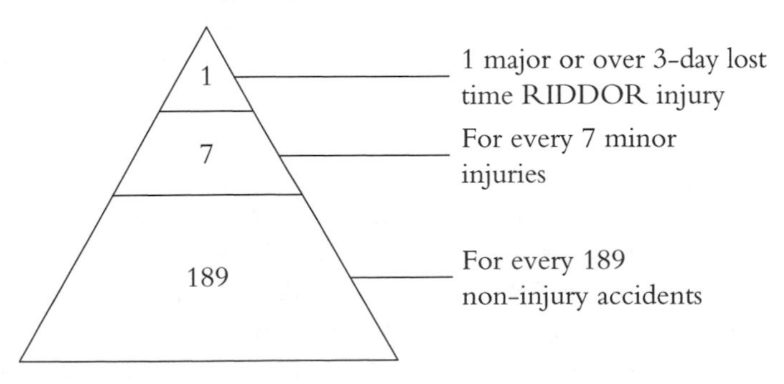

Source: HSE (1997a)

Figure 10.2: Accident triangle (2)

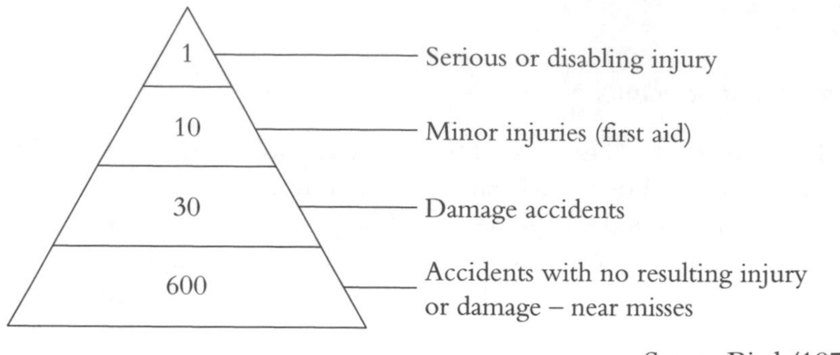

Source: Bird (1974)

ascertained the ratios between different accident severities and near misses (see Figure 10.2).

Incidence rates

10.9 This is a means of making more accurate comparisons over time, with changing numbers of staff. One can break the data down into classes of accidents by severity or type, and other factors such as location. Some regard incidence rates as unsatisfactory for detailed statistical analysis in that they do not take account of part-time, non-permanent workers, or those who do extensive overtime.

HSE's formula for annual incidence rate:

$$\frac{\text{Number of injuries in year}}{\text{Average number employed during year}} \times 100{,}000$$

There are numerous variations on this approach. For example Associated British Ports Holdings Ltd (ABP), a major ports group, uses as its main performance indicator the measurement of *RIDDOR* reportable injuries per thousand employees.

Frequency rates

10.10 This can be a more useful measure for many organisations because it measures the number of injuries with reference to the exposure to workplace risks on the basis of hours worked. Often the rate is calculated per million hours worked, but provided a consistent approach is taken over time there is no reason why a different figure cannot be used. For example 200,000 hours would equate to the time worked in one year by 100 people and is often used in American organisations.

HSE formula for frequency rate:

$$\frac{\text{Number of injuries in the period}}{\text{Total number of hours worked}} \times 1{,}000{,}000$$

The frequency rate is flexible and can be used also to measure particular types of accidents in large populations of workers. (One international industry association, for example, includes in its published statistics the fatal accident rate (FIFR) per 100 million hours worked.)

Lost time injury rates

10.11 The lost time injury rate (LTIR) is very a commonly used measurement. This involves an accident which results in time off work, although the precise definitions vary: It may be simply any time off taken as a result of an accident or injury that prevents a person returning to work on the following day or shift. Some organisations also include work-related illness in this measurement. LTIR mitigates one of the weaknesses of other measures in that it is less prone to inaccuracy through under-reporting of incidents – absence from work is highly visible and is usually recorded in sick-leave records. The information is usually expressed as a frequency rate, by reference to 100,000; 200,000 or 1,000,000 hours worked, and is sometimes expressed as well as a twelve-month rolling average.

A variation on the LTIR is the "significant injury rate" (SIR). This captures in total the less significant accidents which might simply prevent employees from doing their normal jobs (sometimes referred to as "restricted work cases" or RWCs). A variety of other approaches may be taken. There are organisations which measure and compare shorter and larger firm absences from work to gain a better understanding of the severity of accident outcomes.

Accident severity rate

10.12 This is sometimes used in conjunction with LTIR data. It measures the days (or shifts) lost from injury, often per 1,000 employees but sometimes per employee. This can sometimes be slightly misleading as it may not reveal the full extent of severity, for example, hearing loss or a back injury which might not result in a significant period of absence but which could have a long lasting and profound effect on the victim.

Exercising caution with accident data

10.13 Much has been written about limitations that are inherent in focusing on accident data, and sometimes even the descriptions used – that it is "backward-looking" and "reactive" – have pejorative tones. This is unfortunate because while there is no doubt that there are limitations as we have seen it is still valuable for directors, employees and other stakeholders to know how many accidents are happening, to understand the trends and to set realistic targets for improving performance.

The HSE's "*Guide to measuring health and safety performance*" (HSE, 2001b) suggests that accident data should form part of a larger set of information for the purposes of monitoring "outcomes" (or failures). As well as injuries and work-related ill health this should include:

- property damage;
- incidents (including near misses) with potential to cause harm or loss;
- hazards and faults;
- staff complaints about health and safety;
- weaknesses or omissions in performance standards and systems.

If a thorough system is in place for accident investigation (see Chapter 9) and the resulting information is recorded in accessible sources this information is not difficult to collect for monitoring purposes. What useful information can be derived from these "outcomes"? The answers will vary greatly depending on the nature and size of an organisation's activities, but the following questions are ones which managers can usefully ask themselves in most situations:

- What kind of failures have occurred?
  If they are injuries, how serious? If not serious, or no injury at all, what could have been the worst result? What trends and patterns emerge for example, if there are back-injuries rising, why is this?

- Where/when are the failures occurring?
  Are there discrepancies between different plant or shifts? Why might this be? Could it be plant related or down to differences between managers and supervisors or in training?

- Who is at risk?
  Are there particular problems with contractor's staff? Is there evidence of endangering the public, leading to serious loss of reputation and trust?

- Are common or linked underlying reasons emerging from accident investigation?

This could raise questions about the adequacy of experience of those who carry out risk assessments, the understanding of senior management of the principles of safety, whether there is a lack of supervision or investment in training, or something as fundamental as whether changes in the organisation are happening faster than new appropriate precautions can be put into place.

- How much is this costing?
  This may appear an insensitive issue – especially where people have been seriously injured – but as an adjunct to the organisation's budgeting financial risk management it is relevant to know what the drain on resources is from accidents, and how this would justify capital expenditure or other spending on safety-related projects that would yield returns.

- What actions should be prioritised?
  These questions will have been addressed to some extent by accident investigations, but it may be necessary to identify the most urgent matters from different reports or to prioritise certain areas of the business based for example on a high level assessment of the risks with the highest severity and largest numbers of people exposed to it.

- Are things improving?
  Quite simply, if not why not? Other questions are: Are improvements happening within the expected range and timeframe? Are they affecting the whole organisation, or are there areas where certain types of accidents still crop up that warrant special attention?

- Hazards, contraventions, poor standards, complaints
  Why has the system – or confidence in the system – broken down? Are regular reviews being properly carried out? Has there been a loss of corporate memory or are lessons from previous failures not being learned? Are managers consulting safety representatives and staff, and listening to their concerns?

Benchmarking

10.14 As interest in benchmarking business performance has grown generally, and "best-in-class" or "top quartile" are increasingly seen among organisations' strategy goals, opportunities have begun to develop to apply this approach to health and safety performance. Some proponents have seen this as one more way in which to engage interest in safety at senior management level by integrating it into management techniques that appeal to them and have parallels with financial management. The HSE has encouraged the practice in a number of ways and has published guidance (HSE, 1999a).

It must be pointed out that health and safety benchmarking is by no means confirmed to monitoring accident and ill health data. The HSE guidance describes benchmarking in broad terms as "*a planned process by which an organisation compares its health and safety performance with others to learn how to: reduce accidents and ill health; improve compliance with health and safety law; and/or cut compliance costs*". In fact much of the HSE guidance is arguably concerned not so much with analysis of comparative data (as occurs with other forms of benchmarking in the stricter sense) but with joint partnerships, sharing problems and learning from more successful practices in other organisations. In large organisations it may be possible to achieve many of the same benefits through internal benchmarking.

A simple form of benchmarking is to compare one's own accident and ill health incidence and frequency data with the average for those in the same industry. The HSE's website (*www.hse.gov.uk/statistics*) provides rolling annual data under the headings of agriculture, business/finance, construction, education, extractive, health services, hotel/catering, manufacturing, public administration, retail/wholesale and transport (Figure 10.3).

Figure 10.3: Benchmarking process. Based on HSE Guidance (HSE, 1999)

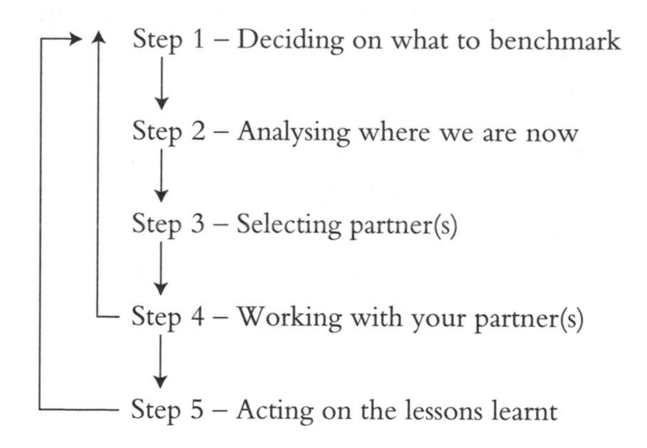

Benchmarking against organisations with similar risk profiles, markets or from the same sector are the main options available. Competitors may be the obvious partners but this can give rise to problems with commercial confidentiality and could in some circumstances create the appearance of inappropriate or unlawful anti-competitive practices. As with financial benchmarking anonymised benchmarking information can be dealt with appropriately via a third party who "moderates" the data.

CASE STUDY 13

10.15

A leisure centre operator became aware that two minor injuries had occurred with users of its gyms in the last eighteen months. Its accident data was routinely compared with that of other similar organisations nationally, and it was comfortably among the best performers. However, after making enquiries among other benchmarking partners it emerged that a significant number of the overall injuries to gym users involved a particular item of equipment. The partners were able to suggest improvements to the manufacturers which were welcomed, the design was modified, and a simple retrofit pack was made available for gym managers. Subsequent monitoring of the partners' accident data showed the injuries had been entirely eliminated.

Health and safety benchmarking schemes

10.16 A number of bodies have established schemes that include health and safety benchmarking.

SHEiiBA: The Safety Health and Environment Intra Industry Benchmarking Associates, *www.sheiiba.com*. This is aimed at health and safety professionals. Its main services are web based and available to subscriber members including tailored league tables which use various measures to compare accident performance, with facilities for anonymous data sharing. (Its website has links to health and safety reporting data from around twenty four large organisations.)

CHaSPI: Corporate Health and Safety Performance Indicator, www.chaspi.info-exchange.com. This is another web-based service. It is for private and public sector organisations with over 250 employees and is sponsored by the HSE. It aims to measure overall health and safety performance by including accident rates. Index results for registered organisations are publicly available on the website, with a CHaSPI overall weighted score on a scale of zero to ten.

Contour: *www.cbi.org*. This is the Confederation of British Industry's environment, health and safety benchmarking tool.

Measuring performance: the bigger picture

10.17 Accident data – if used carefully and presented effectively – can assist greatly in measuring health and safety performance and developing a strong safety culture. If measurement depends wholly on accident data it will result in a distorted view of what has gone wrong, not what is being done right or should be done more. There will be inevitably a significant lag effect if only accident and ill health data is relied upon to identify and drive improvements. Some commentators treat this form of reactive monitoring as inferior and present other active monitoring as being more important. This is misconceived. In truth, both forms of measurement have their place and should be used in combination – a "balanced scorecard" approach which has gained much favour in the world of financial measurement (Kaplan and Norton, 1996) (Table 10.4).

Table 10.4: Advantages and disadvantages of monitoring performances through accident/ill health statistics

Advantages	*Disadvantages*
Easy to obtain data	Under-reporting of accidents can near misses distort the data
Large body of external data for comparison	Individual organisations may have too few data points to provide meaningful analysis
Senior managers and employees readily understand and are motivated by measurement of actual accident numbers	Accident numbers do not tell the whole story – whether an event resulted in injury may have been chance
Clear trends emerge in incidence or frequency of type of accidents/ill health	Injury incidence or frequency rates may disguise severity of outcomes
Verifiable absence of injuries/ ill health is a true measurement of success	Low injury/ill health rates can lead to complacency
Ease of benchmarking	There may well be no correlation between good control of risks and control of major hazards, rendering comparisons ineffective
Lessons are easily learned and real-life case studies can boost training provision	Statistics reflect outcomes of past failures, not underlying causes which may still be affecting operations

The subject of active monitoring is addressed in general terms in Chapter 5 of HS(G)65 (HSE, 1997a) which suggests that it comprises monitoring of:

- the achievement of specific plans and objectives;
- the operation of the health and safety management system;
- compliance with performance standards.

Table 10.5: Active monitoring

1. *Routine procedures to monitor-specific objectives* This could consist of monthly, quarterly or yearly reports on for example, upgraded health and safety training programmes or the elimination of certain hazardous substances from a process.
2. *Periodic examination of documents* For example, checks on whether risk assessments are being reviewed regularly or whether toolbox talks for certain activities are properly scripted and recorded as taking place.
3. *Inspectors of premises and equipment* Are they being maintained in good condition? Could deterioration introduce new risks to safety? How many checks are carried out annually?
4. *Safety sampling* This is an under-used direct observation technique, capable of directly predicting potential accidents, that consist of randomly selecting and measuring the accident potential of particular activities or workplaces. It resembles a risk assessment but, while a desktop exercise, it scores actual hazards and defects, which can then be re-measured after improvement plans have been implemented.
5. *Safety tours* In essence these are short-duration spot checks, made by a party of managers, supervisors, safety representatives or other employers, following a predetermined route or area, looking for unsafe conditions, signs of poor house-keeping, careless treatment of personal protective equipment or other unsatisfactory behaviour. Reports are made and remedial actions have to be closed out. The numbers and frequency of safety tours can be monitored.
6. *Environmental monitoring and health surveillance* Traditionally aimed at picking up signs of contamination or breakdowns in barriers to exposures, measurements can be derived which allow inferences

to be drawn. For example, if back injuries are prevalent, the question arises whether manual handling or display screen equipment risk assessments and training are operating effectively.

7. *Absence monitoring*
   This is particularly important when absence has become subject to greater self-certification by staff and injury/ill health reports when input from doctors and other healthcare professionals is less readily available. Trends may emerge in particular groups of employees, shift patters or locations.

8. *Staff consultation/complaints/confidential reporting lines*
   An early warning measure that can pick up concerns or trends well before accidents/absence measure do so.

9. *Independent audits*
   These may be external auditors or trained auditors from a separate part of the same organisation. Findings may be collated from other audits and investigations for example, insurance surveys, inspections by regulators, safety representatives' inspectors, or other outside consultants such as for noise or water treatment purposes.

10. *Learning and development*
    Is safety awareness being re-enforced through training at all levels? Is there sufficient refresher training at all levels? Is there sufficient refresher training in place? Are directors being kept up to date with regular reports and up to date with industry safety issues, forthcoming law changes and their understanding of directors' responsibilities?

11. *Regulatory interventions*
    Enforcement notices, or "softer" forms of action such as formal warnings or advice on remedial actions provide evidence of management awareness of conditions and risk controls being eroded over time.

12. *Safety climate surveys*

    Surveys which "take the temperature" of the safety culture can take many forms. Knowing what the staff perceive to be the organisation's safety performance, the attitude and leadership of its directors, and the ease with which they feel they can raise concerns or make direct contributions to improvements can be revealing. (The HSE offers a safety climate survey tool on CD (HSE, 1997b) and further useful information is contained in a summary guide for survey tools (HSE, 2001c).) Alternatively, a human factors specialist will be able to develop a bespoke measure, designed to track changes over time in a single industry or organisation.

The composition of a programme of active monitoring has to be considered carefully and tailored to the organisation. Taken to extremes amassing too many measures could become a disproportionate risk "performance culture" rather than underlying safety objectives. It can also work against attempts to focus on priority areas and aiming at key targets. Table 10.5 draws together forms of active monitoring from HS(G)65 along with other suggested management tools.

A detailed analysis of health and safety performance monitoring is beyond the scope of this handbook, but additional information can be found in the materials listed as references at the end of this chapter. Further HSE guidance is available (HSE, 2001a) and a useful paper for managers and safety professionals produced by RoSPA (Figure 10.4).

Figure 10.4: A holistic view of corporate OS&H performance

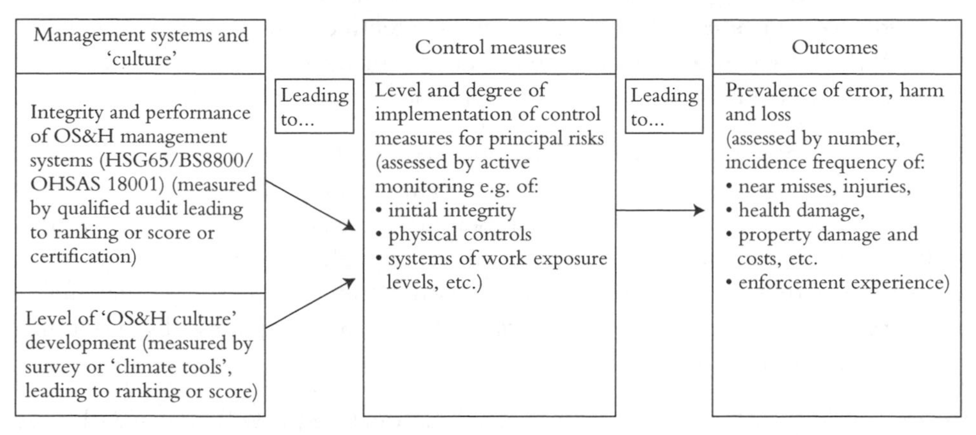

Source: RoSPA (2001)

Setting targets

10.18 It is not uncommon to see policy documents which set health and safety performance targets, for example, stating that it is the goal of an organisation to achieve zero lost time or reportable injuries. There are certain problems with this narrow approach:

1. Without very clear communication it is often unclear if this is a literal objective for which people are individually accountable, or rather part of the rhetorical of management goals intended more than to express a collective aspiration.
2. As a measurement of failure it does not necessarily convey the importance of proactive steps measurable by other KPI's.

3. Damaging tensions can arise if perceived pressure to meet these targets or health and safety reward and recognition schemes result in under-reporting, or blocks the flow of accident and incident data and information.
4. Failure to meet the targets can undermine the credibility of the process for selling them, and that of the system as a whole if accidents are felt to be increasing against all expectations.

These problems are particularly acute if "zero" is expressed as the target, as there is a significant prospect in an organisation of any size that accidents will happen.

> "While many argue that every accident can be prevented, in reality, especially in very large organizations, some level of error leading to harm is probably inevitable. This is particularly so if it is accepted that preventive interventions are always based on incomplete data and understanding and are always likely to involve some degree of non-compliance. Simple calculations suggest, for example, that in a business employing 1,000 people which has managed to achieve one tenth of the national average RIDDOR rate (a substantial achievement in many sectors), there will be a notifiable injury at least once every 18 months" (RoSPA, 2002).

In 2000 the government and the HSC published its *Revitalising Health and Safety*, a ten-year strategy designed to give new momentum to health and safety performance across the sectors. This has been a useful demonstration of the extreme difficulties that arise in setting targets that are achievable in practice (see page 4). Nevertheless there is still HSE guidance recommending that specific measurable targets contributing to the one in *Revitalising Health and Safety* are set by and monitored by senior management.

In this way it is anticipated that the overall national trends will be driven by the continued efforts of (mainly larger) organisations which are more accurately set at central, divisional or site levels.

It has been argued that "*the 'process' of target setting is itself as important as the targets that are eventually set*" (RoSPA, 2002). This is because the process involves extrapolating from historical trends, identifying change, management of obstacles and what can be done to overcome them, close scrutiny of areas of improvement and what can realistically be achieved. The degree of care taken in the process will directly affect the extent of the four problems likely to be encountered which are described above.

As with performance monitoring more generally (see 10.6) a wider perspective that just accident and ill health outcomes of accidents is needed for setting targets. Developing targets for specific objectives, projects, activities which are intended to spot uncontrolled hazards or improvement awareness can be valuable adjuncts to the process (see Figure 10.5). Some examples can be seen from Table 10.5 and the measurements obtained from "active" monitoring. In some cases this will involve measurement of actual numbers – for example, training days undertaken – and in others it may need to be by reference to risk or audit scores or surveys of attitudes and behaviours.

Figure 10.5: Devising and monitoring targets

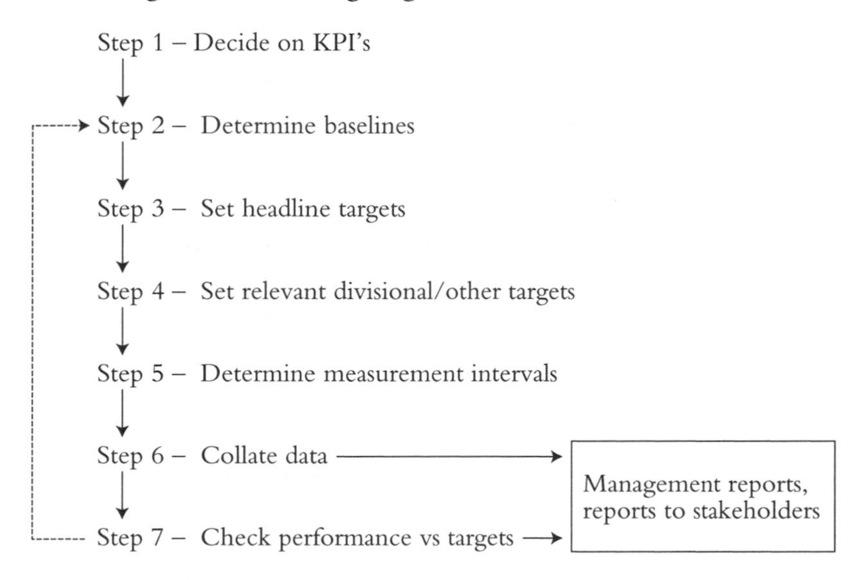

Annual reporting

10.19 One of the approaches supporting the targets set in the *Revitalising Health and Safety* strategy which began in 2000 has been to promote companies and public sector bodies to publish annual reports on their health and safety performance. In part this is to increase the amount of useful information generated, and here the guidance outlined below on the content for such reports. The fundamental policy objective though is to increase awareness of these issues at board level, to encourage their inclusion routinely in high level management reports, and to make evident to shareholders and others the risk management arrangements which the board members have expected the organisation to adhere to.

Having to state publicly on how often these arrangements have failed and resulted in accidents is intended to have a salutary effect.

> "we can think of few things more likely to engage the attention of the chairman and directors of a board than an obligation to furnish regular accounts about how the firm has catered for the safety and health of its employees, and with what results" (Robens, 1990).

Although legislation has existed since 1974 enabling the government to introduce statutory health and safety reporting this has not been used. In 2005, the Companies Act 1985 was amended with a very imprecise requirement on directors of medium sized and large organisations to include in annual reports, information on KPIs relating to environmental and employee matters to the extent this is "necessary for an understanding of the development, performance or position of the business of the company". This stemmed from the EU *Accounts Modernisation Directive (2003/51/EC)* and does not sit easily with UK policies of de-regulation and cutting red tape. A significant practical difficulty is establishing a satisfactory common reporting standard with which to define a statutory obligation to report on health and safety performance.

One view is that by requiring merely the publication of annual statistics for reportable accidents and industrial disease sufficient detail will emerge. It is argued that there would invariably be comment and explanation added by the directors on the statistics, and that this would be as enlightening as the bare numbers prescribed by law.

So far however, the voluntary approach has prevailed, and has had some success in terms of the increasing numbers of organisations adopting annual reporting. Criticisms are still made about the consistency and quality of reporting. While there is no consensus as to common reporting standards there is a certain tendency towards convergence in approach as many larger companies use auditors whose approach is basically the same. The voluntary guidelines in place do at least allow for reports to be adapted to particular organisations and permit an approach which is flexible. Thus many reports seen today are less a statement of annual accident statistics, and more of an operating review and disruption of aims and management arrangements.

HSE annual reporting guidelines (HSE, 2001d)

10.20 When these guidelines were originally published they were sent to the largest 350 listed company boards by the HSE with an exhortation to include health and safety in their criminal reports. There is evidence that this has been a successful exercise in changing corporate behaviour. In 2001 47 per cent of FTSE 100 companies were reporting publicly on health and safety, but the numbers had increased to 96 per cent by 2004. By comparison though reporting by public bodies decreased from 79 per cent in 2002 to 69 per cent in 2004, with local authorities publishing the least.

HSE's guidance on health and safety in annual reports

As a minimum your company's annual report should address key health and safety issues including the effectiveness of your systems for controlling health and safety risks. Reporting should include the following information or give an indication of the steps your company is taking to gather the information for publication in later reports:

- In broad context of your policy on health and safety.
- The significant risks faced by your employees and others and the strategies and systems in place to control them.
- Your health and safety goals. These should relate to your written statement of health and safety policy (and the arrangements for carrying the policy into effect), required by Section 2(3) of the Health and Safety at Work etc Act 1974. Specific and measurable targets – contributing to those in the *Revitalising Health and Safety* Strategy Statement and *Securing Health Together* have a key role.
- Report your progress towards achieving your health and safety goals in the reporting period, and on your health and safety plans for the forthcoming period. There may be specific developments you wish to report on which had an impact on your company's health and safety performance, for example, the introduction of new working practices, technological change or employee training and development. Your company may have significant health and safety plans for the coming years which build on past performance and are noteworthy.
- The arrangements for consulting employees and involving safety representatives.

In addition, your report should provide data on your health and safety performance. Unless it is not available (in which case your report should indicate the steps you are taking to gather the information) the following data should be included:

- The number of injuries, illnesses and dangerous occurrences which should be reported to your health and safety enforcing authority by the *Reporting of Injuries Diseases and Dangerous Occurrences Regulations 1995 (RIDDOR)*. This data should distinguish between fatalities, other injuries, illnesses and dangerous occurrences. More inclusive definitions (lost time injuries, for example) may be used. This may be particularly helpful if you include data from overseas subsidiaries. To help with comparison against the *Revitalising* targets, this data should be presented as the rate of injuries per 100,000 employees.

- Brief details of the circumstances of any fatalities and of the actions taken to prevent any recurrence.
- The number of other cases of physical and mental illness, disability or other health problems that are caused or made worse by someone's work first reported during the period.
- The total number of employee days lost by the company due to all causes of physical and mental illness including injuries, disability or other health problems. You should identify the number of these days thought to be caused or made worse by someone's work and a statement of the main causes of absence.
- The number of health and safety enforcement notices served on the company and information on what the notices required the company to do so.
- The number and nature of convictions for health and safety offences sustained by the company, their outcome in terms of penalty and costs and what has been done to prevent a recurrence.
- The total cost to your company of the occupational injuries and illnesses suffered by your staff in the reporting period.

We encourage companies to go beyond these minimum standards. It can be useful, for example, to include information on the outcome of health and safety audits, and on the extent and effectiveness of health and safety training provided to staff.

An example of an annual report is provided in the Appendix VI.

Source: HSE (2001d)

Performance measurement for SMEs

10.21 Business Link (*www.businesslink.gov.uk*) operates an on-line self-assessment questionnaire which is designed to provide an understanding of how well basic risks (e.g. manual handling) are controlled based on a simple zero–ten scoring system. Comparisons are shown with scores of other similar organisations in the same sector.

The costs of accidents

10.22 The victims of accidents are affected by not just pain, ill health and continuing impairment of activities but also economic losses. The awards of damages by the courts in civil cases attempt to measure all these losses, with conventional ranges of amounts of money set by precedents in personal injury cases in respect of pain and suffering, disability loss of amenity, as well as sums calculated as being the most accurate estimation of loss of earnings and pensions and additional expenses and other outgoings that the victim will incur in the future such as special equipment or care cost. These compensatory awards can be added to other losses such as repair and replacement of damaged equipment to provide quite accurate measures of the *direct* costs of individual accidents but they do not tell the whole story. For companies and other employers the *indirect* costs may be significant but less obvious, and the full cost of accidents has been subject to detailed analysis in the last decade and used by regulators to good effect in promoting the "good health is good business" message.

Numerous measures of accident costs, both direct and indirect, have been identified which are usually hidden amongst other information routinely kept for other purposes such as payroll or other overheads (HSE, 1999b). For example, there is the immediate cost of the employee's absence from work including sick pay arrangements, re-deployment of staff or reliance on agency personnel or contractors and the management time involved in investigations. Then there are the more remote consequences in terms of reduced productivity, staff turnover, recruitment costs and either capital expenditure, repairs or replacement with safer equipment. In terms of the legal aspects, there is the potential for the cost of dealing with regulatory investigations, fines or other sanctions, claims for damages (where not fully insured) and increases in insurance costs.

Figure 10.6 below demonstrates how relatively few of the business costs of accidents fall within the range of losses that are covered by insurance policies. This is because employers and public liability cover only extends to losses and expenses for which there is a legal liability to pay compensation to another party (although some of these other losses may be covered by property insurance or sickness or life cover taken out by the organisations where employees or their dependents are the beneficiaries). Research carried out over a number of years has indicated that the majority of accident costs – in fact about 80 per cent – are indirect costs, and that the proportion of insurance costs to uninsured losses can be very variable but typically between about 12 per cent and 3 per cent; that is to say for each pound of insurance premiums paid the organisations were also covering out of their own resources between eight and thirty six pounds of direct and indirect accident losses. The HSE has estimated that accidents costs in the construction industry for example can be equivalent to 3–6 per cent of a project's costs.

None of these estimates actually provide an accurate measure of the costs to individual organisations, they are merely illustrative, and indeed they must be treated

Figure 10.6

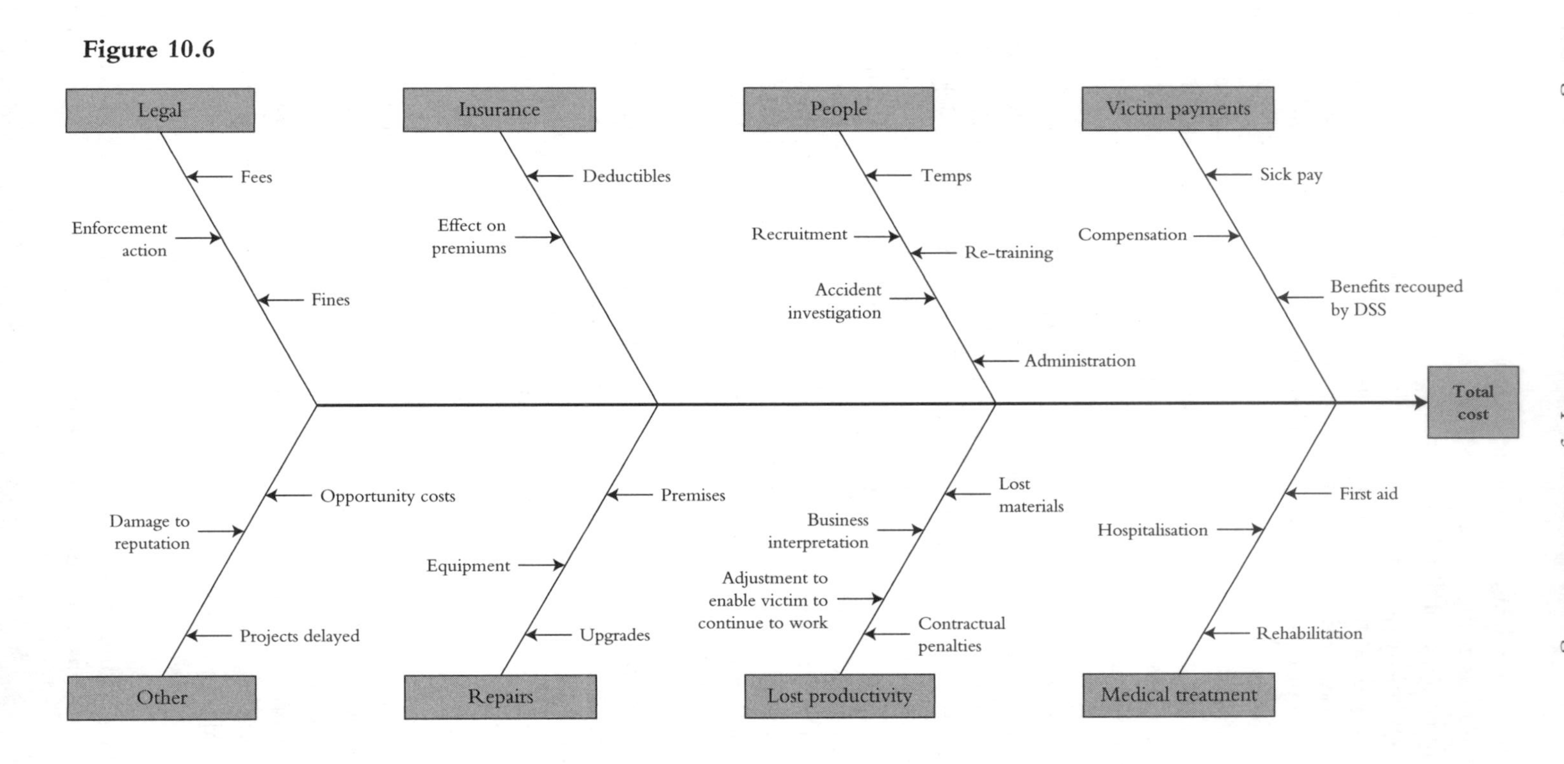

Legal
Fees
Enforcement action
Fines
Insurance
Deductibles
Effect on premiums
People
Temps
Recruitment
Re-training
Accident investigation
Administration
Victim payments
Sick pay
Compensation
Benefits recouped by DSS
Total cost
Opportunity costs
Damage to reputation
Projects delayed
Other
Premises
Equipment
Upgrades
Repairs
Lost materials
Business interpretation
Adjustment to enable victim to continue to work
Contractual penalties
Lost productivity
First aid
Hospitalisation
Rehabilitation
Medical treatment

cautiously as some commentators argue. For instance the total costs can be affected by the amount of retained risk an organisation has under insurance policy excesses, and the potential savings from having fewer accidents may not be as great as is supposed when certain fixed or inelastic costs such as insurance premiums are taken into account. Also some of the costs might offset other losses, for example heavier expenditure on physiotherapy for rehabilitation may reduce lost time from absences. Perhaps more significantly however most accident cost modelling fails to take into account the past and continuing costs *savings* an organisation might have accrued by operating without inadequate safety procedures, equipment and training, or with a lack of supervision. It is therefore necessary to carry out studies on one's own organisation to see exactly what costs are being borne and how these carry through to the bottom line.

Carrying out a costs study

10.23 The purpose of carrying out a costs study (HSE, 2005) is threefold:

- To gain an understanding of the impact on the organisation's financial performance over a set period in terms of the direct and indirect costs of accidents (this also serves as a baseline).
- To identify opportunities to reduce or eliminate these costs by meeting improvement targets.
- To provide a set of data which can be used to attribute typical value to the costs of accidents for measuring purposes in the future.

The process can be begun by looking back at a period to accidents and incidents which have already happened (a *retrospective* study) or by commencing at a certain date and measuring the costs as they arise (a *prospective* study). Some organisations may choose to do both at once to get the biggest possible picture although this does risk including greater fluctuations in some of costs information and comparisons may be deceptive. A prospective study will often be chosen because it has the advantage in that it can be supported with a company-wide campaign to heighten awareness the need to record accidents and also budget holders can be asked in advance of accident events to carefully identify and isolate the costs information needed for the study.

A prospective study might have to last for longer than planned, particularly if it spans a period in which only minor accidents occur. While cumulative costs of these minor accidents are likely to be greater for the organisation in the long run and it is valuable particularly in larger organisations to have the data, it is also desirable to know the costs of over-3-day lost time injuries and major accidents. Even then certain "extreme" costs such as that of responding to enforcement action or a criminal fine would not be counted unless they happened to arise and be costed within the same period, so some costs might have to be projected. Where the underlying causes of an accident have been found to include cost-cutting or under-investment these "savings" should generally not be counted against the costs (or the study would reach perverse findings based on compliance costs being discretionary expenditure) but added as on as representing unbudgeted costs.

Certain limits of scope must be identified before starting a study. Invariably it will aim to capture all serious and other lost time accidents which will be where the greatest expense arises. Will it extend to very minor incidents and near misses as well, and if so will it count events without any actual damage in order to measure lost time or investigation costs? If so there will have to be a limit set to prevent frequent insignificant incidents making the data collection process unmanageable and the study getting bogged down in counting trivial costs. These questions have to be considered in the light of the resources available for monitoring incidents and whether it might be possible in the study to ascribe notional numbers to sums to minor or non-injury accidents using the published accident triangles (see 10.8), and to also give them notional values (say £35 for a first aid only injury or £150 for a damage only incident, values used in the HSE "Ready Reckoner" described below) or a value based on a percentage of injury accidents. (A study by Lawrence Bamber, Ridley (2003) concluded that uninsured costs of accidents averaged £2,097 for an injury causing an absence from work or £141 for every accident causing no injury, but damage to property, plant and equipment. On this basis one might make a rough assumption that the costs of non-injury accidents could be about 7 per cent of the value of that for the injury accidents recorded in the study.)

Values will also have to be established for some costs which do not have an obvious financial measure, for example to cost lost time of workers being absent or being re-deployed to lighter duties, administration costs and management time spent on investigations. This may be done by the appropriate division of the gross annual salary and benefits of the individuals concerned into an equivalent daily or hourly rate. Other costs may be just too intangible to place a value on – loss of employee goodwill, loss of reputation and so on. However where management can identify evidence that suggests that unsatisfactory health and safety performance are potentially a factor in such losses (e.g. through a staff or customer satisfaction surveys, or from low scoring of safety under procurement criteria used in evaluation of a tender) a proportionate notional value may be placed on that loss if it is considered relevant.

It is also worth recording the value of insured losses as well, since this can affect indirect costs in terms of lost potential for reduced insurance premiums and the organisation's insurance brokers or insurance department will be able to compare the effects on renewal premiums. This also enables the study to show a more complete picture of the true economic cost of its accidents.

An outline for a costs study might look like the items shown in Figure 10.6. Care is needed to avoid double counting.

The HSE's "Ready Reckoner"

10.24 The HSE has developed and made available on-line a system for illustrating the approximate costs of accidents (and incidents or "near misses") to support the *Revitalising Health and Safety* strategy. Software can be downloaded from the website at *www.hse.gov.uk/costs* which can be used to process own data by reference to values given by the HSE for typical costs of different accident, incidents, injuries and ill health. The results are not intended to be exact. Three methods of calculation are offered, based on formulae using average values obtained from research into a range of organisations: (1) A multiple of ten times the annual insurance premium; (2) a multiple of the number of employees times the average cost of all accidents; and (3) a multiple of the number of employees times the average cost of different severity levels of accident.

CASE STUDY 14

10.25

1. The Ready Reckoner was used to assess the accidents which were previously given as examples in Table 10.1 of the record for the second quarter of 2007 for (the fictitious) Newco Ltd if this were representative of accidents there over the whole year and there were thirty two lost time injuries. Insurance premiums were £20,000 pa and the company's gross profit margin was 10 per cent. Method (1) was not used because it always indicates the same cost of £200,000 irrespective of the numbers and severity of accidents. The other two methods with their results averaged would indicate uninsured losses as follows:

Method	*Loss pa*	*Revenue to cover this loss*
(2)	£25,125	£251,250
(3)	£71,768	£717,680
Average	**£48,446**	**£484,460**

2. If the Newco Ltd had been able to reduce its accident incidence rate by 50 per cent and eliminate LTIR altogether, its costs, as calculated by the Ready Reckoner, would have reduced on average by 73 per cent and increased the gross profits by £35,534 (not taking into account any further savings through a reduction in insurance premiums).

Method	*Loss pa*	*Revenue to cover this loss*
(2)	£25,125	£251,250
(3)	£700	£7,000
Average	**£12,912**	**£129,120**

3. If only method (3) – probably the most accurate method – is used, the costs savings and improved profitability which flow to Newco Ltd from the better accident performance are seen to be much more significant – 99 per cent and £71,068, respectively.

Checklist for monitoring accident data

10.26

- ✓ Accident and ill health rates are collated consistently and using recognised measures.
- ✓ Data is accessible and can be manipulated to produce reports for the various different purposes needed.
- ✓ Data is used to systematically review performance, plan improvements and set targets.
- ✓ Reactive and active monitoring is combined to provide both lagging and leading indicators.
- ✓ Suitable benchmarking criteria are monitored.
- ✓ Costs of accidents are measured and understood.
- ✓ Senior executives produce and publish high level annual reports on the organisation's health and safety performance.

References

Bird (1974) *Management Guide to Loss Control.* Institute Press, Atlanta, Georgia, USA.

DETR (2000) *Revitalising Health and Safety – Strategy Statement.* www.hse.gov.uk/revitalising/strategy.pdf

HSE (1997) *The Costs of Accidents at Work HS*(G)*96*, Second edition, HSE Books.

HSE (1997a) *Successful Health and Safety Management* (*HS(G)65*). HSE Books.

HSE (1997b) *Health and safety climate tool. Electronic publication*, HSE Books.

HSE (1999a) *Health and Safety benchmarking – Improving together* (*INDG 301*). www.hse.gov.uk/pubns/indg301.pdf

HSE (1999b) *The costs to Britain of workplace accidents and work-related ill health in 1995/96*, Second edition, HSE Books.

HSE (2001a) *Accident Investigations – The drivers, methods and outcomes – HSE Contract Research Report 344/2001.* www.hse.gov.uk/RESEARCH/crr_pdf/2001/crr01344.pdf

HSE (2001b) *A Guide to measuring health and safety performance.* www.hse.gov.uk/opsunit/perfmeas.pdf

HSE (2001c) *Summary guide to safety climate tools, offshore technology report 1999/063.* www.hse.gov.uk/RESEARCH/otopdf/1999/oto99063.pdf

HSE (2001d) *Health and Safety in Annual Reports: Guidance from the Health and Safety Commission.* www.hse.gov.uk/revitalising/tools.htm#annual

HSE (2005) Running a Costs Study. www.hse.gov.uk/costs

Kaplan R. and Norton D. (1996) *The Balanced Scorecard* (Harvard Business School Press).

Ridley (2003) *Safety at Work*, Sixth edition (Butterworth-Heinemann) (includes data provided by Lawrence Bamber)

Robens (1990) *Safety and Health at Work (Robens Report) Cmnd* 5034, HMSO.

RoSPA (2001) *DASH – Director Actions on Safety and Health – Measuring Reporting on Corporate Health and Safety Performance – Towards Best Practice.* www.rospa.co.uk/occupationalsafety/info/dash.pdf

RoSPA (2002) *Targets for change.* www.rospa.co.uk/occupationalsafety/info/targets.pdf

Sheen Report (1987) mv *Herald of Free Enterprise. Formal Investigation. Report of Court No 8074,* HMSO.

Appendices

Appendix I
RIDDOR Schedule 2

Dangerous occurrences

Part I
General

Lifting machinery, etc.

1. The collapse of, the overturning of or the failure of any load-bearing part of any:
   - (a) lift or hoist;
   - (b) crane or derrick;
   - (c) mobile powered access platform;
   - (d) access cradle or window-cleaning cradle;
   - (e) excavator;
   - (f) pile-driving frame or rig having an overall height, when operating, of more than 7 metres; or
   - (g) fork lift truck.

Pressure systems

2. The failure of any closed vessel (including a boiler or boiler tube) or of any associated pipework, in which the internal pressure was above or below atmospheric pressure, where the failure has the potential to cause the death of any person.

Freight containers

3(1) The failure of any freight container in any of its load-bearing parts while it is being raised, lowered or suspended.

(2) In this paragraph, "freight container" means a container as defined in Regulation 2(1) of the Freight Containers (Safety Convention) Regulations 1984.

Overhead electric lines

4. Any unintentional incident in which plant or equipment either:
   (a) comes into contact with an uninsulated overhead electric line in which the voltage exceeds 200 volts; or
   (b) causes an electrical discharge from such an electric line by coming into close proximity to it.

Electrical short circuit

5. Electrical short circuit or overload attended by fire or explosion which results in the stoppage of the plant involved for more than 24 hours or which has the potential to cause the death of any person.

Explosives

6. Any of the following incidents involving explosives:
   (a) any unintentional fire, explosion or ignition at a site:
      (i) where explosives are manufactured by a person who holds a licence, or who does not hold a licence but is required to, in respect of that manufacture under the Manufacture and Storage of Explosives Regulations 2005; or
      (ii) where explosives are stored by a person who holds a licence or is registered, or who is not licensed but is required to be in the absence of any registration, in respect of that storage under those Regulations;
   (aa) the unintentional explosion or ignition of explosives at a place other than a site described in sub-paragraph (1)(a), not being one:
      (i) caused by the unintentional discharge of a weapon where, apart from that unintentional discharge, the weapon and explosives functioned as they were designed to do; or
      (ii) where a fail-safe device or safe system of work functioned so as to prevent any person from being injured in consequence of the explosion or ignition;
   (b) a misfire (other than one at a mine or quarry or inside a well or one involving a weapon) except where a fail-safe device or safe system of work functioned so as to prevent any person from being endangered in consequence of the misfire;
   (c) the failure of the shots in any demolition operation to cause the intended extent of collapse or direction of fall of a building or structure;

(d) the projection of material (other than at a quarry) beyond the boundary of the site on which the explosives are being used or beyond the danger zone in circumstances such that any person was or might have been injured thereby;

(e) any injury to a person (other than at a mine or quarry or one otherwise reportable under these Regulations) involving first aid or medical treatment resulting from the explosion or discharge of any explosives or detonator or from any intentional fire or ignition.

(2) In this paragraph "danger zone" means the area from which persons have been excluded or forbidden to enter to avoid being endangered by any explosion or ignition of explosives; and "explosives" has the same meaning as in the Manufacture and Storage of Explosives Regulations 2005.

Biological agents

7. Any accident or incident which resulted or could have resulted in the release or escape of a biological agent likely to cause severe human infection or illness.

Malfunction of radiation generators, etc.

8(1) Any incident in which:

(a) the malfunction of a radiation generator or its ancillary equipment used in fixed or mobile industrial radiography, the irradiation of food or the processing of products by irradiation, causes it to fail to de-energise at the end of the intended exposure period; or

(b) the malfunction of equipment used in fixed or mobile industrial radiography or gamma irradiation causes a radioactive source to fail to return to its safe position by the normal means at the end of the intended exposure period.

(2) In this paragraph, "radiation generator" means any electrical equipment emitting ionising radiation and containing components operating at a potential difference of more than 5 kilovolt.

Breathing apparatus

9(1) Any incident in which breathing apparatus malfunctions:

(a) while in use; or

(b) during testing immediately prior to use in such a way that had the malfunction occurred while the apparatus was in use it would have posed a danger to the health or safety of the user.

(2) This paragraph shall not apply to breathing apparatus while it is being:

- (a) used in a mine; or
- (b) maintained or tested as part of a routine maintenance procedure.

Diving operations

10. Any of the following incidents in relation to a diving project:

- (a) the failure or the endangering of:
  - (i) any lifting equipment associated with the diving operation; or
  - (ii) life support equipment, including control panels, hoses and breathing apparatus, which puts a diver at risk;
- (b) any damage to, or endangering of, the dive platform, or any failure of the dive platform to remain on station, which puts a diver at risk;
- (c) the trapping of a diver;
- (d) any explosion in the vicinity of a diver; or
- (e) any uncontrolled ascent or any omitted decompression which puts a diver at risk.

Collapse of scaffolding

11. The complete or partial collapse of:

- (a) any scaffold which is:
  - (i) more than 5 metres in height which results in a substantial part of the scaffold falling or overturning; or
  - (ii) erected over or adjacent to water in circumstances such that there would be a risk of drowning to a person falling from the scaffold into the water; or
- (b) the suspension arrangements (including any outrigger) of any slung or suspended scaffold which causes a working platform or cradle to fall.

Train collisions

12. Any unintended collision of a train with any other train or vehicle, other than one reportable under Part IV of this Schedule, which caused, or might have caused, the death of, or major injury to, any person.

Wells

13. Any of the following incidents in relation to a well (other than a well sunk for the purpose of the abstraction of water):
    - (a) a blow-out (that is to say an uncontrolled flow of well-fluids from a well);
    - (b) the coming into operation of a blow-out prevention or diversion system to control a flow from a well where normal control procedures fail;
    - (c) the detection of hydrogen sulphide in the course of operations at a well or in samples of well-fluids from a well where the presence of hydrogen sulphide in the reservoir being drawn on by the well was not anticipated by the responsible person before that detection;
    - (d) the taking of precautionary measures additional to any contained in the original drilling programme following failure to maintain a planned minimum separation distance between wells drilled from a particular installation; or
    - (e) the mechanical failure of any safety critical element of a well (and for this purpose the safety critical element of a well is any part of a well whose failure would cause or contribute to, or whose purpose is to prevent or limit the effect of, the unintentional release of fluids from a well or a reservoir being drawn on by a well).

Pipelines or pipeline works

14. The following incidents in respect of a pipeline or pipeline works:
    - (a) the uncontrolled or accidental escape of anything from, or inrush of anything into, a pipeline which has the potential to cause the death of, major injury or damage to the health of any person or which results in the pipeline being shut down for more than 24 hours;
    - (b) the unintentional ignition of anything in a pipeline or of anything which, immediately before it was ignited, was in a pipeline;
    - (c) any damage to any part of a pipeline which has the potential to cause the death of, major injury or damage to the health of any person or which results in the pipeline being shut down for more than 24 hours;
    - (d) any substantial and unintentional change in the position of a pipeline requiring immediate attention to safeguard the integrity or safety of a pipeline;
    - (e) any unintentional change in the subsoil or seabed in the vicinity of a pipeline which has the potential to affect the integrity or safety of a pipeline;

(f) any failure of any pipeline isolation device, equipment or system which has the potential to cause the death of, major injury or damage to the health of any person or which results in the pipeline being shut down for more than 24 hours; or

(g) any failure of equipment involved with pipeline works which has the potential to cause the death of, major injury or damage to the health of any person.

Fairground equipment

15. The following incidents on fairground equipment in use or under test:

(a) the failure of any load-bearing part;

(b) the failure of any part designed to support or restrain passengers; or

(c) the derailment or the unintended collision of cars or trains.

Carriage of dangerous substances by road

16(1) Any incident involving a road tanker or tank container used for the carriage of … dangerous goods in which:

(a) the road tanker or vehicle carrying the tank container overturns (including turning onto its side);

(b) the tank carrying the dangerous goods is seriously damaged;

(c) there is an uncontrolled release or escape of the dangerous goods being carried; or

(d) there is a fire involving the dangerous goods being carried.

17(1) Any incident involving a vehicle used for the carriage of dangerous goods, other than a vehicle to which paragraph 16 applies, where there is:

(a) an uncontrolled release or escape of the dangerous goods being carried in such a quantity as to have the potential to cause the death of, or major injury to, any person; or

(b) a fire which involves the dangerous goods being carried.

17A

In paragraphs 16 and 17 above, "carriage" and "dangerous goods" have the same meaning as those terms in Regulation 2(1) of the Carriage Regulations.

Dangerous occurrences which are reportable except in relation to offshore workplaces

Collapse of building or structure

18. Any unintended collapse or partial collapse of:
    - (a) any building or structure (whether above or below ground) under construction, reconstruction, alteration or demolition which involves a fall of more than 5 tonnes of material;
    - (b) any floor or wall of any building (whether above or below ground) used as a place of work; or
    - (c) any false work.

Explosion or fire

19. An explosion or fire occurring in any plant or premises which results in the stoppage of that plant or as the case may be the suspension of normal work in those premises for more than 24 hours, where the explosion or fire was due to the ignition of any material.

Escape of flammable substances

20(1) The sudden, uncontrolled release:
- (a) inside a building:
  - (i) of 100 kilograms or more of a flammable liquid;
  - (ii) of 10 kilograms or more of a flammable liquid at a temperature above its normal boiling point;
  - (iii) of 10 kilograms or more of a flammable gas;
- (b) in the open air, of 500 kilograms or more of any of the substances referred to in sub-paragraph (a) above.

(2) In this paragraph, "flammable liquid" and "flammable gas" mean respectively a liquid and a gas so classified in accordance with Regulation 5(2), (3) or (5) of the Chemicals (Hazard Information and Packaging for Supply) Regulations 1994.

Escape of substances

21. The accidental release or escape of any substance in a quantity sufficient to cause the death, major injury or any other damage to the health of any person.

Part II
Dangerous occurrences which are reportable in relation to mines

Fire or ignition of gas

22. The ignition, below ground, of any gas (other than gas in a safety lamp) or of any dust.
23. The accidental ignition of any gas in part of a firedamp drainage system on the surface or in an exhauster house.
24. The outbreak of any fire below ground.
25. An incident where any person in consequence of any smoke or any other indication that a fire may have broken out below ground has been caused to leave any place pursuant to either Regulation 11(1) of the Coal and Other Mines (Fire and Rescue) Regulations 1956 or Section 79 of the Mines and Quarries Act 1954.
26. The outbreak of any fire on the surface which endangers the operation of any winding or haulage apparatus installed at a shaft or unwalkable outlet or of any mechanically operated apparatus for producing ventilation below ground.

Escape of gas

27. Any violent outburst of gas together with coal or other solid matter into the mine workings except when such outburst is caused intentionally.

Failure of plant or equipment

28. The breakage of any rope, chain, coupling, balance rope, guide rope, suspension gear or other gear used for or in connection with the carrying of persons through any shaft or staple shaft.
29. The breakage or unintentional uncoupling of any rope, chain, coupling, rope tensioning system or other gear used for or in connection with the transport of persons below ground, or breakage of any belt, rope or other gear used for or in connection with a belt conveyor designated by the mine manager as a man-riding conveyor.
30. An incident where any conveyance being used for the carriage of persons is overwound; or any conveyance not being so used is overwound and becomes detached from its winding rope; or any conveyance operated by means of the friction of a rope on a winding sheave is brought to rest by the apparatus provided in the headframe of the shaft or in the part of the shaft below the lowest landing for the time being in use, being apparatus provided for bringing the conveyance to rest in the event of its being overwound.

31. The stoppage of any ventilating apparatus (other than an auxiliary fan) which causes a substantial reduction in ventilation of the mine lasting for a period exceeding 30 minutes, except when for the purpose of planned maintenance.
32. The collapse of any headframe, winding engine house, fan house or storage bunker.

Breathing apparatus

33. At any mine an incident where:
    (a) breathing apparatus or a smoke helmet or other apparatus serving the same purpose or a self-rescuer, while being used, fails to function safely or develops a defect likely to affect its safe working; or
    (b) immediately after using and arising out of the use of breathing apparatus or a smoke helmet or other apparatus serving the same purpose or a self-rescuer, any person receives first aid or medical treatment by reason of his unfitness or suspected unfitness at the mine.

Injury by explosion of blasting material, etc.

34. An incident in which any person suffers an injury (not being a major injury or one reportable under Regulation 3(2)) which results from an explosion or discharge of any blasting material or device within the meaning of Section 69(4) of the Mines and Quarries Act 1954 for which he receives first aid or medical treatment at the mine.

Use of emergency escape apparatus

35. An incident where any apparatus is used (other than for the purpose of training and practice) which has been provided at the mine in accordance with Regulation 4 of the Mines (Safety of Exit) Regulations 1988 or where persons leave the mine when apparatus and equipment normally used by persons to leave the mine is unavailable.

Inrush of gas or water

36. Any inrush of noxious or flammable gas from old workings.
37. Any inrush of water or material which flows when wet from any source.

Insecure tip

38. Any movement of material or any fire or any other event which indicates that a tip to which Part I of the Mines and Quarries (Tips) Act 1969 applies, is or is likely to become insecure.

Locomotives

39. Any incident where an underground locomotive when not used for testing purposes is brought to rest by means other than its safety circuit protective devices or normal service brakes.

Falls of ground

40. Any fall of ground, not being part of the normal operations at a mine, which results from a failure of an underground support system and prevents persons travelling through the area affected by the fall or which otherwise exposes them to danger.

Part III
Dangerous occurrences which are reportable in relation to quarries

Collapse of storage bunkers

41. The collapse of any storage bunker.

Sinking of craft

42. The sinking of any water-borne craft or hovercraft.

Injuries

43(1) An incident in which any person suffers an injury (not otherwise reportable under these Regulations) which results from an explosion or from the discharge of any explosives for which he receives first aid or medical treatment at the quarry.

(2) In this paragraph, "explosives" has the same meaning as in Regulation 2(1) of the Quarries Regulations 1999.

Projection of substances outside quarry

44. Any incident in which any substance is ascertained to have been projected beyond a quarry boundary as a result of blasting operations in circumstances in which any person was or might have been endangered.

Misfires

45. Any misfire, as defined by Regulation 2(1) of the Quarries Regulations 1999.

Insecure tips

46. Any event (including any movement of material or any fire) which indicates that a tip [to which the Quarries Regulations 1999 apply] is or is likely to become insecure.

Movement of slopes or faces

47. Any movement or failure of an excavated slope or face which:
    (a) has the potential to cause the death of any person; or
    (b) adversely affects any building, contiguous land, transport system, footpath, public utility or service, watercourse, reservoir or area of public access.

Explosions or fires in vehicles or plant

48(1) Any explosion or fire occurring in any large vehicle or mobile plant which results in the stoppage of that vehicle or plant for more than 24 hours and which affects:
    (a) any place where persons normally work; or
    (b) the route of egress from such a place.

(2) In this paragraph, "large vehicle or mobile plant" means:
    (a) a dump truck having a load capacity of at least 50 tonnes; or
    (b) an excavator having a bucket capacity of at least 5 cubic metres.

Para 46: words "to which the Quarries Regulations 1999 apply" in square brackets substituted by SI 1999/2024, reg 48(2), Sch 5, Pt II.
Date in force: 1 January 2000: see SI 1999/2024, reg 1(1).

Part IV
Dangerous occurrences which are reportable in respect of relevant transport systems

Accidents to passenger trains

49. Any collision in which a passenger train collides with another train.
50. Any case where a passenger train or any part of such a train unintentionally leaves the rails.

Accidents not involving passenger trains

51. Any collision between trains, other than one between a passenger train and another train, on a running line where any train sustains damage as a result of

the collision, and any such collision in a siding which results in a running line being obstructed.

52. Any derailment, of a train other than a passenger train, on a running line, except a derailment which occurs during shunting operations and does not obstruct any other running line.
53. Any derailment, of a train other than a passenger train, in a siding which results in a running line being obstructed.

Accidents involving any kind of train

54. Any case of a train striking a buffer stop, other than in a siding, where damage is caused to the train.
55. Any case of a train striking any cattle or horse, whether or not damage is caused to the train, or striking any other animal if, in consequence, damage (including damage to the windows of the driver's cab but excluding other damage consisting solely in the breakage of glass) is caused to the train necessitating immediate temporary or permanent repair.
56. Any case of a train on a running line striking or being struck by any object which causes damage (including damage to the windows of the driver's cab but excluding other damage consisting solely in the breakage of glass) necessitating immediate temporary or permanent repair or which might have been liable to derail the train.
57. Any case of a train, other than one on a railway, striking or being struck by a road vehicle.
58. Any case of a passenger train, or any other train not fitted with continuous self-applying brakes, becoming unintentionally divided.

59(1) Any of the following classes of accident which occurs or is discovered whilst the train is on a running line:

- (a) the failure of an axle;
- (b) the failure of a wheel or tyre, including a tyre loose on its wheel;
- (c) the failure of a rope or the fastenings thereof or of the winding plant or equipment involved in working an incline;
- (d) any fire, severe electrical arcing or fusing in or on any part of a passenger train or a train carrying dangerous goods;
- (e) in the case of any train other than a passenger train, any severe electrical arcing or fusing, or any fire which was extinguished by a fire-fighting service; or
- (f) any other failure of any part of a train which is likely to cause an accident to that or any other train or to kill or injure any person.

(2) In this paragraph "dangerous goods" has the meaning assigned to it in Regulation 2(1) of the Carriage Regulations.

Accidents and incidents at level crossings

60. Any case of a train striking a road vehicle or gate at a level crossing.
61. Any case of a train running onto a level crossing when not authorised to do so.
62. A failure of the equipment at a level crossing which could endanger users of the road or path crossing the railway.

Accidents involving the permanent way and other works on or connected with a relevant transport system

63. The failure of a rail in a running line or of a rack rail, which results in:
    (a) a complete fracture of the rail through its cross-section; or
    (b) in a piece becoming detached from the rail which necessitates an immediate stoppage of traffic or the immediate imposition of a speed restriction lower than that currently in force.
64. A buckle of a running line which necessitates an immediate stoppage of traffic or the immediate imposition of a speed restriction lower than that currently in force.
65. Any case of an aircraft or a vehicle of any kind landing on, running onto or coming to rest foul of the line, or damaging the line, which causes damage which obstructs the line or which damages any railway equipment at a level crossing.
66. The runaway of an escalator, lift or passenger conveyor.
67. Any fire or severe arcing or fusing which seriously affects the functioning of signalling equipment.
68. Any fire affecting the permanent way or works of a relevant transport system which necessitates the suspension of services over any line, or the closure of any part of a station or signal box or other premises, for a period:
    (a) in the case of a fire affecting any part of a relevant transport system below ground, of more than 30 minutes, and
    (b) in any other case, of more than 1 hour.
69. Any other fire which causes damage which has the potential to affect the running of a relevant transport system.

Accidents involving failure of the works on or connected with a relevant transport system

70(1) The following classes of accident where they are likely either to cause an accident to a train or to endanger any person:

(a) the failure of a tunnel, bridge, viaduct, culvert, station, or other structure or any part thereof including the fixed electrical equipment of an electrified relevant transport system;

(b) any failure in the signalling system which endangers or which has the potential to endanger the safe passage of trains other than a failure of a traffic light controlling the movement of vehicles on a road;

(c) a slip of a cutting or of an embankment;

(d) flooding of the permanent way;

(e) the striking of a bridge by a vessel or by a road vehicle or its load; or

(f) the failure of any other portion of the permanent way or works not specified above.

Incidents of serious congestion

71. Any case where planned procedures or arrangements have been activated in order to control risks arising from an incident of undue passenger congestion at a station unless that congestion has been relieved within a period of time allowed for by those procedures or arrangements.

Incidents of signals passed without authority

72(1) Any case where a train, travelling on a running line or entering a running line from a siding, passes without authority a signal displaying a stop aspect unless:

(a) the stop aspect was not displayed in sufficient time for the driver to stop safely at the signal; …

(b) …

(2) …

Part V

Dangerous occurrences which are reportable in respect of an offshore workplace

Release of petroleum hydrocarbon

73. Any unintentional release of petroleum hydrocarbon on or from an offshore installation which:

(a) results in:

(i) a fire or explosion; or

(ii) the taking of action to prevent or limit the consequences of a potential fire or explosion; or

(b) has the potential to cause death or major injury to any person.

Fire or explosion

74. Any fire or explosion at an offshore installation, other than one to which paragraph 73 above applies, which results in the stoppage of plant or the suspension of normal work.

Release or escape of dangerous substances

75. The uncontrolled or unintentional release or escape of any substance (other than petroleum hydrocarbon) on or from an offshore installation which has the potential to cause the death of, major injury to or damage to the health of any person.

Collapses

76. Any unintended collapse of any offshore installation or any unintended collapse of any part thereof or any plant thereon which jeopardises the overall structural integrity of the installation.

Dangerous occurrences

77. Any of the following occurrences having the potential to cause death or major injury:

    (a) the failure of equipment required to maintain a floating offshore installation on station;

    (b) the dropping of any object on an offshore installation or on an attendant vessel or into the water adjacent to an installation or vessel; or

    (c) damage to or on an offshore installation caused by adverse weather conditions.

Collisions

78. Any collision between a vessel or aircraft and an offshore installation which results in damage to the installation, the vessel or the aircraft.

79. Any occurrence with the potential for a collision between a vessel and an offshore installation where, had a collision occurred, it would have been liable to jeopardise the overall structural integrity of the offshore installation.

Subsidence or collapse of seabed

80. Any subsidence or local collapse of the seabed likely to affect the foundations of an offshore installation or the overall structural integrity of an offshore installation.

Loss of stability or buoyancy

81. Any incident involving loss of stability or buoyancy of a floating offshore installation.

Evacuation

82. Any evacuation (other than one arising out of an incident reportable under any other provision of these Regulations) of an offshore installation, in whole or part, in the interests of safety.

Falls into water

83. Any case of a person falling more than 2 metres into water (unless the fall results in death or injury required to be reported under sub-paragraphs (a) to (d) of Regulation 3(1)).

Appendix II
RIDDOR Schedule 3

Reportable diseases – Regulation 5(1), (2)

Part I
Occupational diseases

Column 1 **Diseases**	*Column 2* **Activities**
Conditions due to physical agents and the physical demands of work	
9. Inflammation, ulceration or malignant disease of the skin due to ionising radiation	Work with ionising radiation
10. Malignant disease of the bones due to ionising radiation	

(*Continued*)

Column 1 ***Diseases***	*Column 2* ***Activities***
11. Blood dyscrasia due to ionising radiation	
12. Cataract due to electromagnetic radiation	Work involving exposure to electromagnetic radiation (including radiant heat)
13. Decompression illness	Work involving breathing gases at increased pressure (including diving)
14. Barotrauma resulting in lung or other organ damage	
15. Dysbaric osteonecrosis	
16. Cramp of the hand or forearm due to repetitive movements	Work involving prolonged periods of handwriting, typing or other repetitive movements of the fingers, hand or arm
17. Subcutaneous cellulitis of the hand (beat hand)	Physically demanding work causing severe or prolonged friction or pressure on the hand Physically demanding work causing severe or prolonged friction or pressure at or about the knee
18. Bursitis or subcutaneous cellulitis arising at or about the knee due to severe or prolonged external friction or pressure at or about the knee (beat knee)	Physically demanding work causing severe or prolonged friction or pressure at or about the elbow
19. Bursitis or subcutaneous cellulitis arising at or about the elbow due to severe or prolonged external friction or pressure at or about the elbow (beat elbow)	Physically demanding work, frequent or repeated movements, constrained postures or extremes of extension or flexion of the hand or wrist
20. Traumatic inflammation of the tendons of the hand or forearm or of the associated tendon sheaths	Work involving the use of hand-held vibrating tools
21. Carpal tunnel syndrome	Work involving prolonged periods of handwriting, typing or other repetitive movements of the fingers, hand or arm
22. Hand–arm vibration syndrome	Work involving: (a) the use of chain saws, brush cutters or hand-held or

(*Continued*)

Column 1 ***Diseases***	*Column 2* ***Activities***
	hand-fed circular saws in forestry or woodworking; (b) the use of hand-held rotary tools in grinding material or in sanding or polishing metal; (c) the holding of material being ground or metal being sanded or polished by rotary tools; (d) the use of hand-held percussive metal-working tools or the holding of metal being worked upon by percussive tools in connection with riveting, caulking, chipping, hammering, fettling or swaging; (e) the use of hand-held powered percussive drills or hand-held powered percussive hammers in mining, quarrying or demolition, or on roads or footpaths (including road construction); or (f) the holding of material being worked upon by pounding machines in shoe manufacture
Infections due to biological agents	
23. Anthrax	(a) Work involving handling infected animals, their products or packaging containing infected material; or (b) work on infected sites
24. Brucellosis	Work involving contact with: (a) animals or their carcasses (including any parts thereof) infected by brucella or the untreated products of same; or (b) laboratory specimens or vaccines of or containing brucella

(*Continued*)

Column 1 ***Diseases***	*Column 2* ***Activities***
25. (a) Avian chlamydiosis	Work involving contact with birds infected with *Chlamydia psittaci*, or the remains or untreated products of such birds
(b) Ovine chlamydiosis	Work involving contact with sheep infected with *Chlamydia psittaci*, or the remains or untreated products of such sheep
26. Hepatitis	Work involving contact with: (a) human blood or human blood products; or (b) any source of viral hepatitis
27. Legionellosis	Work on or near cooling systems which are located in the workplace and use water; or work on hot water service systems located in the workplace which are likely to be a source of contamination
28. Leptospirosis	(a) Work in places which are or are liable to be infested by rats, fieldmice, voles or other small mammals; (b) work at dog kennels or involving the care or handling of dogs; or (c) work involving contact with bovine animals or their meat products or pigs or their meat products
29. Lyme disease	Work involving exposure to ticks (including in particular work by forestry workers, rangers, dairy farmers, game keepers and other persons engaged in countryside management)
30. Q fever	Work involving contact with animals, their remains or their untreated products

(*Continued*)

Column 1 **Diseases**	*Column 2* **Activities**
31. Rabies	Work involving handling or contact with infected animals
32. *Streptococcus suis*	Work involving contact with pigs infected with *Streptococcus suis*, or with the carcasses, products or residues of pigs so affected
33. Tetanus	Work involving contact with soil likely to be contaminated by animals
34. Tuberculosis	Work with persons, animals, human or animal remains or any other material which might be a source of infection
35. Any infection reliably attributable to the performance of the work specified in the dead entry opposite hereto	Work with micro-organisms; work with live or human beings in the course of providing any treatment or service or in conducting any investigation involving exposure to blood or body fluids
Conditions due to substances	
36. Poisonings by any of the following: acrylamide monomer; arsenic or one of its compounds; benzene or a homologue of benzene; beryllium or one of its compounds; cadmium or one of its compounds; carbon disulphide; diethylene dioxide (dioxan); ethylene oxide; lead or one of its compounds; manganese or one of its compounds; mercury or one of its compounds; methyl bromide; nitrochlorobenzene, or a nitro- or amino- or chloro-derivative of benzene or of a homologue of benzene; oxides of nitrogen; phosphorus or one of its compounds	Any activity

(*Continued*)

Column 1	*Column 2*
Diseases	***Activities***
37. Cancer of a bronchus or lung	(a) Work in or about a building where nickel is produced by decomposition of a gaseous nickel compound or where any industrial process which is ancillary or incidental to that process is carried on; or (b) work involving exposure to bis(chloromethyl) ether or any electrolytic chromium processes (excluding passivation) which involve hexavalent chromium compounds, chromate production or zinc chromate pigment manufacture
38. Primary carcinoma of the lung where there is accompanying evidence of silicosis	Any occupation in: (a) glass manufacture; (b) sandstone tunnelling or quarrying; (c) the pottery industry; (d) metal ore mining; (e) slate quarrying or slate production; (f) clay mining; (g) the use of siliceous materials as abrasives; (h) foundry work; (i) granite tunnelling or quarrying; or (j) stone cutting or masonry
39. Cancer of the urinary tract	1. Work involving exposure to any of the following substances: (a) beta-naphthylamine or methylene-bis-orthochloroaniline; (b) diphenyl substituted by at least one nitro or primary amino group or by at least one nitro and primary amino group (including benzidine);

(*Continued*)

Column 1 **Diseases**	*Column 2* **Activities**
	(c) any of the substances mentioned in sub-paragraph (b) above if further ring substituted by halogeno, methyl or methoxy groups, but not by other groups; or (d) the salts of any of the substances mentioned in sub-paragraphs (a) to (c) above 2. The manufacture of auramine or magenta
40. Bladder cancer	Work involving exposure to aluminium smelting using the Soderberg process
41. Angiosarcoma of the liver	(a) Work in or about machinery or apparatus used for the polymerisation of vinyl chloride monomer, a process which, for the purposes of this sub-paragraph, paragraph, comprises all operations up to and including the drying of the slurry produced by the polymerisation and the packaging of the dried product; or (b) work in a building or structure in which any part of the process referred to in the foregoing sub-paragraph takes place
42. Peripheral neuropathy	Work involving the use or handling of or exposure to the fumes of or vapour containing *n*-hexane or methyl *n*-butyl ketone
43. Chrome ulceration of: (a) the nose or throat; or (b) the skin of the hands or forearm	Work involving exposure to chromic acid or to any other chromium compound
44. Folliculitis	Work involving exposure to mineral oil, tar, pitch or arsenic
45. Acne	

(*Continued*)

Column 1 ***Diseases***	*Column 2* ***Activities***
46. Skin cancer	
47. Pneumoconiosis (excluding asbestosis)	1(a) The mining, quarrying or working of silica rock or the working of dried quartzose sand, any dry deposit or residue of silica or any dry admixture containing such materials (including any activity in which any of the aforesaid operations are carried out incidentally to the mining or quarrying of other minerals or to the manufacture of articles containing crushed or ground silica rock); or
	(b) the handling of any of the materials specified in the foregoing sub-paragraph in or incidentally to any of the operations mentioned therein or substantial exposure to the dust arising from such operations
	2. The breaking, crushing or grinding of flint, the working or handling of broken, crushed or ground flint or materials containing such flint or substantial exposure to the dust arising from any of such operations
	3. Sand blasting by means of compressed air with the use of quartzose sand or crushed silica rock or flint or substantial exposure to the dust arising from such sand blasting
	4. Work in a foundry or the performance of, or substantial exposure to the dust arising from, any of the following operations:
	(a) the freeing of steel castings from adherent siliceous substance; or

(*Continued*)

Column 1 ***Diseases***	*Column 2* ***Activities***
	(b) the freeing of metal castings from adherent siliceous substance: (i) by blasting with an abrasive propelled by compressed air, steam or a wheel, or (ii) by the use of power-driven tools
	5. The manufacture of china or earthenware (including sanitary earthenware, electrical earthenware and earthenware tiles) and any activity involving substantial exposure to the dust arising therefrom
	6. The grinding of mineral graphite or substantial exposure to the dust arising from such grinding
	7. The dressing of granite or any igneous rock by masons, the crushing of such materials or substantial exposure to the dust arising from such operations
	8. The use or preparation for use of an abrasive wheel or substantial exposure to the dust arising therefrom
	9(a) Work underground in any mine in which one of the objects of the mining operations is the getting of any material
	(b) The working or handling above ground at any coal or tin mine of any materials extracted therefrom or any operation incidental thereto
	(c) The trimming of coal in any ship, barge, lighter, dock or harbour or at any wharf or quay

(*Continued*)

Column 1 **Diseases**	Column 2 **Activities**
	(d) The sawing, splitting or dressing of slate or any operation incidental thereto
	10. The manufacture or work incidental to the manufacture of carbon electrodes by an industrial undertaking for use in the electrolytic extraction of aluminium from aluminium oxide and any activity involving substantial exposure to the dust therefrom
	11. Boiler scaling or substantial exposure to the dust arising therefrom
48. Byssinosis	The spinning or manipulation of raw or waste cotton or flax or the weaving of cotton or flax, carried out in each case in a room in a factory, together with any other work carried out in such a room
49. Mesothelioma	(a) The working or handling of asbestos or any admixture of asbestos
	(b) The manufacture or repair of asbestos textiles or other articles containing or composed of asbestos
	(c) The cleaning of any machinery or plant used in any of the foregoing operations and of any chambers, fixtures and appliances for the collection of asbestos dust
	(d) Substantial exposure to the dust arising from any of the foregoing operations
50. Lung cancer	
51. Asbestosis	

(Continued)

Column 1 ***Diseases***	*Column 2* ***Activities***
52. Cancer of the nasal cavity or associated air sinuses	1(a) Work in or about a building where wooden furniture is manufactured; (b) Work in a building used for the manufacture of footwear or components of footwear made wholly or partly of leather or fibre board; or (c) Work at a place used wholly or mainly for the repair of footwear made wholly or partly of leather or fibre board 2. Work in or about a factory building where nickel is produced by decomposition of a gaseous nickel compound or in any process which is ancillary or incidental thereto
53. Occupational dermatitis	Work involving exposure to any of the following agents: (a) epoxy resin systems (b) formaldehyde and its resins (c) metal-working fluids (d) chromate (hexavalent and derived from trivalent chromium) (e) cement, plaster or concrete (f) acrylates and methacrylates (g) colophony (rosin) and its modified products; (h) glutaraldehyde; (i) mercaptobenzothiazole, thiurams, substituted paraphenylene-diamines and related rubber processing chemicals; (j) biocides, anti-bacterials, preservatives or disinfectants

(*Continued*)

Column 1 **Diseases**	*Column 2* **Activities**
	(k) organic solvents (l) antibiotics and other pharmaceuticals and therapeutic agents (m) strong acids, strong alkalis, strong solutions (e.g. brine) and oxidising agents including domestic bleach or reducing agents (n) hairdressing products including in particular dyes, shampoos, bleaches and permanent waving solutions (o) soaps and detergents (p) plants and plant-derived material including in particular the daffodil, tulip and chrysanthemum families, the parsley family (carrots, parsnips, parsley and celery), garlic and onion, hardwoods and the pine family; (q) fish, shell-fish or meat; (r) sugar or flour; or (s) any other known irritant or sensitising agent including in particular any chemical bearing the warning may cause sensitisation by "skin contact" or "irritating to the skin"
54. Extrinsic alveolitis (including farmer's lung)	Exposure to moulds, fungal spores or heterologous proteins during work in: (a) agriculture, horticulture, forestry, cultivation of edible fungi or malt-working; (b) loading, unloading or handling mouldy vegetable matter or edible fungi whilst same is being stored;

(*Continued*)

Column 1 ***Diseases***	*Column 2* ***Activities***
	(c) caring for or handling birds; or (d) handling bagasse
55. Occupational asthma	Work involving exposure to any of the following agents: (a) isocyanates; (b) platinum salts (c) fumes or dust arising from the manufacture, transport or use of hardening agents (including epoxy resin curing agents) based on phthalic anhydride, tetrachlorophthalic anhydride, trimellitic anhydride or triethylene-tetramine (d) fumes arising from the use of rosin as a soldering flux (e) proteolytic enzymes (f) animals including insects and other arthropods used for the purposes of research or education or in laboratories (g) dusts arising from the sowing, cultivation, harvesting, drying, handling, milling, transport or storage of barley, oats, rye, wheat or maize or the handling, milling, transport or storage of meal or flour made therefrom (h) antibiotics (i) cimetidine (j) wood dust (k) ispaghula (l) castor bean dust (m) ipecacuanha (n) azodicarbonamide

(*Continued*)

Column 1 **Diseases**	*Column 2* **Activities**
	(o) animals including insects and other arthropods (whether in their larval forms or not) used for the purposes of pest control or fruit cultivation or the larval forms of animals used for the purposes of research or education or in laboratories
	(p) glutaraldehyde
	(q) persulphate salts or henna
	(r) crustaceans or fish or products arising from these in the food processing industry
	(s) reactive dyes;
	(t) soya bean;
	(u) tea dust;
	(v) green coffee bean dust;
	(w) fumes from stainless steel welding;
	(x) any other sensitising agent, including in particular any chemical bearing the warning "may cause sensitisation by inhalation"

Part II
Diseases additionally reportable in respect of offshore workplaces

56. Chickenpox
57. Cholera
58. Diphtheria
59. Dysentery (amoebic or bacillary)
60. Acute encephalitis
61. Erysipelas
62. Food poisoning

63. Legionellosis
64. Malaria
65. Measles
66. Meningitis
67. Meningococcal septicaemia (without meningitis)
68. Mumps
69. Paratyphoid fever
70. Plague
71. Acute poliomyelitis
72. Rabies
73. Rubella
74. Scarlet fever
75. Tetanus
76. Tuberculosis
77. Typhoid fever
78. Typhus
79. Viral haemorrhagic fevers
80. Viral hepatitis

Appendix III
Example of Consent Form for a Medical Report

Employee Consent Form

APPLICATION FOR A MEDICAL REPORT

DATE OF APPLICATION: ______________________________

PERSON REQUESTING REPORT: ______________________

NAME OF EMPLOYEE:________________________________

ACCESS TO MEDICAL REPORTS ACT 1988

AN OUTLINE TO YOUR STATUORY RIGHTS

1. The person requesting must indicate to you that they wish to apply for a medical report and obtain your consent to seek such a report. (*If consent is declined any subsequent decision regarding employment will be based on other available information.*)
2. You have the right to see the medical report before it is sent if you so wish. A copy of your consent will be provided for your doctor.
3. If you wish to see the report it is your responsibility to make the necessary arrangements with your doctor. (*Up to 21 days are allowed for you to see the report form the date of application.*)
4. If you choose to see the report and do not agree with some or all of it, then you may ask in writing to amend it. If the doctor providing the report does not agree a noted amendment can be added or you can refuse for it to be sent.
5. You should be aware that the doctor, in certain circumstances, may be obliged to withhold parts of a medical report.
6. Whether or not a report is seen before despatch to the applicant, the doctor is obliged to keep a copy of the report for at least 6 months during which time you can have access. If a copy is requested from the doctor, a fee may be charged to you.

I have been made aware of my statutory rights under the Access to Medical Reports Act 1988 (see overleaf) and consent to the person named overleaf applying for a medical report from a Doctor(s) involved in my clinical care.

MY FULL NAME IS:
(Block Capitals) ______________________________

EMPLOYEE NUMBER: ______________________________

ADDRESS: ______________________________

______________________________ POSTCODE: ______________

DATE OF BIRTH: ______________________________
(Day/Month/Year)

I DO / I DO NOT wish to see the report before it is sent (Delete as appropriate)

SIGNATURE: ______________________ DATE: ______________

DOCTOR/GP
NAME: (Including Initial) ______________________________

ADDRESS ______________________________

____

TELEPHONE No: (Including Area Code) ______________________

HOSPITAL CONSULTANT
(Including Initial) ______________________

HOSPITAL NO. ______________________

ADDRESS: ______________________________

TELEPHONE No: (Including Area Code) ______________________

2ND HOSPITAL CONSULTANT
(Including Initial) ______________________________

HOSPITAL NO. ______________________________

ADDRESS: ______________________________

TELEPHONE No: (Including Area Code) ______________________

Appendix IV
Example of Letter from Employer to GP Requesting Medical Information

Employer's name and address

GP's name and address
Date

Dear Doctor Brown

Mr X Y Smith DoB: 1 January 1950
Home Address: 1 Lane, Bigtown, Major County

X Y Smith is an employee of this company. As you know, he was injured as a result of an accident at work on (date). I have been in contact with him regularly since the accident. I am now considering what options are available to the company and to Mr Smith with regard to his continued employment with us.

It would be very helpful if you could let me know

- Is he likely to recover to the point where he can return to work?
- If so, how soon is he likely to be able to return to work?
- In your opinion are there any workplace adjustments that might assist achieving an earlier and/or a more successful return to work? (*Offer liaison with your occupational health adviser, who will be able to provide this information if they have the relevant medical information. If you have no occupational health adviser, provide the GP with a basic description of the work that Mr X Y Smith does.)*
- In your opinion what are the obstacles that may prevent a successful return to work?

Please note that I do not need to know the details of his medical case. I need information to assist me in deciding how best to manage Mr Smith's return to work, assuming that his recovery will allow this.

I enclose a copy of Mr Smith's signed consent allowing you to release this information. (*Add if appropriate*: Please note that he has asked to see the report before it is sent, as is his right. I have informed him that I have requested a report from you and suggested that he contact your secretary to make the necessary arrangements. *And note that if Mr Smith has asked to see the report, you are required to also write to him, enclosing a copy of the completed consent form and informing him that you have written to the GP.*)

(*Optional:* A fee is payable for your report if you enclose your invoice. *Expect a fee whether or not you state this*)

Yours sincerely

Appendix V

Searle & Co

Newco Ltd
London Road
Bradfield
Berks

40 Upper Street
Reading
RG1 4DD

Tel 0118 9063000
Fax 0118 9062000

Our Ref: MIT1.49b/AHEN.1

30 July 2006

Dear Sirs

Personal Injury Claim: Mrs Aretha Hennie

We are instructed by your employee Mrs Hennie to claim damages in connection with an accident at her workplace at approximately 2.30 p.m. on 13 June 2006.

Please confirm the identity of your insurers. Please note that the insurers need to see this letter as soon as possible and it may affect your insurance cover and/or the conduct of any subsequent legal proceedings if you do not send this letter to them.

The circumstances of the accident are that in the absence on this day of the usual operator Mrs Hennie was required to operate a power press to cut and form produce base plates for constructing a variety of the company's office chairs. She was given a short demonstration of the task by the supervisor but no previous experience with the equipment.

Shortly after commencing the afternoon's work Mrs Hennie found that a jam occurred in the area where the metal sheets were fed into the equipment preventing its operation. She attempted to clear the jam by pulling out the metal sheet which was in the feeder mechanism. It was necessary for Mrs Hennie to lean over and extend her arm into the equipment to reach the area and in doing so she inadvertently leant on the controls, which activated the equipment. The press fortunately missed her hand but before she could pull her hand free from danger the metal sheet crushed her hand as it bent under the force of the press.

The reason why we are alleging fault is you owed Mrs Hennie a duty of care as her employer to provide safe systems of work. This duty was not met in that that the power press was not fitted with suitable guards, there was inadequate training given to her for the task assigned and the supervision was not adequate for an inexperienced operator.

A description of our client's injuries is as follows:

- Broken thumb and forefingers.
- Damaged tendons, requiring an operation.
- Severe lacerations to the affected areas likely to leave marked scarring.
- Mrs Hennie is in constant pain as a result of her injuries and has suffered from inability to sleep at night.
- She has been advised that it will take around three months of physiotherapy to regain the mobility and grip in her upper arm.

Our client received treatment for her injuries at Royal Berks Hospital. Her hospital reference number is H 19098768.

Mrs Hennie was absent from work for four weeks due to her injuries. She required help with domestic tasks and transport and she is still been to drive.

At this stage of our enquiries we would expect the documents contained in the attached schedule to be relevant to this claim.

Please note that we are acting for our client under a conditional fee arrangement dated into on 24 July 2006 which provides for a success fee within the meaning of Section 58(2) of the Courts and legal Services Act 1990. Our client has taken out an insurance policy with Assurance Co to which Section 29 of the Access to Justice Act 1999 applies. The policy number is 101222/06 and the policy is dated 30 July 2006.

A copy of this letter is attached for you to send to your insurers. Finally we expect an acknowledgement of this letter within twenty one days by yourselves or your insurers.

Yours faithfully

Searle & Co

Relevant Documents For Disclosure Under The Personal Injury Pre-Action Protocol

1. Accident book entries
2. First abider report
3. Foreman/supervisor accident report
4. Safety representatives accident report

5. RIDDOR report
6. Other communications with the HSE
7. Report to DSS
8. Documents listed above relative to any previous similar incident
9. Earnings information
10. Pre-accident risk assessment required by the Management of Health and Safety at Work Regulations 1999
11. Post-accident risk assessment required by the Management of Health and Safety at Work Regulations 1999
12. Accident investigation report prepared in implementing the requirements of MHSWR, Regulations 4, 6 and 9
13. Information provided to employees under MHSWR, Regulation 8
14. Documents relating to employees' training required by MHSWR, Regulation 11
    - (i) Manufacturers' specifications and instructions in respect of relevant work equipment establishing its suitability to comply with Regulation 5.
    - (ii) Maintenance log/maintenance records required to comply with Regulation 6.
    - (iii) Documents providing information and instructions to employees to comply with PUWER Regulation 8.
    - (iv) Documents provided to the employee in respect of training for use to comply with PUWER Regulation 9.
    - (v) Any notice, sign or document relied upon as a defence to alleged breaches of PUWER Regulations 14 to 18 dealing with controls and control systems.
    - (vi) Instruction/training documents issued to comply with the requirements of PUWER Regulation 22 insofar as it deals with maintenance operations where the machinery is not shut down.
    - (vii) Copies of markings required to comply with PUWER Regulation 23.
    - (viii) Copies of warnings required to comply with V Regulation 24.

Appendix VI Associated British Ports Holdings plc Corporate Social Responsibility Report 2005 (Health and Safety Extracts)

Introduction

Given the nature of our operations, health and safety remains a key area of focus for us. Our approach to the management of health and safety matters continues to be based on the belief that accidents and injuries suffered at work are preventable and is backed by a strong commitment from our board. To achieve continuous performance improvement in this area, we are working hard to ensure that our employees are made aware of the potential risks and are equipped with the relevant training. Further information on our health and safety policies and practices can be found on our CSR website www.csr.abports.co.uk. In this section, we provide an update on our performance in 2005, along with future initiatives that we are looking to implement during 2006.

Performance summary: UK

A summary of our UK performance in 2005 is provided in Table 1. Fatalities are the worst thing that can happen in our business. There were no fatalities relating to our employees or contractors during 2005. Tragically, during the year, a fatality did occur at Immingham that was related to operations there. This incident involved a crew member of a visiting vessel and is currently being investigated by the appropriate authority. Under UK regulations, certain incidents are required to be reported as dangerous occurrences (see glossary). The increase in our reportable dangerous occurrences in 2005 is being addressed. Five of these reported incidents involved cargo-handling equipment. Even though our follow-up investigations did not highlight any common weaknesses in our existing procedures, the Group Health & Safety Manager increased his focus on this area in the last quarter of 2005. Methods used to raise awareness on this subject have included the issuing of a safety alert and discussions about these types of risks in regular briefings known as "tool-box talks". Operationally, reportable injuries per thousand employees remain the prime indicator of our performance on health and safety. Figure 1 below provides a comparison of our performance on reportable injuries per thousand employees against the equivalent figure for the membership of Ports Skills and Safety Limited (PSSL) for the 5 years to 31 December 2005.

Table 1: Health and safety performance UK

*Performance indicator**	*2005*	*2004*
Fatal accidents to employees/contractors	–	–
Reportable dangerous occurrences	6	3
Reportable injuries	23	35
Reportable injuries per thousand employees	9.3	14.0
Lost days per employee due to injuries at work	0.26	0.51
Improvements notices issued by the HSE	–	–
Prohibition notices issued by the HSE	1	–
HSE fines	£60,000	–

* See glossary for explanation of our key performance indicators, performance indicators exclude data relating to agency staff

Figure 1: UK reportable injuries per thousand employees

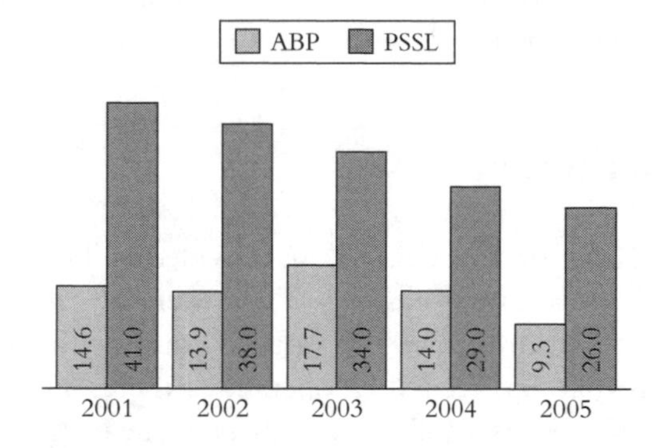

We had significant success in improving our reportable injury rate in 2005 – it fell to 9.3 per thousand employees (2004: 14.0 per thousand employees). As such, we achieved our target of 12 or fewer reportable injuries per thousand employees for the first time.

Figure 2 provides an analysis of our 2005 and 2004 reportable injuries by type. Figure 3 on page 411 provides an analysis of our 2005 and 2004 reportable injuries by severity (assessed by the number of days lost).

A significant proportion of our reportable injuries continue to relate to slips, trips and falls. Our experience is consistent with the nature of our industry and the type of operations in which we are involved. Our initiatives and safety processes continue to focus on addressing these areas. The reduction in our reportable injuries

Figure 2: UK reportable injuries by type

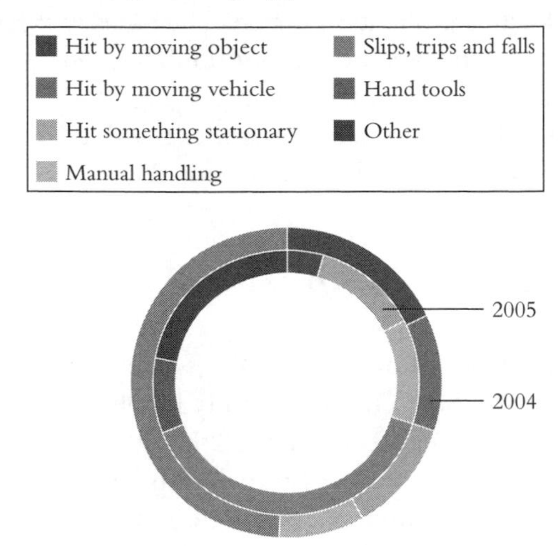

Figure 3: UK reportable injuries by days lost

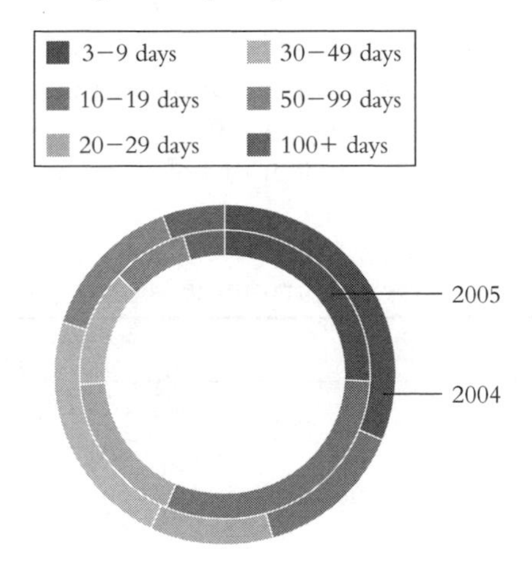

was one of the factors contributing to the decrease in lost days per employee due to injuries at work from 0.51 in 2004 to 0.26 in 2005.

Notwithstanding our performance on reportable injuries, we realise that there is no room for complacency and that the focus of our efforts now needs to be on sustaining this improvement. In recognition of this, we have reduced our target from 12 to 9 or fewer reportable injuries per thousand employees.

We were not served with any improvement notices during 2005 or 2004. In February 2005, we were served with a prohibition notice. The notice was served to prevent the stand-by electricity from being used by a tenant until the specified condition was met. The position was rectified immediately upon serving of the notice. In 2005, we received a fine from the Health and Safety Executive (HSE) of £60,000 (plus £35,000 costs) in relation to a fatality at Hull that occurred in 2003.

Performance summary: USA

A summary of our USA performance in 2005 is provided in Table 2. Reporting requirements differ in the USA to those we apply in the UK and we use recordable injuries, as defined by the Occupational Safety & Health Administration (OSHA), as the primary indicator of our performance on health and safety for our USA business. During 2005, our USA business achieved a much-improved performance as recordable injuries decreased by 23% to 49 compared with 64 for 2004. This performance was marginally better than our 2005 target of reducing recordable injuries to 50 or less. Figure 4 provides an analysis of our USA recordable injuries by type and Figure 5 provides an analysis of our USA recordable injuries by severity. Even though our USA operations saw a reduction in the number of recordable injuries from 2004, the greater severity of accidents meant that the number of lost days per employee increased over the same period. We remain committed to maintaining recordable injuries for our USA business at or below 50 in 2006.

Table 2: Health and safety performance USA

*Performance indicator**	*2005*	*2004*
Fatal accidents to employees/contractors	–	–
Recordable injuries	49	64
Recordable injuries per thousand employees	90.5	113.8
Reportable injuries to employees	–	–
Lost days per employee due to injuries at work	1.46	1.00

* See glossary for explanation of our key performance indicators

Risk assessments

Risk assessments and the subsequent development of safe systems of work continue to underpin the development of our policies, procedures and initiatives on health and safety matters. Our existing activities are maintained under regular review and we undertake detailed health and safety assessments in relation to all new major

Figure 4: USA recordable injuries by type

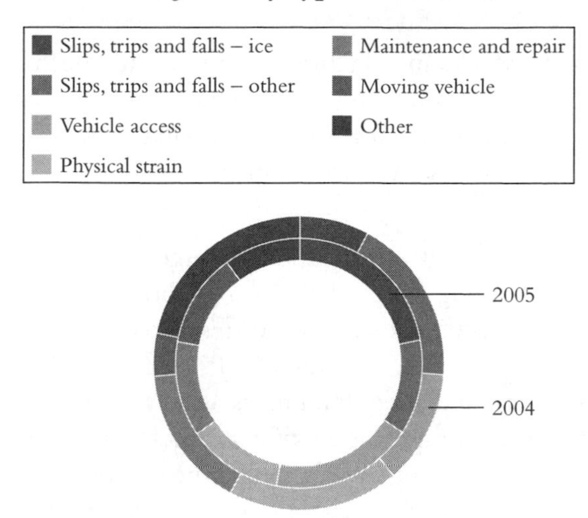

Figure 5: USA recordable injuries by days lost

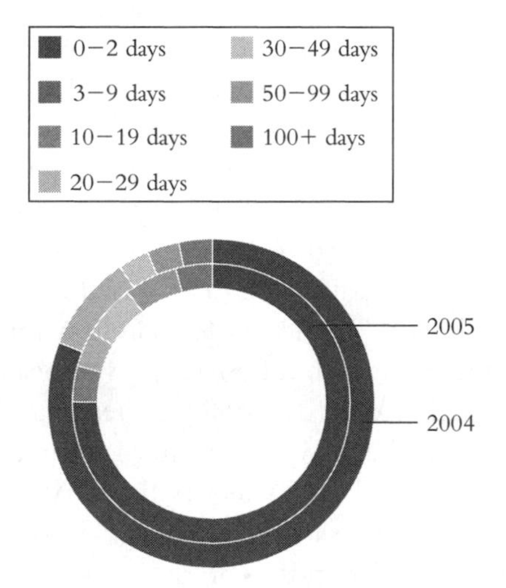

schemes. This includes a multi-disciplinary approach to assessing risk in the planning of major projects, working closely with contractors to manage safety during the construction and commissioning phases, through to developing operational safe systems of work for implementation when new facilities are handed over for use.

This approach is undertaken on all projects. During the development of our two new terminals at the Port of Immingham, our on-site safety manager delivered site safety induction training to almost 2,000 contractors, employees and site visitors.

Training and education

We operate our business within a culture committed to the highest standards of health and safety. Our commitment to health and safety matters is promoted through employee training, awareness campaigns and by linking a proportion of employees' bonus payments to health and safety performance.

We performed satisfactorily against our training targets in 2005. Although we fell short of our refresher training target, with 77% of those requiring training having received it by the end of 2005, we completed this initiative in early 2006. Overall, our performance improved, with 89% (2004: 80%) of our UK employees having received accredited health and safety training as at 31 December 2005, against our long-term target of 95%. Our goal of providing cargo-handling training to all relevant UK employees is on track for completion by the end of 2006. Our ambition to provide training to meet the specific needs of junior managers and those with supervisory responsibilities has been postponed. We had intended to achieve Institution of Occupational Safety and Health (IOSH) accreditation for an internally developed "Leading Teams Safely" course, until the IOSH subsequently announced its intention to review its own suite of training courses. This may result in the IOSH producing an equivalent version of "Leading Teams Safely", in which case we will consider the merits of adopting its training material, as we do currently with the majority of the training that we provide. Consequently, the situation remains under review, although it remains our intention to supplement the training we already offer by introducing a course tailored to the needs of junior managers and supervisors. Our health and safety campaign in 2005 focused on the theme of "reporting and analysis of potential hazards". The 2006 awareness campaign centres on the need to maintain the momentum behind our improved health and safety performance. We recognise the need to keep the profile of health and safety issues high and these campaigns are firmly embedded in the internal communications techniques that we use to promote our health and safety policies. As such, we have ceased to make the annual execution of an awareness campaign an ongoing target. However, we will continue to provide regular updates on the ways in which we raise awareness amongst our employees on health and safety matters.

Management of contractors

We engage contractors to undertake a variety of work within our UK and USA businesses. We continue to believe that it is important for third parties and contractors to obtain appropriate health and safety training. Whilst the responsibility for this lies with the employer, we are increasing our efforts in encouraging them to do so. One of the key selection criteria that we use for contractors is an assessment of the health and safety performance of the company in question and their relevant experience in the type of work involved.

Want to know more?

Our health and safety policy:
http://www.csr.abports.co.uk/workplace/healthsafety/policy.htm
Performance monitoring:
http://www.csr.abports.co.uk/workplace/healthsafety/mg_procedures/perf_monitoring.htm
PSSL:www.portskillsandsafety.co.uk
HSE: www.hse.gov.uk
IOSH: www.iosh.co.uk
OSHA: www.osha.gov

Monitoring and improving performance

During 2005, our internal audit function performed reviews on our health and safety processes and procedures at Garston, Hams Hall, Hull and Southampton. Findings from these reviews were reported to our audit committee and the associated action points were implemented in line with the agreed timescales. In 2006, we shall be conducting in-house health and safety audits at each of our UK business locations. Each site will be thoroughly audited to check the following:

- Adherence to group health and safety policies and procedures, broken down within each department by activity/task
- Compliance with recommendations made in previous audit reports
- Robustness of risk assessments
- Application of safe systems of work
- Maintenance and inspection standards
- Provision of training
- Control of contractors.

We acknowledge the weaknesses identified by ERM in relation to our processes for recording the outcome of decisions that injuries are not work related and for monitoring occupational diseases or chronic injuries. During 2006, we intend to address the recommendations made by ERM and in addition, we are focusing our efforts on meeting our new challenge of reducing our UK reportable injuries per thousand employees to 9 or less and of maintaining our USA recordable injuries at 50 or less.

Summary of progress against targets and initiatives

Table 3 provides a summary of our current targets and initiatives and the status of our progress as at 31 December 2005.

Table 3:

Initiative	*Status*
Key performance indicators and monitoring	
A. Reduce UK reportable injuries per thousand employees to 12 or less	A. Target achieved during 2005 with a rate of 9.3 (2004: 14.0). Target for 2006 is 9 or less
B. Reduce USA recordable injuries to 50 or less	B. Target achieved during 2005 as recordable injuries fell to 49 (2004: 64). Target for 2006 is 50 or less
C. Conduct internal safety audits at all UK locations	C. New initiative to be completed by the end of 2006
Training and awareness	
A. Provide accredited safety training to all UK employees	A. 89% or relevant employees had received accredited training by the end of 2005 (2004: 80%). Long-term target is for 95% of employees to have been provided with relevant IOSH-accredited training
B. Provide refresher courses to all employees whose accredited safety training took place five or more years ago	B. During 2005, 77% or relevant employees were provided with refresher training. This target will not be reported during 2006 as the above target includes employees who have received refresher training
C. Provide IOSH-accredited training course "Cargo Handling in Ports" to all ABP employees involved in cargo handling	C. On track to be completed by the end of 2006
D. Introduce "Leading Teams Safely" training course once IOSH accreditation has been received	D. Target suspended pending the outcome of the IOSH's internal review of training (see page 9)
E. Conduct campaign on the theme of "Reporting and Analysis of Potential Hazards"	E. Completed during 2005
Other	
A. Implement workplace transport policy and procedures	A. Completed and embedded within our overall health and safety procedures

Table of Cases

Table of Cases

Table of Statutes

Table of Statutory Instruments

Index

D

G

J

K

L

M